T0236494

Lecture Notes in Computer Science

Lecture Notes in Bioinformatics 13883

The series Lecture Notes in Bioinformatics (LNBI) was established in 2003 as a topical subseries of LNCS devoted to bioinformatics and computational biology.

The series publishes state-of-the-art research results at a high level. As with the LNCS mother series, the mission of the series is to serve the international R & D community by providing an invaluable service, mainly focused on the publication of conference and workshop proceedings and postproceedings.

Katharina Jahn · Tomáš Vinař

Editors

Comparative Genomics

20th International Conference, RECOMB-CG 2023
Istanbul, Turkey, April 14–15, 2023
Proceedings

 Springer

Editors
Katharina Jahn
Freie Universität Berlin
Berlin, Germany

Tomáš Vinař
Comenius University
Bratislava, Slovakia

ISSN 0302-9743   ISSN 1611-3349 (electronic)
Lecture Notes in Bioinformatics
ISBN 978-3-031-36910-0   ISBN 978-3-031-36911-7 (eBook)
https://doi.org/10.1007/978-3-031-36911-7

LNCS Sublibrary: SL8 – Bioinformatics

This Springer imprint is published by the registered company Springer Nature Switzerland AG
The registered company address is: Gewerbestrasse 11, 6330 Cham, Switzerland

Preface

This volume contains the papers presented at the 20th RECOMB Satellite Conference on Comparative Genomics (RECOMB-CG), held on April 14–15, 2023 in Istanbul, Turkey. RECOMB-CG, founded in 2003, brings together leading researchers in the mathematical, computational and life sciences to discuss cutting edge research in comparative genomics, with an emphasis on computational approaches and novel experimental results. RECOMB-CG was co-located with the annual RECOMB conference.

There were 25 submissions from authors from 15 countries. The submissions were reviewed by 3–5 (average 3.84) members of the program committee and invited sub-reviewers. Based on the reviews, 15 submissions were selected for presentation at the conference and are included in these proceedings. The RECOMB-CG contributed talks were complemented by keynotes from two invited speakers, Rute da Fonseca (University of Copenhagen, Denmark) and Mehmet Somel (Middle East Technical University, Turkey). The program also included six poster presentations.

The best paper award was awarded to Ali Osman Berk Şapcı for the presentation of paper "CONSULT-II: Taxonomic Identification Using Locality Sensitive Hashing." Askar Gafurov was awarded the best poster award for the presentation of the poster "Efficient Analysis of Annotation Colocalization Accounting for Genomic Contexts."

We would like to thank the members of the steering committee for guidance and helpful discussions and local organizers Can Alkan, A. Ercüment Çiçek from Bilkent University, and Arzucan Özgür from Boğaziçi University for organizing the RECOMB satellites and securing financial support for the conference. We would also like to thank the members of the program committee and sub-reviewers. Finally, we would like to thank all researchers who submitted papers and posters and attended RECOMB-CG 2023.

April 2023

Katharina Jahn
Tomáš Vinař

Organization

Program Chairs

Katharina Jahn   Freie Universität Berlin, Germany
Tomáš Vinař   Comenius University in Bratislava, Slovakia

Steering Committee

Marília Braga   Bielefeld University, Germany
Dannie Durand   Carnegie Mellon University, USA
Jens Lagergren   KTH Royal Institute of Technology, Sweden
Aoife McLysaght   Trinity College Dublin, Ireland
Luay Nakhleh   Rice University, USA
David Sankoff   University of Ottawa, Canada

Program Committee

Max Alekseyev   George Washington University, USA
Lars Arvestad   Stockholm University, Sweden
Mukul S. Bansal   University of Connecticut, USA
Anne Bergeron   Université du Québec à Montréal, Canada
Sèverine Bérard   Université de Montpellier, France
Paola Bonizzoni   Università di Milano-Bicocca, Italy
Marilia Braga   Bielefeld University, Germany
Broňa Brejová   Comenius University in Bratislava, Slovakia
Cedric Chauve   Simon Fraser University, Canada
Miklós Csűrös   University of Montreal, Canada
Daniel Doerr   Heinrich Heine University Düsseldorf, Germany
Mohammed El-Kebir   University of Illinois at Urbana-Champaign, USA
Nadia El-Mabrouk   University of Montreal, Canada
Oliver Eulenstein   Iowa State University, USA
Guillaume Fertin   University of Nantes, France
Martin Frith   University of Tokyo, Japan
Pawel Gorecki   University of Warsaw, Poland
Wataru Iwasaki   University of Tokyo, Japan
Asif Javed   University of Hong Kong
Xingpeng Jiang   Central China Normal University, China
Lingling Jin   University of Saskatchewan, Canada

Jaebum Kim	Konkuk University, South Korea
Manuel Lafond	Universitè de Sherbrooke, Canada
Kevin Liu	Michigan State University, USA
Istvan Miklos	Renyi Institute, Hungary
Siavash Mirarab	University of California, San Diego, USA
Aïda Ouangraoua	Universitè de Sherbrooke, Canada
Fabio Pardi	LIRMM - CNRS, France
Teresa Przytycka	National Center of Biotechnology Information, USA
Aakrosh Ratan	University of Virginia, USA
Michael Sammeth	Coburg University, Germany
Mingfu Shao	Carnegie Mellon University, USA
Sagi Snir	University of Haifa, Israel
Giltae Song	Pusan National University, South Korea
Yanni Sun	City University of Hong Kong
Wing-Kin Sung	National University of Singapore, Singapore
Krister Swenson	CNRS, Universitè de Montpellier, France
Olivier Tremblay-Savard	University of Manitoba, Canada
Tamir Tuller	Tel Aviv University, Israel
Jean-Stèphane Varrè	Universitè de Lille, France
Fábio Martinez	Federal University of Mato Grosso do Sul, Brazil
Yong Wang	Academy of Mathematics and Systems Science, China
Tandy Warnow	University of Illinois at Urbana-Champaign, USA
Yufeng Wu	University of Connecticut, USA
Xiuwei Zhang	Georgia Institute of Technology, USA
Louxin Zhang	National University of Singapore, Singapore
Fa Zhang	Chinese Academy of Science, China
Jie Zheng	ShanghaiTech University, China

Sub-Reviewers

Eloi Araujo	Herui Liao
Ke Chen	Xueheng Lyu
Simone Ciccolella	Alexey Markin
Rei Doko	Diego P. Rubert
Askar Gafurov	Jiayu Shang
Meijun Gao	Clarence Todd
Stephane Guindon	Sanket Wagle
Thulani Hewavithana	Jiale Yu
Aleksandar Jovanovic	Julia Zheng

Previous RECOMB-CG conferences

The 19th RECOMB-CG, La Jolla, USA, in 2022,
The 18th RECOMB-CG held virtually in 2021,
The 17th RECOMB-CG, Montpellier, France, in 2019,
The 16th RECOMB-CG, Quebec, Canada, in 2018,
The 15th RECOMB-CG, Barcelona, Spain, in 2017,
The 14th RECOMB-CG, Montreal, Canada, in 2016,
The 13th RECOMB-CG, Frankfurt, Germany, in 2015,
The 12th RECOMB-CG, New York, USA, in 2014,
The 11th RECOMB-CG, Lyon, France, in 2013,
The 10th RECOMB-CG, Niterói, Brazil, in 2012,
The 9th RECOMB-CG, Galway, Ireland, in 2011,
The 8th RECOMB-CG, Ottawa, Canada, in 2010,
The 7th RECOMB-CG, Budapest, Hungary, in 2009,
The 6th RECOMB-CG, Paris, France, in 2008,
The 5th RECOMB-CG, in San Diego, USA, in 2007,
The 4th RECOMB-CG, Montreal, Canada, in 2006,
The 3rd RECOMB-CG, Dublin, Ireland, in 2005,
The 2nd RECOMB-CG, Bertinoro, Italy, in 2004,
The 1st RECOMB-CG, Minneapolis, USA, in 2003.

Contents

Classifying the Post-duplication Fate of Paralogous Genes 1
  Reza Kalhor, Guillaume Beslon, Manuel Lafond, and Celine Scornavacca

Inferring Clusters of Orthologous and Paralogous Transcripts 19
  Wend Yam Donald Davy Ouedraogo and Aida Ouangraoua

On the Class of Double Distance Problems 35
  *Marília D. V. Braga, Leonie R. Brockmann, Katharina Klerx,
  and Jens Stoye*

The Floor Is Lava - Halving Genomes with Viaducts, Piers and Pontoons 51
  Leonard Bohnenkämper

Two Strikes Against the Phage Recombination Problem 68
  Manuel Lafond, Anne Bergeron, and Krister M. Swenson

Physical Mapping of Two Nested Fixed Inversions in the X Chromosome
of the Malaria Mosquito *Anopheles messeae* 84
  *Evgenia S. Soboleva, Kirill M. Kirilenko, Valentina S. Fedorova,
  Alina A. Kokhanenko, Gleb N. Artemov, and Igor V. Sharakhov*

Gene Order Phylogeny via Ancestral Genome Reconstruction Under Dollo 100
  Qiaoji Xu and David Sankoff

Prior Density Learning in Variational Bayesian Phylogenetic Parameters
Inference 112
  Amine M. Remita, Golrokh Vitae, and Abdoulaye Baniré Diallo

The Asymmetric Cluster Affinity Cost 131
  *Sanket Wagle, Alexey Markin, Paweł Górecki, Tavis Anderson,
  and Oliver Eulenstein*

The K-Robinson Foulds Measures for Labeled Trees 146
  Elahe Khayatian, Gabriel Valiente, and Louxin Zhang

Bounding the Number of Reticulations in a Tree-Child Network
that Displays a Set of Trees 162
  Yufeng Wu and Louxin Zhang

Finding Agreement Cherry-Reduced Subnetworks in Level-1 Networks 179
    Kaari Landry, Olivier Tremblay-Savard, and Manuel Lafond

CONSULT-II: Taxonomic Identification Using Locality Sensitive Hashing 196
    Ali Osman Berk Şapcı, Eleonora Rachtman, and Siavash Mirarab

MAGE: Strain Level Profiling of Metagenome Samples 215
    Vidushi Walia, V. G. Saipradeep, Rajgopal Srinivasan,
    and Naveen Sivadasan

MoTERNN: Classifying the Mode of Cancer Evolution Using Recursive
Neural Networks 232
    Mohammadamin Edrisi, Huw A. Ogilvie, Meng Li, and Luay Nakhleh

Author Index 249

Classifying the Post-duplication Fate
of Paralogous Genes

Reza Kalhor[1]([envelope]), Guillaume Beslon[2], Manuel Lafond[1],
and Celine Scornavacca[3]

[1] Department of Computer Science, Université de Sherbrooke, Sherbrooke, Canada
`reza.kalhor@usherbrooke.ca`
[2] Université de Lyon, INSA-Lyon, INRIA, CNRS, LIRIS UMR5205, Lyon, France
[3] Institut des Sciences de l'Evolution de Montpellier (Université de Montpellier, CNRS, IRD, EPHE), Montpellier, France

Abstract. Gene duplication is one of the main drivers of evolution. It is well-known that copies arising from duplication can undergo multiple evolutionary fates, but little is known on their relative frequency, and on how environmental conditions affect it. In this paper we provide a general framework to characterize the fate of duplicated genes and formally differentiate the different fates. To test our framework, we simulate the evolution of populations using aevol, an *in silico* experimental evolution platform. When classifying the resulting duplications, we observe several patterns that, in addition to confirming previous studies, exhibit new tendencies that may open up new avenues to better understand the role of duplications.

Keywords: Gene duplication · Duplication fates · Classification · Paralogy · Simulation

1 Introduction

Gene duplication is largely responsible for boosting the innovation and function variation of genomes [1–3], and plays a central role in the evolution of gene families [4]. Copies of genes arising from duplication can undergo multiple evolutionary fates [5]. For instance, the copies may perform the same role, share functions, or one of them could accumulate mutations while the other maintains the original function [6]. The more commonly-studied fates, described in detail in the following section, are pseudogenization (one gene is lost), (double)-neofunctionalization (both/one gene diverges in function), conservation (both genes preserve functions), subfunctionalization (genes split the functions) and specialization (genes split functions and acquire novel ones).

Still, little is known on whether some of these fates are more frequent than others, and on how environmental conditions affect their relative frequency. Inferring the fate of paralogous genes is a difficult task for two main reasons. First, the functions of their lowest common ancestor is usually unknown, making it difficult to predict how the roles of each gene evolved. Second, even if the ancestral functions were known, their evolution may not fit perfectly into one of the

K. Jahn and T. Vinař (Eds.): RECOMB-CG 2023, LNBI 13883, pp. 1–18, 2023.
https://doi.org/10.1007/978-3-031-36911-7_1

established classes. Several works have focused on understanding the role of duplications (see e.g. [7]), but to our knowledge, no rigorous framework has been developed to classify these roles. Here, we aim at providing a general framework to formally characterize the possible fates of duplicated genes to be able to discriminate them using phylogenetic data. Our approach is based on comparison of the biological functions of the original gene and the duplicated ones, and provides a continuum between the different fates.

Most research works on the topic are theoretical and propose statistical fate models to make predictions. For example, Lynch et al. [8,9] model genes as discrete sets of functions and propose a population-based model of subfunctionalization that considers mutation rates at regulatory regions. They notably show that the probability of subfunctionalization tends to 0 as population sizes increase. Using similar ideas, Walsh [10] compares pseudogenization against other fates, showing that predictions depend on mutation rates. In [11], the authors also compare subfunctionalization and pseudogenization using a mechanistic model based on Markov chains, which allows for data fitting and improved characterizations of hazard rates of pseudogenization. Markov chains were also used in [12] to predict the evolution of gene families undergoing duplications, loss, and partial gain/loss of function. Also, the theoretical impacts of neofunctionalization on orthology prediction were discussed in [13]. Classification tools based on gene-species reconciliation have also been proposed, e.g. for xenologs [14], which are pairs of genes whose divergence includes a horizontal gene transfer.

In more practical settings, perhaps the closest work to ours is that of Assis and Bachtrog [15]. Based on the ideas of [16], they used Euclidean distances between gene expression profiles to distinguish between the fates neofunctionalization, subfunctionalization, conservation and specialization. Using drosophilia data, they show that neofunctionalization is the dominant fate, followed by conservation and specialization, and they find very few cases of subfunctionalization. In [17], the authors use d_N/d_S ratios and expression data to distinguish subfunctionalization and neofunctionalization. They notably conclude that such dichotomic fate models are insufficient to explain the variety of functional patterns of duplicate genes. This motivates the need to develop classification methods that account for hybrid fates. Several works have also focused on pseudogenization, based on sequence comparisons and homology detection, showing that this fate is very likely in certain species [18,19]. For instance in Zebrafish, it is estimated that up to 20% of duplicated genes are retained and the rest are non-functional [20]. Neofunctionalization has also been studied in practice. This fate can occur through changes in the biological processes of a copy, but also in the expression at the transcriptional level. The latter was argued to play an important role in evolution [21–23]. Functional changes can occur at the enzymatic level [24] and, more recently, were shown to also occur at the post-translational level [25]. This was achieved by comparing one fate against another for three species in which short regulatory motifs were identified and statistically correlated with observed post-translational changes.

Our framework aims at generalizing the approaches developed in these experimental studies. To test our framework, we use an *in silico* experimental evolution platform that enable to simulate the evolution of a population of individuals under the combined effect of selection and variation [26,27]. Specifically, we used the aevol platform [28], a computing platform where populations of digital organisms can evolve under various conditions, enabling to experimentally study the effect of the different evolutionary forces on genomes, gene repertoire and phenotypes. Aevol has already been used to study the direct and indirect effect of segmental duplications/deletions, showing that their mutational effect is likely to regulate the amount of non-coding sequences due to robustness constraints [29,30]. The platform has also been used to show that genetic association can help maintaining cooperative behaviour in bacterial populations [31]. More recently, aevol has been used to study the "complexity ratchet", showing that epistatic conflicts between genes duplication-divergence (i.e. neofunctionalization or double-neofunctionalization fates) and local events (i.e. allelic variation of a single gene) opens the route to biological complexity even in situations where simple phenotypes would easily thrive [32]. However, although it as been shown that gene duplications is a rather frequent event in aevol, (almost half of the gene families being created by a segmental event [33]), the precise fate of gene duplicates has never been specifically studied in the model.

In this paper, we fill this gap by simulating the evolution of populations of individuals via aevol and classifying the resulting duplications using our framework. Our tests on aevol confirm the experimental studies on drosophilia data [15] and show that conservation of the original function in both copies is rather unlikely, the general trend being that the more frequent fates are those exhibiting a higher level of function acquisition.

2   Post-duplication Fates

Several classes and sub-classes of post-duplication fates have been proposed in the literature; here we recall the main ones that we model in our framework. These fates have been chosen because they are generally agreed upon, as discussed in various surveys (see e.g. [34,35]); each class is assigned an acronym that we shall use in the following of the paper.

Pseudogenization (P): one copy retains its functions, while the other diverges and becomes non-functional [5]. Pseudogenization is believed to be very likely, since losing one copy can repair an "accidental" duplication. In this study, we consider only a type of pseudogenization, called *compensatory drift*, in which the expression level of at least one of the duplicated genes is too low to supply the function [36,37]. Note that a gene could be lost by a deletion event or by a mutation that would, e.g., inactivate its promoter. However, these fates are not considered here as we focus on gene duplication leading to paralogy in extant genomes.

Neofunctionalization (N): when one copy diverges as above, it may acquire novel functions instead of pseudogeneizing [38]. This is often believed to be a major mechanism of function acquisition, as neofunctionalization can use a copy of a functional gene as a template to favor adaptation [39].

Double-neofunctionalization (DN): both copies acquire distinct functions that are different from the original gene (hence, the original function is not performed by any of the two copies). To our knowledge, there is no established name for this fate, although this phenomenon occurs frequently in our experiments. Double-neofunctionalization can arise when a gene is not required for survival, for instance when a copy of a duplicated gene undergoes a second duplication. In this case, both sub-copies are free to develop new functions.

Conservation (C): this process is such that neither of the duplicated copies changes, both performing the same functions as the original gene, potentially doubling its expression level. One could argue that this provides no advantage to an adapted organism (it could even be harmful due to dosage effect). However, conservation can also be advantageous when increased gene dosage is required for adaptation [40], or when one copy needs to be kept as a "backup" [36].

Subfunctionalization (SF): the copies partition the original functions and are thus complementary and necessary to perform them [41]. This is sometimes called duplication-degeneration-complementation (DDC) [40]. Subfunctionalization has also been associated with changes in expression patterns [36], especially in cases where the copies become expressed less but, together, still produce the same amount of proteins as before. The latter is sometimes distinguished as hypofunctionalization [42]. In this paper, we consider both situations as mere subfunctionalization.

Specialization (SP): this fate occurs when the genes copies are able to perform the original functions, but *also* both develop novel functions. This differs from DN, since the original function is still performed, but also differs from SF because of the novel functions. The term was introduced in [16] and described as a mix of SF and N. In this work, we consider that this fate occurs as long as the original function exists (whether it is by SF or not) and both copies acquire a significant amount of new functions.

3   Methods

We first describe our theoretical model of fate classification, and then proceed to describe our experiments.

We assume the existence of a set of possible biological functions that we denote by $\mathcal{F}$. We allow any representation of functions as a set and $\mathcal{F}$ can be discrete or continuous (for instance, Gene Ontology terms, or coordinates in a multidimensional functional universe). A *gene* g expresses some functions of $\mathcal{F}$ to some degree. For this purpose, we model a gene as a (mathematical) function $g : \mathcal{F} \to \mathbb{R}$, where $g(\zeta)$ represents the activation level of function $\zeta \in \mathcal{F}$. If $g(\zeta) = 0$, then g does not contribute to performing function ζ. Importantly, notice that $g(\zeta)$ can be negative, which models the fact that g *inhibits* function ζ. These concepts are illustrated in Fig. (1.a), which shows a gene whose expression pattern has a triangular shape (note that this shape is merely for illustration, as our model applies to any shape). This gene expresses functions in the range $[0, 25, 0.75]$, and the expression of each function ζ in this range is the height of the triangle at x-coordinate ζ (for instance, $g(0.5) = 1, g(0.75) = 0$).

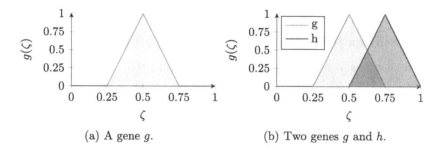

(a) A gene g. (b) Two genes g and h.

Fig. 1. An illustration of genes expressing functions in a triangle pattern.

We define the following comparative tools for two genes g and h:

– $[g + h]$ represents function addition, which can be seen as a gene described by the functional landscape that g and h accomplish together (note that they may cancel each other in case of inhibition). For each $\zeta \in \mathcal{F}$, it is defined as

$$[g + h](\zeta) = g(\zeta) + h(\zeta)$$

– $[g \cap h]$ represents function intersection and, for each $\zeta \in \mathcal{F}$, is defined as

$$[g \cap h](\zeta) = \begin{cases} \min(g(\zeta), h(\zeta)) & \text{if } g(\zeta) \geq 0, h(\zeta) \geq 0 \\ \max(g(\zeta), h(\zeta)) & \text{if } g(\zeta) < 0, h(\zeta) < 0 \\ 0 & \text{otherwise} \end{cases}$$

– for gene g, we define $contrib(g)$ as the total functional contribution of the gene, i.e. as the sum of absolute values of its expression levels. If $\mathcal{F}$ is discrete, we define $contrib(g) = \sum_{\zeta \in \mathcal{F}} |g(\zeta)|$, and if $\mathcal{F}$ is continuous, we define $contrib(g) = \int_{\mathcal{F}} |g(\zeta)| d\zeta$.

– $i_{g|h}$ represents the function coverage of g by h, i.e. the proportion of functions of g that can be performed by h, and is defined as

$$i_{g|h} = \frac{contrib([g \cap h])}{contrib(g)}$$

We may write $g + h$ and $g \cap h$ without brackets when no confusion can arise. Note that $[g + h] = [h + g]$ and $[g \cap h] = [h \cap g]$, but $i_{g|h}$ might differ from

$i_{h|g}$. These notions can be visualized from Fig. (1.b): $[g + h]$ can be seen as the points on the leftmost diagonal edge of the g triangle, on the top edge of the light gray area, and on the rightmost diagonal edge of the h triangle; $[g \cap h]$ can be seen as the points on the diagonal edges of the triangle formed by the overlap of the g and h triangles, which is another triangle with height 0.5 and area $0.5 \cdot 0.25/2 = 0.0625 = 1/16$. Hence, $i_{g|h} = i_{h|g} = (1/16)/(1/4) = 1/4$.

3.1 Classifying the Fates of Paralogs

Suppose that a and b are two extant paralogs and that their least common ancestor is g. For each fate described in Sect. 2, i.e. for each fate $X \in \{P, N, DN, C, SF, SP\}$, we quantify how much a and b appear to have undergone X, using appropriate $i_{g|h}$ proportions as defined above. The main challenge in developing a continuum between fates is to ensure that each fate has a distinguishing feature against the others. In our design, each pair of fates has a factor that contributes conversely to the two fates (while also correctly modeling them, of course). For example, N expects exactly one of $i_{a|g}$ or $i_{b|g}$ to be 1, whereas DN expects both to be 0, and values in-between have opposite effects. It was also necessary to include thresholds to model some of the fates properly, as follows:

- $\delta_\tau(x) = \max(0, \frac{x - \tau}{1 - \tau})$ is a generic *threshold function* with respect to a parameter τ. It equals 0 for $x \leq \tau$, and then increases linearly from 0 to 1 in the interval $x \in [\tau, 1]$. This is useful to model fates that require a threshold.
- $\rho \in [0, 1]$ is a *pseudogene threshold*, used to determine how much functionality a copied gene must lose to be considered a pseudogene. For example, if $\rho = 0.2$, the amount of P of a gene linearly increases from 0 to 1 as its coverage of its parent drops between one fifth and 0.
- $\nu \in [0, 1]$ is a *novelty threshold* that determines how much a copy must dedicate to the parental functions to be considered as "not too new". For instance if $\nu = 0.25$, the fates C, SF require the copied genes to dedicate a quarter or more of their functions to the parental functions, and otherwise they are excluded as possible fates. Conversely, $1 - \nu$ could be interpreted as "new enough", and determines how much novelty is needed for SP.

The formulas for computing the proportion of each fate are detailed in Table 1.

Using the triangular gene illustrations, Fig. 2 shows that each canonical fate has an inferred proportion of 1 in our model. It can also be verified that when this occurs, the other fates have proportion 0. Also note that P and N are the only fates to use a maximum of two values. This is because there are two ways in which P can occur (either gene loses functions), and in which N can occur (either gene diverges). In the other fates (DN, C, SF, SP), the two genes behave in a similar manner instead. Although it is difficult to validate the formulas formally, we provide the rationale behind each of them:

- *Pseudogenization*: P_a should be close to 1 when a has not developed novel functions *and* has lost most of g's functions. The $i_{a|g}$ factor ensures the first condition by checking that a is covered by g. The $(1 - \frac{i_{g|a}}{\rho})$ factor implements

Table 1. The formulas used to compute the proportion of each fate.

Fate	Formula
Pseudogenization (P)	$P_a = i_{a\|g} \cdot (1 - \frac{i_{g\|a}}{\rho})$
	$P_b = i_{b\|g} \cdot (1 - \frac{i_{g\|b}}{\rho})$
	$P = \max(0, P_a, P_b)$
Neofunc. (N)	$N_a = (1 - i_{a\|g}) \cdot \delta_\nu(i_{b\|g}) \cdot i_{g\|b}$
	$N_b = (1 - i_{b\|g}) \cdot \delta_\nu(i_{a\|g}) \cdot i_{g\|a}$
	$N = \max(N_a, N_b) \cdot (1 - P)$
Double-neo. (DN)	$DN = (1 - i_{a\|g})(1 - i_{b\|g})(1 - i_{g\|b})(1 - i_{g\|b})(1 - P)$
Conservation (C)	$C = \delta_\nu(i_{a\|g}) \cdot \delta_\nu(i_{b\|g}) \cdot i_{g\|a+b} \cdot (1 - \delta_{0.5}(i_{a+b\|g})) \cdot (1 - P)$
Subfunc. (SF)	$SF = \delta_\nu(i_{a\|g}) \cdot \delta_\nu(i_{b\|g}) \cdot i_{g\|a+b} \cdot \delta_{0.5}(i_{a+b\|g}) \cdot (1 - P)$
Specialization (SP)	$SP = i_{g\|a+b} \cdot (1 - \delta_\nu(i_{a\|g})) \cdot (1 - \delta_\nu(i_{b\|g})) \cdot (1 - P)$

our threshold idea for the second condition, as this factor increases linearly as a covers less functions of g, but only once the threshold ρ is crossed. The same applies to b and P_b, and P is the maximum of P_a and P_b.

Note that all further fates consider the level of pseudogenization P by multiplying them by $(1 - P)$. This is because the more a gene has pseudogeneized, the less it should be considered for other fates.

- *Neofunctionalization*: N_a should be close to 1 when a acquires entirely new functions. Since a is novel, $1 - i_{a|g}$ should equal 1, and since b should only perform g, $\delta_\nu(i_{b|g})$ should be 1 (and $i_{g|b}$ should equal 1 because b covers g). The same applies to b and N_b when b neofunctionalizes.
- *Double-neo*: neither of a and b should intersect with g, and thus each of $i_{a|g}, i_{b|g}, i_{g|b}, i_{g|a}$ should be close to 0.
- *Conservation*: a and b should be identical to g, and thus a, b should be dedicated to g without "too much" novelty (i.e. $\delta_\nu(i_{a|g}), \delta_\nu(i_{b|g})$ should be 1), and g should be covered by $[a + b]$ ($i_{g|a+b}$ should be 1). Moreover, $[a + b]$ should double each of g's functions. The $1 - \delta_{0.5}(i_{g|a+b})$ factor hence expects $[a + b]$ to be covered by g by a proportion of 0.5 or less, and penalizes the fate if the coverage is higher. This factor separates C from SF.
- *Subfunctionalization*: a and b should be dedicated to performing g without too much novelty ($\delta_\nu(i_{a|g}), \delta_\nu(i_{b|g})$ should be 1), and $a + b$ should perform g together ($i_{g|a+b}$ should be 1). Unlike conservation, $[a + b]$ should be entirely covered by g since a and b have split the functions of g. The $\delta_{0.5}(i_{a+b|g})$ factor increases linearly from 0 to 1 for $i_{a+b|g} \in [0.5, 1]$, which is the opposite of conservation.
- *Specialization*: g should be performed by a and g, and thus $i_{g|a+b}$ should be 1. Moreover, a and b should both develop enough novel functions. For a, the amount of novelty is expressed as $1 - i_{a|g}$. We cannot expect this term to be 1 in the SP fate, since a portion of a performs g. Using $1 - \delta_\nu(i_{a|g})$ instead tolerates a to dedicate a proportion of up to ν to perform g without penalty, as long as a has enough novelty. The same holds for b.

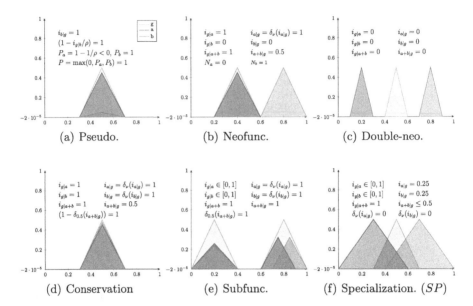

Fig. 2. The canonical fates using the triangle representation (note that two possible ways in which SF can occur are shown in the same subfigure). We assume thresholds $\rho = 0.2$ (relevant for P) and $\nu = 0.25$ (mostly relevant for SP).

If one considers our formulas as a probability distributions on fates, the sum of values of each fate should sum to 1 (i.e. $P + N + C + SF + SP + DN = 1$). However, the six categories presented here may not cover all the possible fates of genes after a duplication. Indeed, in our experiments, we regularly observed situations where $P + N + C + SF + SP + DN < 1$. Note however that we never observed situations where the sum of fate values is larger than 1 (see Table 4). Since we studied thousands of duplications, we conjecture that the sum of fate values should be bounded by 1, leaving the proof as an open problem.

3.2   Computing the Fate Between All Paralogs in a Gene Tree

The previous section describes how to compute the fate of a gene g and two of its paralogous descendants a and b. However, in the case of successive duplications, g may have multiple pairs of such paralogous descendants. In Algorithm 1, we describe how to compute the fate proportions between all paralogs in a gene tree G, in which leaves are extant genes and internal nodes are ancestral genes. For the purposes of our algorithm, we assume that the functions of both extant and ancestral genes are known. We also assume knowledge of a set of duplication nodes D, which can be inferred through reconciliation [43,44]. Then for each gene $g \in D$ affected by a duplication, the algorithm looks at its two child copies g_1 and g_2. It then finds the extant descendants $a_1, \ldots, a_n$ of g_1 (left leaves of g) and $b_1, \ldots, b_m$ of g_2 (right leaves of g), and calculates each fate for each triple of the form g, a_i and b_j. In our results, we report the average proportion of each fate, taken over all pairs of paralogs analyzed, as computed in Algorithm 1.

```
 1  function ClassifyDuplicationFates(G)
 2      // D is the set of duplication nodes;
 3      Fates ← array of 6 values, initialized to 0;
 4      NbParalogies ← 0;
 5      for each g ∈ D do
 6          Let g₁, g₂ be two children of g in G;
 7          Let A = {a₁, a₂, ..., aₙ} be extant descendants of g₁;
 8          Let B = {b₁, b₂, ..., bₘ} be extant descendants of g₂;
 9          for each X ∈ {P, N, DN, C, SF, SP} do
10              for each aᵢ ∈ A do
11                  for each bⱼ ∈ B do
12                      Fates[X] + = ComputeFate[X](g, aᵢ, bⱼ);
13                      NbParalogies + = 1;
14      for each X ∈ {P, N, DN, C, SF, SP}  do  Fates[X] = Fates[X]/NbParalogies;
```

Algorithm 1: Algorithm to classify duplication events. $ComputeFate[X](g, a_i, b_j)$ calculates the average proportion of each fate for each triple g, a_i and b_j.

3.3   Simulations

As already mentioned, to test our method, we used simulated data generated using the aevol platform. Aevol is an *in silico* experimental evolution platform that simulates the evolution of a population or digital organisms [45]. In aevol, each organism owns a genome (double-stranded circular sequence inspired from bacterial chromosome, see Fig. 3, upper part) and the model simulates transcription and translation to identify genes on the sequence. Each gene is then decoded into a $[0, 1] \rightarrow [-1, 1]$ mathematical kernel function (a "protein") and all the kernels are linearly combined to compute the phenotype (a $[0, 1] \rightarrow [0, 1]$ function – Fig. 3, bottom). A population of such organisms replicate through a Wright-Fisher scheme. At each generation, the fitnesses of all the organisms are computed by comparing the phenotypic function with a target function that indirectly represents the environment (see Figs. 3 and 4) and, during replication, organisms may undergo various kinds of sequence mutations, including substitutions, Indels and chromosomal rearrangements (including inversions, duplications and deletions). Organisms are thus embeded into an evolutionary loop, enabling to study the relative effects of the different evolutionary forces on genome structure, genome sequence and gene repertoire.

As aevol has already been extensively described elsewhere [27,28,30,32,46], we will not describe it in more details here. Now, given our objective, there are a number of advantages of using aevol. First, the platform enables both variation of gene content and genes sequences, a mandatory property to study the fate of duplicated genes. Second, in aevol, each gene is decoded into a mathematical function representing the genes function and the sum of all genes functions enables computing the organisms phenotype. The reproductive success (or the extinction) of an organism then depends on the adequacy of its phenotype function and the target function representing the environmental conditions. This enables a formal characterisation of genes functions, hence of the different possible fates of gene duplicates. Finally, the aevol platform has already – and successfully – be used as a benchmark to test bioinformatics methods [47]. Furthermore, it has not been designed specifically to test our framework, hence providing an independent test-bed.

We now discuss our simulation framework. As briefly described above, in aevol the environment is represented by a $[0, 1] \rightarrow [0, 1]$ target function that the phenotypes must fit. We considered four different environments shown in Fig. 4. We used environment (a) to generate the initial genomes, which means we let a population evolve for 1.1 million generations in this environment[1], extracted ancestor individual of final population at generation 1 million, and used it as the initial genome for further simulations. These initial genomes are called *wild-types*, and are well adapted to their environment (this "pre-evolution" step is required since evolution is heavily random in naive populations). In aevol a specific parameter ($0 < w_{max} \leq 1$) enables tuning the maximum pleiotropy in the model (the higher w_{max} the higher the pleiotropy level – $w_{max} = 1$ representing the maximum, where a gene can have an effect on all functions). As pleiotropy level is suspected to influence the fate of duplicated genes [48], we generated wild-types with four different values of $w_{max} \in \{0.01, 0.1, 0.5, 1\}$ in environment (a). These four different wild-types enable us to test whether the pleiotropy of an organism has an impact on duplication fates. Figure 3 shows the sequence level (top) and functional level (bottom) of a wild-type evolved for 1 million generations with a minimal pleiotropy level ($w_{max} = 0.01$). Note the gene highlighted in red on the bottom-left figure. Though not active enough to reach the target, it exists in three copies on the genome, hence increasing its effect (red triangle on the bottom right). This results from two successive duplication events with fate C.

We used each generated wild-type as an initial genome for further 1 million generations of evolution in our four different environments. Note that, since wild-types are already adapted to environment (a), we expect very few

---

[1] All evolutionary simulations were conducted with a population size of 1024 individuals and a mutation rate of 10^{-6} mutations per base pair per generation for each kind of mutational event. Previous experiments with the model showed that this parameter set leads to genomic structures akin to prokaryotic ones, though globally smaller [29]. For instance, the wild-type presented on Fig. 3 has a 10,541 bp-long genome carrying 118 genes located on 50 mRNAs with a coding fraction of 77%.

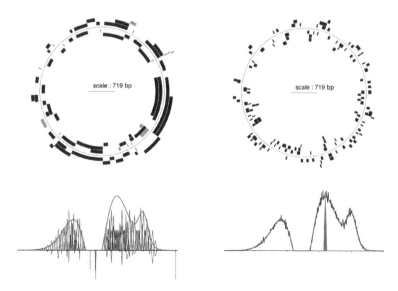

Fig. 3. Overview of an aevol wild-type. Top: sequence level (genome, RNAs and genes). The double-stranded genome is represented by a circle (thin line). Black arcs represent RNAs (left) and genes (right) on each strand (grey arcs represent non-coding RNAs and non-functional genes respectively). Note the presence of polycistronic sequences. Bottom: environmental target (red curve) and functional levels (genes and phenotype, in black) with one specific function highlighted in red (see main text for details). Left: each triangle corresponds to a mathematical kernel which parameters are decoded from a gene sequence. Note the presence of function-activating/repressing genes (positive/negative triangles respectively). Right: organism's phenotype resulting from the sum of all kernels. (Color figure online)

duplications to occur in this environment. The other three environments range from mild, medium, and heavy change with respect to the original environment; the intent of these simulations is to evaluate how individuals respond to different degrees of changes in their environment. Therefore, we expect the genomes that evolve under (d) to undergo more duplications. For each wild-type and each environment, we then performed 20 independent simulations.

Finally, we collected the most fit individuals at the end of each simulation. The extant paralogs that we analyzed were those found in their genome at the end of the process. As explained above, this procedure does not consider genes lost after duplication (either through sequence deletion or inactivation of transcription/translation initiation sequences). Thus, the pseudogenization fate here only considers extant genes whose activity has been strongly reduced[2].

---

[2] The source code is available at https://github.com/r3zakalhor/Post-Duplication-Fate-Framework.

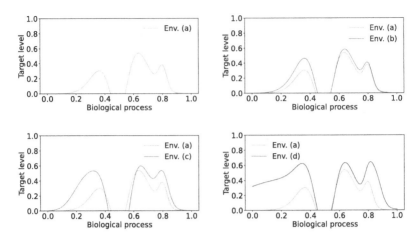

Fig. 4. The four different environments used in the simulations. On the x-axis, we assume that the set of functions (biological processes) is the interval $[0, 1]$. The y-axis depicts the target level which, for each function, indicates the ideal amount of expression to survive in the environment.

4   Results

As explained above, starting from wild-types evolved in environment (a) with different maximum pleiotropic levels w_{max}, we simulated the evolution 20 populations in 4 environments (ordered by increased variation compared to the environment of the wild-type) and for 1 million generations. We first verified that our phylogenies contain enough fixed duplications to enable studying the fate of duplicated genes with a reasonable precision. Table 2 shows the number of duplications per million generations observed for each environment. Recall that observed duplications are only those that result in extant paralogs, i.e. we do not consider duplications in intergenic regions, or in which a copy is lost.

Table 2. Rate of gene duplications for each environment (number of gene duplications fixed per million generations, averaged over every possible w_{max}).

	Env. (a)	Env. (b)	Env. (c)	Env. (d)
Gene dup. rate	3.559	15.825	40.712	55.326

Table 3. Rate of gene duplications for each pleiotropy level (number of gene duplications fixed per million generations, averaged over every environment).

	$w_{max} = 0.01$	$w_{max} = 0.1$	$w_{max} = 0.5$	$w_{max} = 1$
Gene dup. rate	53.735	28.220	17.641	15.825

Not surprisingly, the rate of fixed duplications is minimum when the organisms evolve in the constant environment (a) and it increases with the amount of change in the environments. Since each dataset comprises a million generations, the number of duplications is large enough to observe a large variety of fates. Interestingly, the number of gene duplications not only depends on the amount of environmental variation but also on the degree of pleiotropy. Indeed, Table 3 clearly shows that the lower the pleiotropy (i.e. the smaller w_{max}), the higher the number of fixed gene duplications (hence the higher the number of paralogs at the end of the simulation). One explanation is that a smaller w_{max} implies that genes have a narrower function spectrum. Thus, having more genes may increase the chance of adding new functions, thus improving fitness.

Table 4. Average fate proportions. Most frequent fates are boldfaced.

w_{max}	P	N	DN	C	SF	SP	Total	Dup. rate
			Environment (a)					
0.01	0.079	0.269	**0.357**	0.068	0.001	0.006	0.780	4.720
0.1	**0.395**	0.166	0.192	0.000	0.000	0.031	0.784	2.150
0.5	**0.239**	0.213	0.197	0.040	0.007	0.066	0.762	4.567
1	0.240	**0.434**	0.130	0.067	0.010	0.037	0.918	2.800
			Environment (b)					
0.01	0.080	0.286	**0.459**	0.049	0.002	0.010	0.886	34.510
0.1	0.133	0.241	**0.343**	0.044	0.000	0.069	0.830	13.850
0.5	0.183	0.227	**0.261**	0.102	0.004	0.064	0.841	8.117
1	0.112	0.232	**0.244**	0.069	0.017	0.100	0.774	6.825
			Environment (c)					
0.01	0.075	0.290	**0.465**	0.070	0.001	0.011	0.912	76.000
0.1	0.131	**0.265**	0.254	0.068	0.004	0.053	0.775	35.550
0.5	0.106	**0.293**	**0.293**	0.071	0.011	0.087	0.861	26.950
1	0.097	**0.283**	0.280	0.102	0.012	0.074	0.848	24.350
			Environment (d)					
0.01	0.089	0.273	**0.482**	0.052	0.002	0.011	0.909	99.713
0.1	0.110	0.266	**0.340**	0.063	0.007	0.053	0.839	61.333
0.5	0.156	**0.266**	0.260	0.072	0.008	0.080	0.842	30.933
1	0.117	**0.306**	0.233	0.119	0.014	0.071	0.860	29.325

Table 4 show the proportions of the different fates estimated on the aevol simulations (for each wild-type we simulated 4 environments × 20 parallel repetitions evolved for 1 million generations[3]). Except in rare situations, all fates are

---

[3] We note here that for some w_{max} we were able to generate and summarize statistics for several wild types: for $w_{max} = 0.01$ we have five wild types, for $w_{max} = 0.1$ one, for $w_{max} = 0.5$ three and for $w_{max} = 1$ two, leading to a total of 880 experiments.

observed and classified by our classification rules. The column "Total" reports the sum of proportions for each row. The gap between these values and 1 can be interpreted as the amount of fates that remained "unclassified". It would be easy to turn our predictions into a probability distribution by normalizing them, but we prefer to emphasize the fact that paralogs underwent fates that, on average, had between 10–25% of their behavior that did not fit any of the canonical fates.

Several notable results can be observed from this table. When the organisms must adapt to a new environment (b, c and d), the most frequent fate of duplications is N or DN while P is more frequent when the organisms face the same environment as the wild-type. Moreover, while the rate of N seems independent of the mean level of pleiotropy imposed by w_{max}, we observe that the rate of DN decreases as pleiotroty increases. This emphasizes the need to differentiate both classes as we do. Note that this phenomenon is not surprising. Indeed, as pleiotropy increases, the range of functions performed by an individual gene increases, hence the probability that both duplicates lose the ancestral function and acquire a new one decreases. A most striking result is the very low percentage of SF fate. However, this result is coherent with the theoretical predictions of [8] and the experimental results of [15], and probably results from the fact that SF provides no fitness advantage (since the extant function is the same as the ancestral one) but requires a transitory loss of fitness (when both copies have not yet diverged). Notably, the proportion of SF consistently increases with w_{max}. This may be explained by the fact that a higher pleiotropy level allows for alternative adaptive pathways (by adapting either genes with a high/low pleiotropy) which can compensate each others. The situation is slightly more favorable for C, especially when the environment has changed, in agreement with findings in [15]. This is probably due to the fact that the new environments may require dosage adaptation for genes function (see Fig. 3 for an example such effect). In that case, duplicating a gene enables a rapid adaptation. A similar reasoning applies to SP, which has low frequency. This confirms that the conservation of the original function in both copies is rather unlikely, the general trend being $SF < C, SP < DN, N$, sorted by increasing level of function acquisition.

5   Discussion

In this paper, we proposed a methodology to formally classify the fate of gene duplicates depending on the functions of the extant paralogs and of the ancestral gene. The objective isto provide the community with clear definitions as well as a mathematical toolbox to discriminate the different fates. Indeed, in the absence of such a toolbox it is almost impossible to compare experimental and/or theoretical studies limiting the possibility of developing a global understanding of gene duplication, even though this mechanism is considered central in molecular evolution. Our framework has been extensively tested on simulated data provided by aevol, an independently designed platform. Our tests confirmed several tendencies reported in the literature [15,48], showing the relevance of our classification. Further work will permit to study a broader set of parameters and environments to confirm these trends. Incidentally, our results also confirm the interest of using aevol as benchmark to test bioinformatics tools.

Our work opens several fields of research, in comparative genomics and phylogeny, in simulation and, of course, in evolutionary biology. Indeed, although the fraction of gene duplicates classified is high (> 0.75 in all situations), it also shows that further work is required to analyze the remaining fates. Also, even though these data are not reported here, we observed a small fraction of "hybrid fates" which deserve a specific study. Finally, as our methodology is based on the analysis of extant paralogs, it cannot account for the whole diversity of pseudogeneization fates. Indeed, in our results P is always lower than 20% (except in constant environment – see Table 4), which is much lower than the 80% observed in the Zebrafish [16]. We conjecture that the difference is due to the way we selected gene duplicates in our study. Extending P class to account for the whole variety of pseudogeneization fates is an exciting direction of research. Finally, we could used real data available in published datasets such as [49] to further test our approach. While aevol simulations enabled testing the continuous version of our framework, other datasets could enable testing the discrete version, e.g. by classifying paralogs annotated with Gene Ontology [50].

In this study, we used aevol to test our framework, showing that it generates data similar to real observations. This motivates us to further study the gene duplications in the simulator. In particular, aevol not only provides the final organisms, but also the past individuals and the exact gene phylogeny, making it possible to know the exact fate of each gene along each branch (including gene loss). We plan to use this information to study how the fate of duplicated genes evolves in time after the founding duplication event. Indeed, it is straightforward that the fate changes in time (being almost necessarily C just after the duplication event). However, while this question is almost impossible to study in vivo, aevol could reveal whether some fates open the path to others. This would enable studying whether old duplications (or successive duplications) are more likely to be associated to (double-)neofunctionalization or the evolution of the pseudogeneisation probability depending on the time elapsed after a duplication. Such information could be used to predict the evolution of a specific gene branch following recent duplications. The model also enables "*in silico* genetic engineering". We plan to construct a series of mutants in which genes are manually duplicated and let evolve. This will open the route to a systematic study of gene duplication in the model. Another interesting line of investigation could be to understand the impact of regulation on the fates frequency and especially on subfunctionalization. While regulation was not included in the current version of aevol, an extension is under development to account for the evolution of transcription factors that will allow addressing this question.

Finally, it would also be interesting to study how specific biological duplication mechanisms, for instance unequal crossing over, tandem duplication or retrotransposition [51], are associated with fates. Such investigations would probably require to analyse not only gene functions but also gene genealogies. Combining our framework to tools such as PAINT [49] or PANTHER [52], that predict the functions of ancestral genes given a gene phylogeny and the functions of the extant genes, would enable us to analyse real data. We leave this for future work.

References

1. Carvalho, C.M., Zhang, F., Lupski, J.R.: Genomic disorders: A window into human gene and genome evolution. Proc. Natl. Acad. Sci. **107**, 1765–1771 (2010)
2. Kuzmin, E., Taylor, J.S., Boone, C.: Retention of duplicated genes in evolution. Trends Genet. (2021)
3. Vosseberg, J., et al.: Timing the origin of eukaryotic cellular complexity with ancient duplications. Nat. Ecol. Evol. **5**, 92–100 (2021)
4. Demuth, J.P., Hahn, M.W.: The life and death of gene families. Bioessays **31**, 29–39 (2009)
5. Ohno, S.: Evolution by Gene Duplication. Springer Science & Business Media (2013)
6. Ohno, S.: Gene duplication and the uniqueness of vertebrate genomes circa 1970–1999. Semin. Cell Develop. Biol. **10**, 517–522 (1999)
7. Ascencio, D., et al.: Expression attenuation as a mechanism of robustness against gene duplication. Proc. Natl. Acad. Sci. **118**, e2014345118 (2021)
8. Lynch, M., Force, A.: The probability of duplicate gene preservation by subfunctionalization. Genetics **154**, 459–473 (2000)
9. Lynch, M., O'Hely, M., Walsh, B., Force, A.: The probability of preservation of a newly arisen gene duplicate. Genetics **159**, 1789–1804 (2001)
10. Walsh, B.: Origin and Evolution of New Gene Functions, pp. 279–294. Springer (2003)
11. Stark, T.L., Liberles, D.A., Holland, B.R., O'Reilly, M.M.: Analysis of a mechanistic Markov model for gene duplicates evolving under subfunctionalization. BMC Evolution. Biol. **17**, 1–16 (2017)
12. Diao, J., Stark, T.L., Liberles, D.A., O'Reilly, M.M., Holland, B.R.: Level-dependent QBD models for the evolution of a family of gene duplicates. Stochast. Models **36**, 285–311 (2020)
13. Lafond, M., Meghdari Miardan, M., Sanko, D.: Accurate prediction of orthologs in the presence of divergence after duplication. Bioinformatics **34**, i366–i375 (2018)
14. Darby, C.A., Stolzer, M., Ropp, P.J., Barker, D., Durand, D.: Xenolog classification. Bioinformatics **33**, 640–649 (2017)
15. Assis, R., Bachtrog, D.: Neofunctionalization of young duplicate genes in Drosophila. Proc. Natl. Acad. Sci. **110**, 17409–17414 (2013)
16. Otto, S.P., Yong, P.: The evolution of gene duplicates. Adv. Genet. **46**, 451–483 (2002)
17. He, X., Zhang, J.: Rapid subfunctionalization accompanied by prolonged and substantial neofunctionalization in duplicate gene evolution. Genetics **169**, 1157–1164 (2005)
18. Jaillon, O., et al.: Genome duplication in the teleost fish Tetraodon nigroviridis reveals the early vertebrate proto-karyotype. Nature **431**, 946–957 (2004)
19. Brunet, F.G., et al.: Gene loss and evolutionary rates following wholegenome duplication in teleost fishes. Molecul. Biol. Evol. **23**, 1808–1816 (2006)
20. Woods, I.G., et al.: The zebrafish gene map defines ancestral vertebrate chromosomes. Genome Res. **15**, 1307–1314 (2005)

21. Gu, Z., Rifkin, S.A., White, K.P., Li, W.-H.: Duplicate genes increase gene expression diversity within and between species. Nat. Genet. **36**, 577–579 (2004)
22. Huminiecki, L., Wolfe, K.H.: Divergence of spatial gene expression profiles following species-specific gene duplications in human and mouse. Genome Res. **14**, 1870–1879 (2004)
23. Gu, X., Zhang, Z., Huang, W.: Rapid evolution of expression and regulatory divergences after yeast gene duplication. Proc. Natl. Acad. Sci. **102**, 707–712 (2005)
24. Conant, G.C., Wolfe, K.H.: Turning a hobby into a job: How duplicated genes find new functions. Nat. Rev. Genet. **9**, 938–950 (2008)
25. Nguyen Ba, A.N., et al.: Detecting functional divergence after gene duplication through evolutionary changes in posttranslational regulatory sequences. PLoS Comput. Biol. **10**, e1003977 (2014)
26. Hindré, T., Knibbe, C., Beslon, G., Schneider, D.: New insights into bacterial adaptation through in vivo and in silico experimental evolution. Nat. Rev. Microbiol. **10**, 352–365 (2012)
27. Batut, B., Parsons, D.P., Fischer, S., Beslon, G., Knibbe, C.: In silico experimental evolution: A tool to test evolutionary scenarios. BMC Bioinformatics **14**, 1–11 (2013)
28. Knibbe, C.: Structuration des génomes par sélection indirecte de la vari-abilité mutationnelle: une approche de modélisation et de simulation Ph.D. thesis. INSA de Lyon (2006)
29. Knibbe, C., Coulon, A., Mazet, O., Fayard, J.-M., Beslon, G.: A long-term evolutionary pressure on the amount of noncoding DNA. Molecul. Biol. Evol. **24**, 2344–2353 (2007)
30. Rutten, J.P., Hogeweg, P., Beslon, G.: Adapting the engine to the fuel: Mutator populations can reduce the mutational load by reorganizing their genome structure. BMC Evolution. Biol. **19**, 1–17 (2019)
31. Frénoy, A., Taddei, F., Misevic, D.: Genetic architecture promotes the evolution and maintenance of cooperation. PLoS Comput. Biol. **9**, e1003339 (2013)
32. Liard, V., Parsons, D.P., Rouzaud-Cornabas, J., Beslon, G.: The complexity ratchet: Stronger than selection, stronger than evolvability, weaker than robustness. Artificial Life **26**, 38–57 (2020)
33. Knibbe, C.: What happened to my genes? Insights on gene family dynamics from digital genetics experiments. In: ALIFE 14: The Fourteenth International Conference on the Synthesis and Simulation of Living Systems, pp. 33–40 (2014)
34. Zhang, J.: Evolution by gene duplication: An update. Trends Ecol. Evol. **18**, 292–298 (2003)
35. Hahn, M.W.: Distinguishing among evolutionary models for the maintenance of gene duplicates. J. Heredity **100**, 605–617 (2009)
36. Birchler, J.A., Yang, H.: The multiple fates of gene duplications: Deletion, hypofunctionalization, subfunctionalization, neofunctionalization, dosage balance constraints, and neutral variation. The Plant Cell (2022)
37. Thompson, A., Zakon, H.H., Kirkpatrick, M.: Compensatory drift and the evolutionary dynamics of dosage-sensitive duplicate genes. Genetics **202**, 765–774 (2016)
38. Force, A., et al.: Preservation of duplicate genes by complementary, degenerative mutations. Genetics **151**, 1531–1545 (1999)
39. Lynch, M., Conery, J.S.: The evolutionary fate and consequences of duplicate genes. Science **290**, 1151–1155 (2000)
40. Panchy, N., Lehti-Shiu, M., Shiu, S.-H.: Evolution of gene duplication in plants. Plant Physiol. **171**, 2294–2316 (2016)

41. Conrad, B., Antonarakis, S.E.: Gene duplication: A drive for phenotypic diversity and cause of human disease. Annu. Rev. Genom. Hum. Genet. **8**, 17–35 (2007)
42. Veitia, R.A.: Gene duplicates: Agents of robustness or fragility? Trends Genet. **33**, 377–379 (2017)
43. Chauve, C., El-Mabrouk, N.: New perspectives on gene family evolution: Losses in reconciliation and a link with supertrees in Research. In: Proceedings of the Computational Molecular Biology: 13th Annual International Conference, RECOMB 2009, Tucson, 18–21 May 2009, vol. 13, pp. 46–58 (2009)
44. Jacox, E., Chauve, C., Szöllösi, G.J., Ponty, Y., Scornavacca, C.: ecceTERA: Comprehensive gene tree-species tree reconciliation using parsimony. Bioinformatics **32**, 2056–2058 (2016)
45. http://www.aevol.fr and https://gitlab.inria.fr/aevol/aevol
46. Knibbe, C., Mazet, O., Chaudier, F., Fayard, J.-M., Beslon, G.: Evolutionary coupling between the deleteriousness of gene mutations and the amount of non-coding sequences. J. Theoret. Biol. **244**, 621–630 (2007)
47. Biller, P., Knibbe, C., Beslon, G., Tannier, E.: Comparative genomics on artificial life. In: Proceedings of the Pursuit of the Universal: 12th Conference on Computability in Europe, CIE 2016, Paris, France, June 27–July 1 2016, vol. 12, pp. 35–44 (2016)
48. Guillaume, F., Otto, S.P.: Gene functional trade-o s and the evolution of pleiotropy. Genetics **192**, 1389–1409 (2012)
49. Gaudet, P., Livstone, M.S., Lewis, S.E., Thomas, P.D.: Phylogeneticbased propagation of functional annotations within the Gene Ontology consortium. Briefings Bioinform. **12**, 449–462 (2011)
50. Zhao, Y., et al.: A literature review of gene function prediction by modeling gene ontology. Front. Genet. **11**, 400 (2020)
51. Reams, A.B., Kofoid, E., Kugelberg, E., Roth, J.R.: Multiple pathways of duplication formation with and without recombination (RecA) in Salmonella enterica. Genetics **192**, 397–415 (2012)
52. Mi, H., et al.: PANTHER version 11: Expanded annotation data from Gene Ontology and Reactome pathways, and data analysis tool enhancements. Nucl. Acids Res. **45**, D183–D189 (2017)

Inferring Clusters of Orthologous and Paralogous Transcripts

Wend Yam Donald Davy Ouedraogo and Aida Ouangraoua[✉]

Université de Sherbrooke, Sherbrooke, QC J1K2R1, Canada
{wend.yam.donald.davy.ouedraogo,aida.ouangraoua}@usherbrooke.ca

Abstract. The alternative processing of eukaryote genes allows producing multiple distinct transcripts from a single gene, thereby contributing to the transcriptome diversity. Recent studies suggest that more than 90% of human genes are concerned, and the transcripts resulting from alternative processing are highly conserved between orthologous genes.

In this paper, we first present a model to define orthology and paralogy relationships at the transcriptome level, then we present an algorithm to infer clusters of orthologous and paralogous transcripts. Gene-level homology relationships are used to define different types of homology relationships between transcripts and a Reciprocal Best Hits approach is used to infer clusters of isoorthologous and recent paralogous transcripts.

We applied the method to transcripts of gene families from the Ensembl-Compara database. The results are agreeing with those from previous studies comparing orthologous gene transcripts. The results also provide evidence that searching for conserved transcripts beyond orthologous genes will likely yield valuable information. The results obtained on the Ensembl-Compara gene families are available at https://github.com/UdeS-CoBIUS/TranscriptOrthology. Supplementary material can be found at https://doi.org/10.5281/zenodo.7750949.

Keywords: Transcriptome · Orthology and paralogy · Isoorthology · Evolution

1 Introduction

Alternative splicing is one of the most important mechanisms whose extent was revealed in the post-genomic era [8]. It allows distinct transcripts to be produced from the same gene. Over the past decade, the number of alternatively spliced genes and alternative transcripts annotated in eukaryote organisms has increased dramatically [21]. It has now been established that alternative splicing was likely a feature of the eukaryotes' common ancestor.

Over the past decade, several methods have been developed to study the conservation of sets of alternative transcripts annotated in orthologous genes. They identify splicing orthologous transcripts between genes, defined as alternative transcripts of orthologous genes composed of orthologous exons [3,7,9,16,20]. Other studies have proposed various models of transcript evolution with associated algorithms to reconstruct transcript phylogenies using parsimony-based tree

K. Jahn and T. Vinař (Eds.): RECOMB-CG 2023, LNBI 13883, pp. 19–34, 2023.
https://doi.org/10.1007/978-3-031-36911-7_2

search methods [1,5,6] or supertree methods [11,12]. However, several questions about the evolution of sets of alternative transcripts in a gene family remain open [10]. For example, are alternative transcripts more conserved between orthologous genes than paralogous genes? How do new alternative transcripts arise during evolution? Is an alternative transcript preserved between multiple homologous genes and species? Where was it gained or lost in the evolution? Moreover, beyond identifying orthologous transcripts between orthologous genes, no method exists to compare all transcripts of a gene family to provide measures of similarity between the transcripts of all the genes. Furthermore, no frameworks, such as those developed for classifying gene homology types, currently exist for classifying transcript homology types. Beyond allowing a better understanding of alternative transcripts evolution, the prediction of orthologous and paralogous transcripts has other important potential applications. It can be useful for gene orthology inference and gene tree correction, as well as gene function prediction.

In this paper, we present a model to define orthology and paralogy relations between transcripts of a gene family. This model is mainly inspired by the reconciliation model between gene trees and species trees which allows defining orthologs, paralogs and isoorthologs at the gene-level. Isoorthologous genes are the least divergent orthologs that have retained the function of their lowest common ancestor [15,19]. We present an algorithm associated to our model to infer groups of isoorthologous and paralogous transcripts. The algorithm uses transcript pairwise similarity scores as conservation measures to identify pairs of recent paralogs and isoorthologs through a Reciprocal Best Hit (RBH) approach. These relations are then used to infer ortholog groups in which the pairs of transcripts are isoorthologs, recent paralogs or related through a path of isoorthology and recent paralogy relations. The paper is organized as follows. Section 2 provides the definitions and notations required for the remaining of the paper. Section 3 describes our graph-based algorithm to infer ortholog groups. Section 4 contains the results of the application on gene families and sets of transcripts from the Ensembl-Compara database [21].

2   Preliminaries: Phylogenetic Trees, Reconciliation, Orthology, Paralogy

$\mathbb{S}$ denotes a set of species. $\mathbb{G}$ denotes a set of homologous genes from a gene family. $\mathbb{T}$ denotes a set of transcripts descending from the same ancestral transcript. The three sets are related by two functions $s : \mathbb{G} \rightarrow \mathbb{S}$ that maps each gene to its corresponding species, and $g : \mathbb{T} \rightarrow \mathbb{G}$ that maps each transcript to its corresponding gene such that $\{g(\mathbf{t}) : \mathbf{t} \in \mathbb{T}\} = \mathbb{G}$ and $\{s(\mathbf{g}) : \mathbf{g} \in \mathbb{G}\} = \mathbb{S}$. The induced set function g^{-1} associates each gene to its set of corresponding transcripts.

All trees are considered rooted and binary. Given a tree P, $v(P)$ the set of nodes of P, and $l(P)$ its leafset. Given a node x of P, $P[x]$ denotes the subtree of P rooted in x. A node x is an ancestor of a node y if y is a node of $P[x]$. If x is an internal node, x_l and x_r denote its two children. Given a subset L' of $l(P)$, $lca_P(L')$ denotes the lowest common ancestor (LCA) in P of L', defined as the ancestor common to all the nodes in L' that is the most distant from the root.

A tree on a set Σ is a tree whose leafset is Σ. S denotes a species tree on $\mathbb{S}$ whose internal nodes represent a partially ordered set of speciation events that have led to $\mathbb{S}$. G denotes a gene tree on $\mathbb{G}$ whose internal nodes represent speciation and gene duplication events that have led to $\mathbb{G}$. T denotes a transcript tree on $\mathbb{T}$ whose internal nodes represent speciation, gene duplication, and transcript creation events that have led to $\mathbb{T}$. We extend the mapping functions s from $v(G)$ to $v(S)$, and g from $v(T)$ to $v(G)$ as follows: for any node g in $v(G)$, $s(\mathsf{g}) = lca_S(\{s(\mathsf{g}') : \mathsf{g}' \in l(G[\mathsf{g}])\})$, and for any node t in $v(T)$, $g(\mathsf{t}) = lca_G(\{g(\mathsf{t}') : \mathsf{t}' \in l(T[\mathsf{t}])\})$.

In the Duplication-Loss (DL) model of gene evolution, genes undergo speciation events when their corresponding species do, but also duplication events when a new gene copy is created and loss events when a gene is lost in a species. Likewise, in the Creation-Loss (CL) model of transcript evolution, transcripts undergo speciation and duplication events when their corresponding genes do, but also creation events when a new transcript is created and loss events when a transcript is lost in a gene. The labeling of the internal nodes of a gene tree as speciation or gene duplication events is obtained through the reconciliation with a species tree. The labeling of the internal nodes of a transcript tree is obtained through the reconciliation with a gene tree. Several definitions of reconciliation between a gene tree and a species tree exist. Here, we consider the LCA-reconciliation based on the LCA mapping.

Definition 1 (LCA-reconciliation at the gene and transcript levels).
The LCA-reconciliation of G and S is a function $rec_G : v(G) - \mathbb{G} \to \{Spe, Dup\}$ that labels any internal node g of G as a duplication (Dup) if $s(\mathsf{g}) = s(\mathsf{g}_l)$ or $s(\mathsf{g}) = s(\mathsf{g}_r)$, and as a speciation (Spe) otherwise.
Similarly, the LCA-reconciliation of T and G is a function $rec_T : v(T) - \mathbb{T} \to \{Spec, Dup, Cre\}$ that labels any internal node t of T as a creation (Cre) if $g(\mathsf{t}) = g(\mathsf{t}_l)$ or $g(\mathsf{t}) = g(\mathsf{t}_r)$, otherwise as a duplication (Dup) if $rec_G(g(\mathsf{t})) = Dup$, and as a speciation (Spe) if $rec_G(g(\mathsf{t})) = Spe$.

Figure 1 shows an illustration for Definition 1. rec_G provides a reconciliation between G and S that minimizes the number of gene duplications and losses [4]. rec_T also provides a reconciliation between T and G that minimizes the number of transcript creations and losses [12].

Orthology and paralogy are relations defined over pairs of homologous genes based on the gene tree - species tree reconciliation. Orthology is the relation between two genes for which any two distinct ancestors taken at a point of time always appear in distinct ancestral species (i.e. apart, orthogonal). Paralogy is the relation between two genes having two distinct ancestors at a point of time that are in the same ancestral species (i.e. beside, parallel). Based on the co-occurrence of transcript ancestors in the same ancestral genes, we can extend orthology and paralogy definitions over pairs of homologous transcripts.

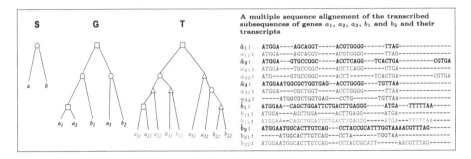

Fig. 1. A species tree S on $\mathbb{S} = \{a, b\}$, a gene tree G on $\mathbb{G} = \{a_1, a_2, a_3, b_1, b_2\}$, and a transcript tree T on $\mathbb{T} = \{a_{11}, a_{21}, a_{21}, a_{31}, a_{32}, b_{11}, b_{12}, b_{21}, b_{22}\}$ such that for any species $x \in \mathbb{S}$, gene $x_i \in \mathbb{G}$, and transcript $x_{ij} \in \mathbb{T}$, $s(x_i) = x$ and $g(x_{ij}) = x_i$. Round nodes represent speciations, square nodes gene duplications, and triangle nodes transcript creations in the LCA-reconciliation of G and S, and the LCA-reconciliation of T and G. Divergence edges after creation nodes are represented as dashed lines. The isoortholog groups of $\mathbb{T}$ are displayed using different colors.

Definition 2 (Orthology, paralogy at the gene and transcript levels).
Two distinct genes g_1 and g_2 of $\mathbb{G}$ are:

– *orthologs if their LCA in G is a speciation, i.e. $rec_G(lca_G(\{g_1, g_2\})) = Spe$;*
– *recent paralogs if $rec_G(lca_G(\{g_1, g_2\})) = Dup$ and $s(lca_G(\{g_1, g_2\}) = s(g_1)$*
– *ancient paralogs otherwise.*

Likewise, two distinct transcripts t_1 and t_2 of $\mathbb{T}$ are [12]:

– *ortho-orthologs if their LCA in T is a speciation,*
  i.e $rec_T(lca_T(\{t_1, t_2\})) = Spe$;
– *para-orthologs if $rec_T(lca_T(\{t_1, t_2\})) = Dup$;*
– *recent paralogs if $rec_T(lca_T(\{t_1, t_2\})) = Cre$ and $g(lca_T(\{t_1, t_2\})) = g(t_1)$;*
– *ancient paralogs otherwise.*

For example in Fig. 1, a_{11} and b_{12} are ortho-orthologs, a_{11} and b_{21} are para-orthologs, b_{21} and b_{22} are recent paralogs, and a_{11} and a_{22} are ancient paralogs. Notice that if all gene pairs are orthologs then all pairs of orthologous transcripts are ortho-orthologs.

Lemma 1 (Link between homology relationships at the gene and transcript levels). *If two transcripts t_1 and t_2 are ortho-orthologs then the genes $g(t_1)$ and $g(t_2)$ are orthologs, and if t_1 and t_2 are para-orthologs then $g(t_1)$ and $g(t_2)$ are paralogs. If t_1 and t_2 are recent paralogs then $g(t_1) = g(t_2)$. None of the converse statements are true.*

Proof. Let $\mathsf{t} = lca_T(\{t_1, t_2\})$. If t_1 and t_2 are ortho-orthologs or para-orthologs then $g(\mathsf{t}) \neq g(t_l)$ and $g(\mathsf{t}) \neq g(t_r)$. Therefore, $lca_G(g(t_1), g(t_2)) = g(\mathsf{t})$, and then $rec_G(lca_G(g(t_1), g(t_2))) = rec_G(g(\mathsf{t}))$ which is a speciation if t_1 and t_2

are ortho-orthologs, and a duplication otherwise. If t_1 and t_2 are recent paralogs then any leaf t' of the subtree $T[t]$ must satisfy $g(t') = g(lca_T(\{t_1, t_2\}))$. Therefore, $g(t_1) = g(t_2) = g(lca_T(\{t_1, t_2\}))$.

Figure 1 shows an example where a_3 and b_2 (resp. a_1 and a_2) are orthologous (resp. paralogous) genes but their transcripts a_{31} and b_{21} (resp. a_{11} and a_{22}) are not ortho-orthologs (resp. para-orthologs). Likewise, $g(a_{21}) = g(a_{22}) = a_2$ but a_{21} and a_{22} are not recent paralogs.

Lemma 2 (Link between recent paralogy and orthology). *(1) If three transcripts t_1, t_2 and t_3 are such that t_1 and t_2 are recent paralogs, and t_1 and t_3 are ortho-orthologs (resp. para-orthologs), then t_2 and t_3 are ortho-orthologs (resp. para-orthologs). (2) If t_1 and t_3 are recent paralogs, and t_2 and t_3 are recent paralogs, then t_1 and t_2 are also recent paralogs.*

Proof. If t_1, t_2 and t_3 are in the configuration of (1) then they must satisfy $lca_T(\{t_2, t_3\}) = lca_T(\{t_1, t_3\})$. Therefore, t_2 and t_3 have the same relation as t_1 and t_3. (2) is trivial.

The key assumption of our method is that after a transcript creation event, the newly created transcript tends to diverge from the original transcript from which it was modified, whereas the original transcript tends to remain conserved, like for the inference of gene ortholog groups using graph-based methods where isoorthologous gene pairs are considered [2,13,15,19]. It should be noted that the conservation/divergence between two transcripts is based on the comparison of their content in exons, and not of their nucleotide sequences. Therefore, given a creation node t in the LCA-reconciliation of T with G, one of its edges descending to its children, say (t, t_l) without loss of generality, corresponds to the original transcript conserved, whereas the other edge (t, t_r) corresponds to the newly created divergent transcript. In this case, we call (t, t_l) a conservation edge, whereas (t, t_r) is called a divergence edge. For example, in Fig. 1, the divergence edges after creation nodes appear in dashed lines. Distinguishing conservation edges and divergence edges after a creation node allows to define a particular type of orthology relation between transcripts.

Definition 3 (Isoorthology at the transcript level). *Two ortho- (resp. para-) orthologous transcripts t_1 and t_2 of $\mathbb{T}$ are ortho- (resp. para-) isoorthologs if there are no divergence edges on the path between t_1 and t_2 in T.*

Notice that the isorthology relation is transitive, which allows $\mathbb{T}$ to be partitioned into ortholog groups.

Definition 4 (Ortholog groups at the transcript level). *An ortholog group $\mathbb{O}$ of $\mathbb{T}$ is a subset of $\mathbb{T}$ such that any two distinct transcripts t_1 and t_2 belonging to $\mathbb{O}$ are isoorthologs (i.e. ortho- or para-isoorthologs), recent paralogs, or there exist two transcripts t'_1 and t'_2 in $\mathbb{O}$ such that $t_1 = t'_1$ or t_1 and t'_1 are recent paralogs, $t_2 = t'_2$ or t_2 and t'_2 are recent paralogs, and t'_1 and t'_2 are isoorthologs.*

For example, in Fig. 1, $\{a_{11}, a_{21}, b_{11}, a_{31}\}$, $\{a_{22}\}$, $\{b_{12}\}$, $\{a_{32}, b_{21}\}$, $\{b_{22}\}$ are the maximum inclusive-wise ortholog groups of $\mathbb{T}$.

3   A Graph-Based Algorithm to Infer Isoorthology and Recent Paralogy Relations Between Transcripts

In this section, we present a graph-based method to infer isoorthology and recent paralogy relations in a set of homologous transcripts $\mathbb{T}$. The method relies on a pairwise similarity measure between transcripts to infer ortholog groups.

3.1   Pairwise Similarity Score Between Transcripts

A gene $\mathbf{g} \in \mathbb{G}$ is a DNA sequence on the alphabet of nucleotides $\Sigma = \{A, C, G, T\}$. A transcript $\mathbf{t}$ of $\mathbf{g}$ (i.e $\mathbf{t} \in g^{-1}(\mathbf{g})$) is a subsequence of $\mathbf{g}$ obtained by concatenating an ordered set of substrings of $\mathbf{g}$ such that each substring is an exon of $\mathbf{g}$ that is present in $\mathbf{t}$. The transcribed subsequence of a gene $\mathbf{g} \in \mathbb{G}$, denoted by $\hat{\mathbf{g}}$, is the subsequence of $\mathbf{g}$ obtained by deleting from $\mathbf{g}$ any nucleotide that is absent from all transcripts of $\mathbf{g}$. $\hat{\mathbb{G}}$ denotes the set of transcribed subsequences of all genes in $\mathbb{G}$. Note that $| \hat{\mathbb{G}} | = | \mathbb{G} |$. Figure 2 shows an illustration.

A multiple sequence alignment $\mathbb{A}$ of all the transcribed subsequences in $\hat{\mathbb{G}}$ and all the transcripts in $\mathbb{T}$ is obtained by first computing a multiple sequence alignment M of the transcribed subsequences in $\hat{\mathbb{G}}$, and then mapping each transcript $\mathbf{t} \in \mathbb{T}$ on its corresponding transcribed subsequence within M to obtain the resulting alignment $\mathbb{A}$.

In the sequel, $\mathbb{A}$ denotes a multiple sequence alignment of the all transcribed subsequences in $\hat{\mathbb{G}}$ and all the transcript sequences in $\mathbb{T}$, represented as a $n \times m$ matrix such that $n = | \mathbb{T} | + | \hat{\mathbb{G}} |$ and m is the number of columns of the alignment.

Following the block-based model used in [16] to represent transcripts, a multiple sequence alignment $\mathbb{A}$ of $\mathbb{T}$ and $\hat{\mathbb{G}}$ is partitioned into a set of non-overlapping blocks of columns as follows:

Definition 5 (Decomposition of multiple sequence alignment). *Let $\mathbb{A}$ be a multiple sequence alignment of $\mathbb{T}$ and $\hat{\mathbb{G}}$. $\mathbb{A}_b$ denotes the binary matrix of same dimension as $\mathbb{A}$ such that each nucleotide A, C, G, or T in $\mathbb{A}$ is replaced by 1 in $\mathbb{A}_b$, and each gap character '-' is replaced by 0. A block of $\mathbb{A}$ is a set of consecutive columns of $\mathbb{A}$ which correspond to a maximum inclusive-wise set of consecutive columns of $\mathbb{A}_b$ which are equal.*

For any block B of $\mathbb{A}$, $\alpha(B)$ denotes a positive number representing the weight of the block B.

For example, in Fig. 2, the alignment is decomposed into 9 blocks.

Lemma 3 (Aligned sequences in blocks). *For any aligned sequence t' in $\mathbb{A}$ and any block B of $\mathbb{A}$, t' contains either only nucleotides, or only gaps in B.*

Proof. Trivial, by definition of blocks.

(A) Two genes g1 and g2 and their transcripts t11, t12, t21, t22

```
g1  :  ***ATGGAATGC************AAGCAGGTCTGG****ACGTGG****GGTGATTGA***
t11:       ATGC            AAGCAGGTCTGG              GGTGA
t12:       ATGGAATGC       AAGCAG          ACGTGG    GGTGATTGA

g2  :  *******ATGA****ATGCCG****GTAACG**************GACTTTGAATAA***
t21:          ATGA    ATGCCG    GTAACG            GACTTTGA
t22:                  ATGCCG                      GACTTTGAATAA
```

(B) The transcribed subsequences ĝ1, ĝ2 of genes g1 and g2

```
ĝ1:  ATGGAATGCAAGCAGGTCTGGACGTGGGGTGATTGA
ĝ2:  ATGAATGCCGGTAACGGACTTTGAATAA
```

(C) A multiple sequence alignement of the transcribed subsequences ĝ1, ĝ2 and transcripts t11, t12, t21, t22, decomposed into 9 blocks

```
ĝ1:  ATGGAATGCAAGCAGGTCTGGACGTGGGG---TGATTGA 1 2 3 4 5 6    8 9
t11: -----ATGCAAGCAGGTCTGG------GG---TGA---- 2 3 4    6    8
t12: ATGGAATGCAAGCAG------ACGTGGGG---TGATTGA 1 2 3    5 6   8 9
ĝ2:  -----ATGAATGCCGGTAACG------GACTTTGAATAA  2 3 4     6 7 8 9
t21: -----ATGAATGCCGGTAACG------GACTTTGA----  2 3 4     6 7 8
t22: ---------ATGCCG------------GACTTTGAATAA     3       6 7 8 9
     =====---=======------======--===---====
      1   2   3    4    5 6 7 8  9
```

Fig. 2. A multiple sequence alignment of the transcribed subsequences of two genes g1 and g2 with their transcripts, decomposed into 9 blocks. Non-coding nucleotides in the gene sequence (i.e. introns, untranscribed and untranslated regions) are represented with the character '*'.

Definition 6 (Block-based representation of transcripts and genes). *Given the ordered set of blocks defined by the partition of* $\mathbb{A}$, *for each transcript* $t \in \mathbb{T}$, $\mathbb{B}(t)$ *denotes the ordered subset of blocks in which the aligned sequence* t' *corresponding to* t *contains nucleotides.*

Likewise, for each gene $g \in \mathbb{G}$, $\mathbb{B}(g)$ *denotes the ordered subset of blocks in which the aligned sequence* g' *corresponding to* $\hat{g}$ *contains nucleotides.*

Lemma 4 (Link between representations of transcripts and genes). *For any gene* $g \in \mathbb{G}$, $\mathbb{B}(g)$ *contains all blocks in which at least one aligned sequence* t' *corresponding to a transcript* t *of* g *contains nucleotides.*

Proof. Let B be a block contained in $\mathbb{B}(g)$. Then, the transcribed subsequence $\hat{g}$ contains a segment of nucleotides in B. Therefore, there exists at least one transcript t of g which contains this segment, and then the corresponding aligned sequence t' contains nucleotides in B. Conversely, any block containing nucleotides from a transcript of g belongs necessarily to $\mathbb{B}(g)$.

We are now ready to give the definition of the transcript similarity measure.

Definition 7 (Pairwise transcript similarity). *Let t_1 and t_2 be two distinct transcripts in $\mathbb{T}$. Consider the sets of blocks shared by t_1 and t_2, $\mathbb{BI}(t_1, t_2) = \mathbb{B}(t_1) \cap \mathbb{B}(t_2)$, and the set of blocks $\mathbb{BU}(t_1, t_2) = \mathbb{B}(t_1) \cup \mathbb{B}(t_2)$ and $\mathbb{BU}_+(t_1, t_2) = \mathbb{BU}(t_1, t_2) \cap (\mathbb{B}(g(t_1)) \cap \mathbb{B}(g(t_2)))$ which contains the blocks of t_1 shared with $g(t_2)$ and the blocks of t_2 shared with $g(t_1)$. The similarity score between t_1 and t_2 equals:*

$$\text{tsm}(t_1, t_2) = \frac{\sum_{B \in \mathbb{BI}(t_1, t_2)} \alpha(B)}{\sum_{B \in \mathbb{BU}(t_1, t_2)} \alpha(B)}$$

The corrected similarity score between t_1 and t_2 equals:

$$\text{tsm}_+(t_1, t_2) = \frac{\sum_{B \in \mathbb{BI}(t_1, t_2)} \alpha(B)}{\sum_{B \in \mathbb{BU}_+}(t_1, t_2)\alpha(B)}$$

If the weights associated to the blocks are unitary, the similarity score between two transcripts t_1 and t_2 is the ratio between the number of blocks shared by the two transcripts and the number of blocks in at least one of the two transcripts. However, for the corrected similarity score, the blocks which are contained in the symmetric difference of $\mathbb{B}(t_1)$ and $\mathbb{B}(t_2)$ are only counted in the denominator if they belong to both $\mathbb{B}(g_1)$ and $\mathbb{B}(g_2)$. This correction allows to account for differences at the transcript-level only, and to avoid those at the gene-level. For instance, in the example provided in Fig. 2, $tsm_+(t11, t21) = |\{2, 3, 4, 6, 8\}| / |\{2, 3, 4, 6, 8\}| = 1$ even if they do not share the block 7. If the weight associated to a block is its length, the similarity score corresponds to the case where each column of the multiple sequence alignment $\mathbb{A}$ is considered as a block.

3.2   Orthology Graph Construction and Ortholog Groups Inference

Using the pairwise similarity scores between transcripts, the method identifies pairs of recent paralogs and putative pairs of isoorthologs through an RBH. The key idea is that between two homologous genes $\mathbf{g}_1$ and $\mathbf{g}_2$, the pairs of isoorthologous transcripts (i.e., para- or ortho-isoorthologs) should be the most conserved. Moreover, two recent paralogous transcripts within a gene $\mathbf{g}_1$ should share more similarities than to any transcript in another gene $\mathbf{g}_2$, because their LCA which is a creation node is lower than the node from which any pair of transcripts from $\mathbf{g}_1$ and $\mathbf{g}_2$ diverged.

Definition 8 (Inferred recent paralogs). *Two distinct transcripts t_1 and t_2 of $\mathbb{T}$ such that $g(t_1) = g(t_2)$ are inferred as recent paralogs if:*

– *$tsm(t_1, t_2) > max\{tsm(t_1, t) : t \in \mathbb{T} - g^{-1}(g(t_2))\}$ and $tsm(t_1, t_2) > max\{tsm(t_2, t) : t \in \mathbb{T} - g^{-1}(g(t_1))\}$;*
– *or there exists a third transcript t_3 of $\mathbb{T}$ such that t_1 and t_3 are recent paralogs and, t_2 and t_3 are also recent paralogs.*

Definition 9 (Putative isoorthologs). *Two transcripts* t_1 *and* t_2 *of* $\mathbb{T}$ *of two distinct genes are inferred as putative isoorthologs if:*
$tsm(t_1, t_2) = max\{tsm(t_1, t) : t \in g^{-1}(g(t_2))\} = max\{tsm(t_2, t) : t \in g^{-1}(g(t_1))\}.$

Using RBH to define putative isoorthologs makes the method more robust to transcript loss or incomplete transcript annotation in some genes. The next step is to define an orthology graph whose set of vertices is $\mathbb{T}$, edges represent inferred recent paralogs or putative isoorthologs, and connected components define ortholog groups.

Definition 10 (Orthology graph). *An orthology graph for* $\mathbb{T}$ *is a graph* $G = (V, E)$ *whose set of vertices* $V = \mathbb{T}$ *and for any two distinct transcripts* t_1 *and* t_2 *of* $\mathbb{T}$:

- *(1) if* $(t_1, t_2) \in E$ *then* t_1 *and* t_2 *are either inferred recent paralogs or putative isoorthologs, and;*
- *(2) if* $g(t_1) = g(t_2)$, *then* t_1 *and* t_2 *belong to the same connected component of* G *if and only if* t_1 *and* t_2 *are recent paralogs.*

Algorithm 1: Construction of the orthology graph

Data: $\mathbb{RP}$, the set of all pairs of inferred recent paralogs
  $\mathbb{PO}$, the ordered set of all pairs of putative isoorthologs
Result: $G = (V, E)$, an orthology graph of $\mathbb{T}$
$V \leftarrow \mathbb{T}$; $E \leftarrow \emptyset$;
foreach $((t_1, t_2) \in \mathbb{RP}$ **do**
  $E \leftarrow E + \{(t_1, t_2)\}$;
end
foreach $(t_1, t_2) \in \mathbb{PO}$ **do**
  $cc_1 \leftarrow cc_{(V,E)}(t_1)$; $cc_2 \leftarrow cc_{(V,E)}(t_2)$;
  if $cc_1 == cc_2$ **then**
    $E \leftarrow E + \{(t_1, t_2)\}$;
  else
    $merge \leftarrow True$;
    foreach $(t'_1, t'_2) \in cc_1 \times cc_2$ **do**
      if $g(t'_1) == g(t'_2)$ *and* $(t'_1, t'_2) \notin E$ **then**
        $merge \leftarrow False$;
      end
    end
    if $merge == True$ **then**
      $E \leftarrow E + \{(t_1, t_2)\}$;
    end
  end
end
$G \leftarrow (V, E)$;
return G;

The objective is to construct an orthology graph for $\mathbb{T}$ that contains a minimum number of connected components. Given a graph (V, E), $CC(V, E)$ denotes the set of all connected components of the graph. We define the function $cc_{(V,E)} : V \rightarrow CC(V, E)$ which associates each vertex $x \in V$ to the connected component to which it belongs. Algorithm 1 uses a progressive heuristic approach to construct an orthology graph for $\mathbb{T}$. It takes as input the set $\mathbb{RP}$ of all the pairs of inferred recent paralogs, and the ordered set $\mathbb{PO}$ of all the pairs of putative isoorthologs ordered by decreasing similarity. It starts with an empty set of edges, then edges corresponding to recent paralogs are added. Next, edges that correspond to putative isoorthologs are considered progressively and added if their addition preserves the property that the graph is an orthology graph.

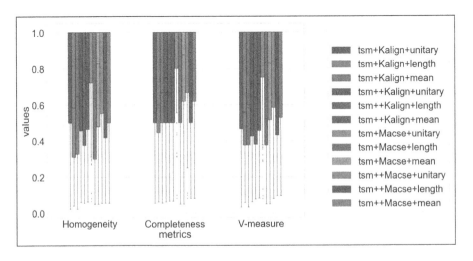

Fig. 3. Homogeneity, Completeness and V-measure scores of our predictions for the 253 triplets of orthologous genes from Guillaudeux et al. [7]

Lemma 5. (Correctness of Algorithm 1). *Given $\mathbb{RP}$ the set of all the pairs of inferred recent paralogs, and $\mathbb{PO}$ the ordered set of all the pairs of putative isoorthologs, Algorithm 1 computes an orthology graph for $\mathbb{T}$.*

Proof. In Algorithm 1, after the first `for` loop, the set of edges contains only recent paralogy edges. Therefore, at this point the graph is an orthology graph. In the remaining of the algorithm, an isoorthology edge is added if and only it preserves the property that the graph is an orthology graph.

Given a multiple sequence alignment $\mathbb{A}$ of dimension $n \times m$ such that $n = | \mathbb{T} | + | \hat{\mathbb{G}} |$ and m is the number of columns, the decomposition of the multiple sequence alignment $\mathbb{A}$ into blocks is computed in $O(n \times m)$ time complexity. The pairwise transcript similarity scores are computed in $O(n^2 \times b)$ time complexity where b is the number of blocks in the decomposition of $\mathbb{A}$ and $b \ll m$. The pairs of inferred recent paralogs and putative isoorthologs are computed in $O(n^2)$ time

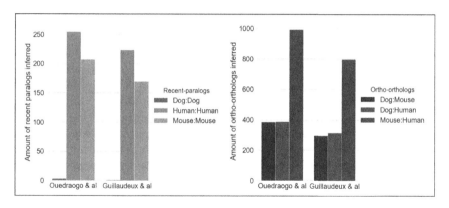

Fig. 4. Comparison of the number of recent paralogs and the number of ortho-isoorthologs inferred by the methods.

complexity. Finally, Algorithm 1 runs in $O(n^2)$ time complexity. Therefore, given $\mathbb{A}$, the whole method runs in $O(n \times m + n^2)$ time complexity.

4   Results and Discussion

4.1   Comparison with Ortholog Groups Predicted in Human, Mouse and Dog One-to-one Orthologous Genes

Since this paper introduces the notion of orthology and paralogy at the transcript level, no previous work exists on the computation of transcript ortholog groups. However, many previous works have studied the problem of identifying conserved transcripts between one-to-one orthologous genes. Most of this work is limited to comparing two species. An exception is the work of Guillaudeux et al. [7] who proposed a formal definition of the splicing structure orthology and an algorithm that was used to predict transcript orthologs in *human, mouse and dog*. A comparison between the approach of Guillaudeux et al. and that of our method is provided in supplementary material. Their dataset is publicly available and includes 253 triplets of one-to-one orthologous genes from 236 gene families in the Ensembl-Compara database. Their method predicted 879 transcript ortholog groups for a total of 1896 transcripts.

We compared the 879 orthologous groups of Guillaudeux et al. to orthologous groups obtained using 12 different settings of our method using: (1) MACSE [17] or Kalign [14] to compute multiple sequence alignments; (2) unitary weights for alignment blocks, or weights corresponding to their lengths, or the mean of the two similarity scores; (3) a corrected or an uncorrected transcript similarity measure. For each version of our method, we used our results as the prediction and the 784 orthologous groups of Guillaudeux et al. as ground truth. We considered 3 performance measures: 1) the homogeneity score which computes the ratio of the pairs of transcripts which are predicted in the same group and are truly in the same group to the pairs of transcripts which are predicted in the same group; 2) the completeness score which computes the ratio of the pairs of transcripts which are predicted in the same group and are truly in the same group to the pairs of transcripts which are truly in the same group; 3) the V-measure which is the harmonic mean between the scores of homogeneity and completeness. Figure 3 provides the details of the score distributions for all 253 gene triplets. The high scores obtained show that our results are agreeing with those of Guillaudeux et al. In particular, the best scores are achieved with the setting that uses MACSE, the unitary-weights for block alignments, and the uncorrected transcript similarity measure. However, for all versions, our method tends to cluster transcripts more (i.e., the homogeneity score is always less than the completeness score).

The detailed comparison of the groups obtained by the Guillaudeux et al. method and the best performing setting of our method shows that 495 clusters of 166 families are exactly the same for the two methods. Our method predicts 1896 ortho-isoorthologs and 466 recent paralogs compared to Guillaudeux et al. who predicts 1,408 ortho-isoorthologs and 395 recent paralogs. Our method also finds all recent paralogs inferred by Guillaudeux et al. Figure 4 shows the number of recent paralogs per species and the number of ortho-isoorthologs per species pair predicted by the two methods.

4.2 Comparison of the Proportions of Ortho-orthologs, Para-orthologs and Recent Paralogs Predicted

Table 1. Description of the dataset of 20 gene families. The total number of genes, the total number of transcripts and the numbers per species are given.

Gene Families ID	# of genes	# of transcripts
ENSGT00390000000715	$6 = 1 + 1 + 1 + 1 + 1 + 1$	$17 = 6 + 4 + 4 + 1 + 1 + 1$
ENSGT00390000003967	$6 = 1 + 1 + 1 + 1 + 1 + 1$	$13 = 3 + 2 + 4 + 1 + 2 + 1$
ENSGT00390000004965	$6 = 1 + 1 + 1 + 1 + 1 + 1$	$8 = 2 + 1 + 2 + 1 + 1 + 1$
ENSGT00390000005532	$6 = 1 + 1 + 1 + 1 + 1 + 1$	$14 = 5 + 2 + 2 + 1 + 2 + 2$
ENSGT00390000008371	$12 = 2 + 2 + 2 + 2 + 2 + 2$	$32 = 11 + 9 + 3 + 3 + 3 + 3$
ENSGT00530000063023	$23 = 4 + 4 + 4 + 4 + 4 + 3$	$65 = 26 + 11 + 12 + 5 + 6 + 5$
ENSGT00530000063187	$17 = 3 + 3 + 3 + 3 + 3 + 2$	$34 = 8 + 6 + 7 + 6 + 4 + 3$
ENSGT00530000063205	$19 = 3 + 3 + 3 + 3 + 4 + 3$	$54 = 19 + 11 + 6 + 3 + 10 + 5$
ENSGT00940000153241	$14 = 2 + 2 + 3 + 3 + 2 + 2$	$23 = 3 + 4 + 7 + 4 + 3 + 2$
ENSGT00940000157909	$6 = 1 + 1 + 1 + 1 + 1 + 1$	$47 = 17 + 13 + 5 + 3 + 4 + 5$
ENSGT00950000182681	$43 = 8 + 7 + 7 + 7 + 7 + 7$	$104 = 26 + 24 + 19 + 8 + 16 + 11$
ENSGT00950000182705	$39 = 7 + 7 + 6 + 6 + 6 + 7$	$158 = 60 + 41 + 14 + 10 + 19 + 14$
ENSGT00950000182727	$30 = 5 + 5 + 5 + 5 + 5 + 5$	$121 = 23 + 15 + 21 + 26 + 23 + 1$
ENSGT00950000182728	$35 = 6 + 6 + 6 + 6 + 6 + 5$	$112 = 31 + 13 + 15 + 30 + 12 + 11$
ENSGT00950000182783	$29 = 5 + 5 + 4 + 5 + 5 + 5$	$97 = 40 + 22 + 9 + 9 + 7 + 10$
ENSGT00950000182875	$37 = 7 + 7 + 5 + 5 + 5 + 8$	$77 = 26 + 14 + 10 + 6 + 10 + 11$
ENSGT00950000182931	$24 = 4 + 4 + 4 + 4 + 4 + 4$	$116 = 57 + 15 + 18 + 9 + 9 + 8$
ENSGT00950000182956	$23 = 4 + 4 + 4 + 4 + 4 + 3$	$109 = 25 + 29 + 21 + 9 + 15 + 10$
ENSGT00950000182978	$24 = 4 + 4 + 4 + 4 + 4 + 4$	$116 = 34 + 20 + 19 + 11 + 18 + 14$
ENSGT00950000183192	$19 = 4 + 3 + 3 + 3 + 3 + 3$	$85 = 50 + 9 + 7 + 5 + 8 + 6$

Table 2. The number of ortholog groups predicted, the ratio of isoortholog pairs to all transcript pairs, the ratio of recent paralogs pairs to all transcript pairs, the ratio of ortho-isoorthologs pairs to all isoortholog pairs divided by the ratio of gene ortholog pairs to all gene pairs, and the ratio of para-isoortholog pairs to all isoortholog pairs divided by the ratio of gene paralog pairs to all gene pairs.

Gene Family ID	Number of ortholog groups	Ratio of iso-orthologs	Ratio of recent paralogs	Normalized ratio of ortho-isoorthologs	Normalized ratio of para-isoorthologs
ENSGT00390000000715	4	0.4118	0.0882	0.7738	1.3393
ENSGT00390000003967	5	0.4103	0.0513	1.0	0.0
ENSGT00390000004965	2	0.7143	0.0357	1.0	0.0
ENSGT00390000005532	6	0.1758	0.0	0.75	1.125
ENSGT00390000008371	6	0.5423	0.0665	0.9574	1.145
ENSGT00530000063023	12	0.5029	0.0207	0.8293	1.034
ENSGT00530000063187	13	0.303	0.0071	0.9	1.0214
ENSGT00530000063205	5	0.5877	0.0307	0.6227	1.0389
ENSGT00940000153241	6	0.5257	0.0158	0.9431	1.039
ENSGT00940000157909	19	0.0722	0.0213	1.0	0.0
ENSGT00950000182681	11	0.6701	0.0149	1.1272	0.9925
ENSGT00950000182705	12	0.7817	0.0373	0.7965	1.011
ENSGT00950000182727	14	0.6354	0.0216	0.9795	1.0338
ENSGT00950000182728	11	0.772	0.0246	1.0074	0.9992
ENSGT00950000182783	9	0.561	0.0217	0.8978	1.017
ENSGT00950000182875	9	0.6548	0.0154	1.1482	0.9143
ENSGT00950000182931	33	0.2937	0.0127	0.98	1.0039
ENSGT00950000182956	24	0.4392	0.0221	0.9481	1.0557
ENSGT00950000182978	13	0.7217	0.036	0.9302	1.0962
ENSGT00950000183192	22	0.4031	0.0459	0.902	1.1559

We randomly selected 20 gene families composed of genes from 6 species: *human, mouse, dog, dingo, cow, chicken*. Table 1 describes the dataset. We used the best performing setting of our method using MACSE to compute the multiple sequence alignment, the unitary-weights similarity scores, and the uncorrected transcript similarity measure. From a total of 1,402 transcripts, we identified 236 ortholog groups. Table 2 shows the ratio of isoorthologues and recent paralogs found between transcript pairs. In this experiment, we could identify para-isoorthologs because the dataset contains paralogous genes. Table 2 also shows the ratio of ortho-isoorthology relations normalized by the ratio of gene orthology relations and the ratio of para-isoorthology relations normalized by the ratio of gene paralogy relations. When the normalized ratio of ortho-isoorthology (resp. para-isoorthology) is greater than 1, it means more relations were predicted than expected given the ratio of gene orthology (resp. paralogy) relations to all gene pairs. In 14 out of 20 families, the normalized para-isoorthology ratio is greater than 1, and greater than the normalized ortho-isoorthology ratio. Therefore, it seems that isoorthologous transcripts tend to be more present between paralogous genes than between orthologous genes. This is consistent with previous studies that have found evidence against the ortholog conjecture in the context of gene function prediction by transferring annotations between homologous genes [18]. The ortholog conjecture proposes that orthologous genes should be preferred when making such predictions because they evolve functions more slowly than paralogous genes. Our results support that orthologs and paralogs should be considered to provide higher prediction accuracy. However, this interpretation should also be taken with caution as there may be errors in the orthology/paralogy relationships between genes.

5   Conclusion

The ability to classify homology relationships between homologous transcripts in gene families is a fundamental step to study the evolution of alternative transcripts. Identifying groups of transcripts that are isoorthologs helps to identify and study the function of transcripts conserved across multiple genes and species. It also provides a framework for gene tree correction, and to study the impact that the evolution of transcripts through creation and loss events has on the evolution of gene functions.

In this work, we revisit the notions of orthology, paralogy and isoorthology at the transcript-level with an associated algorithm to infer ortholog groups composed of transcripts that are isoorthologs, recent paralogs, or related through a path of isoorthology and recent paralogy relations. The method provides results that are consistent with results from methods that identify conserved transcripts between orthologous genes. The results also show that the proportion of conserved transcripts between paralogous genes is not negligible compared to conserved transcripts between orthologous genes, thus justifying the relevance of further studying the relationship between the evolution of transcripts and genes in entire gene families.

The method offers many possibilities for improvement and extension. First, the quality of inference strongly depends on the quality of the multiple sequence alignment and the definition of the transcript similarity measure. Future works will explore different ways to compare transcripts. We will also explore alternative algorithms for computing ortholog groups given putative pairs of isoorthologs. Finally, the method can be extended to infer all pairwise relations between homologous transcripts of a gene family by using the ortholog groups to compute complete transcript trees which can then be reconciled with the gene tree. The predictions can then be evaluated using the annotated functions of the proteins corresponding to transcripts.

References

1. Ait-Hamlat, A., Zea, D.J., Labeeuw, A., Polit, L., Richard, H., Laine, E.: Transcripts' evolutionary history and structural dynamics give mechanistic insights into the functional diversity of the jnk family. J. Molecular Biol. **432**(7), 2121–2140 (2020)

2. Altenhoff, A.M., Gil, M., Gonnet, G.H., Dessimoz, C.: Inferring hierarchical orthologous groups from orthologous gene pairs. PLoS ONE **8**(1), e53786 (2013)
3. Blanquart, S., Varré, J.-S., Guertin, P., Perrin, A., Bergeron, A., Swenson, K.M.: Assisted transcriptome reconstruction and splicing orthology. BMC Genom. **17**(10), 157 (2016)
4. Chauve, C., El-Mabrouk, N.: New perspectives on gene family evolution: losses in reconciliation and a link with supertrees. In: Batzoglou, S. (ed.) RECOMB 2009. LNCS, vol. 5541, pp. 46–58. Springer, Heidelberg (2009). https://doi.org/10.1007/978-3-642-02008-7_4
5. Christinat, Y., Moret, B.M.E.: Inferring transcript phylogenies. BMC Bioinform. **13**(9), S1 (2012)
6. Christinat, Y., Moret, B.M.E.: A transcript perspective on evolution. IEEE/ACM Trans. Comput. Biol. Bioinf. **10**(6), 1403–1411 (2013)
7. Guillaudeux, N., Belleannée, C., Blanquart, S.: Identifying genes with conserved splicing structure and orthologous isoforms in human, mouse and dog. BMC Genom. **23**(1), 1–14 (2022)
8. Harrow, J., et al.: Gencode: the reference human genome annotation for the encode project. Genome Res. **22**(9), 1760–1774 (2012)
9. Jammali, S., Aguilar, J.-D., Kuitche, E., Ouangraoua, A.: Splicedfamalign: Cds-to-gene spliced alignment and identification of transcript orthology groups. BMC Bioinform. **20**(3), 133 (2019)
10. Keren, H., Lev-Maor, G., Ast, G.: Alternative splicing and evolution: diversification, exon definition and function. Nat. Rev. Genet. **11**(5), 345–355 (2010)
11. Kuitche, E., Jammali, S., Ouangraoua, A.: Simspliceevol: alternative splicing-aware simulation of biological sequence evolution. BMC Bioinform. **20**(20), 640 (2019)
12. Kuitche, E., Lafond, M., Ouangraoua, A.: Reconstructing protein and gene phylogenies using reconciliation and soft-clustering. J. Bioinform. Comput. Biol. **15**(06), 1740007 (2017)
13. Lafond, M., Miardan, M.M., Sankoff, D.: Accurate prediction of orthologs in the presence of divergence after duplication. Bioinformatics **34**(13), i366–i375 (2018)
14. Lassmann, T., Sonnhammer, E.L.L.: Kalign-an accurate and fast multiple sequence alignment algorithm. BMC Bioinform. **6**(1), 1–9 (2005)
15. Li, L., Stoeckert, C.J., Roos, D.S.: Orthomcl: identification of ortholog groups for eukaryotic genomes. Genome Res. **13**(9), 2178–2189 (2003)
16. Ouangraoua, A., Swenson, K.M., Bergeron, A.: On the comparison of sets of alternative transcripts. In: Bleris, L., Măndoiu, I., Schwartz, R., Wang, J. (eds.) ISBRA 2012. LNCS, vol. 7292, pp. 201–212. Springer, Heidelberg (2012). https://doi.org/10.1007/978-3-642-30191-9_19
17. Ranwez, V., Douzery, E.J.P., Cambon, C., Chantret, N., Delsuc, F.: Macse v2: toolkit for the alignment of coding sequences accounting for frameshifts and stop codons. Molecular Biol. Evolut. **35**(10), 2582–2584 (2018)
18. Stamboulian, M., Guerrero, R.F., Hahn, M.W., Radivojac, P.: The ortholog conjecture revisited: the value of orthologs and paralogs in function prediction. Bioinformatics **36**(Supplement_1), i219–i226 (2020)
19. Swenson, K.M., El-Mabrouk, N.: Gene trees and species trees: irreconcilable differences. BMC Bioinform. **13**, 1–9. BioMed Central (2012)
20. Zambelli, F., Pavesi, G., Gissi, C., Horner, D.S., Pesole, G.: Assessment of orthologous splicing isoforms in human and mouse orthologous genes. BMC Genom. **11**(1), 1 (2010)
21. Zerbino, D.R., et al.: Ensembl 2018. Nucleic Acids Res. **46**(D1), D754–D761 (2018)

On the Class of Double Distance Problems

Marília D. V. Braga, Leonie R. Brockmann, Katharina Klerx,
and Jens Stoye[(✉)]

Faculty of Technology and CeBiTec, Bielefeld University, Bielefeld, Germany
jens.stoye@uni-bielefeld.de

Abstract. This work is about comparing two genomes $\mathbb{S}$ and $\mathbb{D}$ over the
same set of gene families, such that $\mathbb{S}$ is *singular* (has one gene per family), while $\mathbb{D}$ is *duplicated* (has two genes per family). Considering some
underlying model, that can be simply the minimization of *breakpoints*
or finding the smallest sequence of mutations mimicked by the *double-
cut-and-join* (DCJ) operation, the *double distance* of $\mathbb{S}$ and $\mathbb{D}$ aims to
find the smallest distance between $\mathbb{D}$ and any element from the set $2\mathbb{S}$,
that contains all possible genome configurations obtained by *doubling*
the chromosomes of $\mathbb{S}$. The *breakpoint* double distance of $\mathbb{S}$ and $\mathbb{D}$ can be
greedily solved in linear time. In contrast, the DCJ double distance of $\mathbb{S}$
and $\mathbb{D}$ was proven to be NP-hard. The complexity space between these
two extremes can be explored with the help of an intermediate family of
problems, the σ_k distances, defined for each $k \in \{2, 4, 6, ..., \infty\}$, in a way
such that the σ_2 distance equals the breakpoint distance and the σ_∞ distance equals the DCJ distance. With this class of problems it is possible
to investigate the complexity of the double distance under the σ_k distance, increasing the value k in an attempt to identify the smallest value
for which the double distance becomes NP-hard, indicating the point in
which the complexity changes. In our more recent work we have proven
that, for the particular case in which genomes can only be composed of
circular chromosomes, both σ_4 and σ_6 double distances can be solved in
linear time. Here we present a non-trivial extension of these results to
genomes including linear chromosomes.

1 Introduction

A *genome* is a multiset of *chromosomes* and each chromosome is a sequence
of *genes*, where the genes are classified into *families*. Considering that an *adjacency* is the oriented neighborhood between two genes in one chromosome of a
genome, a simple *distance* measure between genomes is the *breakpoint* distance,
that consists in quantifying their distinct adjacencies [9]. Other distance models rely on large-scale genome *rearrangements*, such as inversions, translocations,
fusions and fissions, or the general *double-cut-and-join* (DCJ) operation, yielding
distances that correspond to the minimum number of rearrangements required
to transform one genome into another [7,8,10].

Our study relies on the *breakpoint graph*, a structure that represents the
relation between two given genomes [1]. When the two genomes have the same

K. Jahn and T. Vinař (Eds.): RECOMB-CG 2023, LNBI 13883, pp. 35–50, 2023.
https://doi.org/10.1007/978-3-031-36911-7_3

set of n_* genes, their breakpoint graph is a collection of cycles of even length and paths. For even k, let c_k and p_k be respectively the numbers of cycles and of paths of length k. The corresponding breakpoint distance is equal to $n_* - \left(c_2 + \frac{p_0}{2}\right)$ [9]. Similarly, when the considered rearrangements are those modeled by the *double-cut-and-join* (DCJ) operation [10], the rearrangement distance is $n_* - \left(c + \frac{p_e}{2}\right)$, where c is the total number of cycles and p_e is the total number of even paths [2].

Both breakpoint and DCJ distances are basic constituents of a problem related to the event of a *whole genome duplication* (WGD) [6,9]. The *double distance* of a *singular* genome $\mathbb{S}$ and a *duplicated* genome $\mathbb{D}$ aims to find the smallest distance between $\mathbb{D}$ and any element from the set $2\mathbb{S}$, that contains all possible genome configurations obtained by *doubling* the chromosomes of $\mathbb{S}$. Interestingly, it can be solved in linear time under the breakpoint distance, but is NP-hard under the DCJ distance [9]. One way of exploring the complexity space between these two extremes is to consider a σ_k distance [5], defined to be $n_* - \left(c_2 + c_4 + \ldots + c_k + \frac{p_0 + p_2 + \ldots + p_{k-2}}{2}\right)$. Note that the σ_2 distance is the breakpoint distance and the σ_∞ distance is the DCJ distance. We can then increasingly investigate the complexities of the double distance under the σ_4 distance, then under the σ_6 distance, and so on, in an attempt to identify the smallest value for which the double distance becomes NP-hard.

In our recent work, we succeeded in devising efficient algorithms for σ_4 and σ_6 if the input genomes are exclusively composed of circular chromosomes [3]. Here we close the gaps of these results by giving a solution for the double distance under σ_4 and σ_6 even if the input genomes include linear chromosomes. This work has an extended version containing detailed proofs and extra figures [4].

2   Background

A *chromosome* can be either linear or circular and is represented by its sequence of genes, where each *gene* is an oriented DNA fragment. We assume that each gene belongs to a *family*, which is a set of homologous genes. A gene that belongs to a family X is represented by the symbol X itself if it is read in forward orientation or by the symbol $\overline{X}$ if it is read in reverse orientation. For example, the sequences $[1\,\overline{3}\,2]$ and (4) represent, respectively, a linear (flanked by square brackets) and a circular chromosome (flanked by parentheses), both shown in Fig. 1, the first composed of three genes and the second composed of a single gene. Note that if a sequence s represents a chromosome K, then K can be equally represented by the *reverse complement* of s, denoted by $\overline{s}$, obtained by reversing the order and the orientation of the genes in s. Moreover, if K is circular, it can be equally represented by any circular rotation of s and $\overline{s}$. Recall that a gene is an *occurrence* of a family, therefore distinct genes from the same family are represented by the same symbol.

We can also represent a gene from family X referring to its *extremities* X^h (head) and X^t (tail). The *adjacencies* in a chromosome are the neighboring extremities of distinct genes. The remaining extremities, that are at the ends of linear chromosomes, are *telomeres*. In linear chromosome $[1\overline{3}2]$, the adjacencies are $\{1^h3^h, 3^t2^t\}$ and the telomeres are $\{1^t, 2^h\}$. Note that an adjacency has no orientation, that is, an adjacency between extremities 1^h and 3^h can be equally represented by $1^h 3^h$ and by $3^h 1^h$. In the particular case of a single-gene circular chromosome, e.g. (4), an adjacency exceptionally occurs between the extremities of the same gene (here $4^h 4^t$).

A *genome* is then a multiset of chromosomes and we denote by $\mathcal{F}(\mathbb{G})$ the set of gene families that occur in genome $\mathbb{G}$. In addition, we denote by $\mathcal{A}(\mathbb{G})$ the multiset of adjacencies and by $\mathcal{T}(\mathbb{G})$ the multiset of telomeres that occur in $\mathbb{G}$. A genome $\mathbb{S}$ is called *singular* if each gene family occurs exactly once in $\mathbb{S}$. Similarly, a genome $\mathbb{D}$ is called *duplicated* if each gene family occurs exactly twice in $\mathbb{D}$. The two occurrences of a family in a duplicated genome are called *paralogs*. A *doubled* genome is a special type of duplicated genome in which each adjacency or telomere occurs exactly twice. These two copies of the same adjacency (respectively same telomere) in a doubled genome are called *paralogous adjacencies* (respectively *paralogous telomeres*). Observe that distinct doubled genomes can have exactly the same adjacencies and telomeres, as we show in Fig. 1, where we also give an example of a singular genome.

Fig. 1. On the left we show the singular genome $\mathbb{S} = \{[1\bar{3}2](4)\}$ and on the right the two doubled genomes $\{[1\bar{3}2][1\bar{3}2](4)(4)\}$ and $\{[1\bar{3}2][1\bar{3}2](44)\}$.

2.1 Comparing Canonical Genomes

Two genomes $\mathbb{S}_1$ and $\mathbb{S}_2$ are said to be a *canonical pair* when they are singular and have the same gene families, that is, $\mathcal{F}(\mathbb{S}_1) = \mathcal{F}(\mathbb{S}_2)$. Denote by $\mathcal{F}_*$ the set of families occurring in canonical genomes $\mathbb{S}_1$ and $\mathbb{S}_2$. For example, genomes $\mathbb{S}_1 = \{(1\bar{3}2)(4)\}$ and $\mathbb{S}_2 = \{(12)(3\bar{4})\}$ are canonical with $\mathcal{F}_* = \{1, 2, 3, 4\}$.

Breakpoint Graph. A multigraph representing the adjacencies of $\mathbb{S}_1$ and $\mathbb{S}_2$ is the *breakpoint graph* $BG(\mathbb{S}_1, \mathbb{S}_2) = (V, E)$ [1]. The vertex set V comprises, for each family $\mathtt{X}$ in $\mathcal{F}_*$, one vertex for $\mathtt{X}^h$ and one vertex for $\mathtt{X}^t$. The edge multiset E represents the adjacencies: for each adjacency in $\mathbb{S}_1$ (respectively $\mathbb{S}_2$) there exists one $\mathbb{S}_1$-edge (respectively $\mathbb{S}_2$-edge) in E linking its two extremities.

The degree of each vertex can be 0, 1 or 2 and each connected *component* alternates between $\mathbb{S}_1$- and $\mathbb{S}_2$-edges. As a consequence, the components of $BG(\mathbb{S}_1, \mathbb{S}_2)$ can be cycles of even length or paths. An even path has one endpoint in $\mathbb{S}_1$ ($\mathbb{S}_1$-*telomere*) and the other in $\mathbb{S}_2$ ($\mathbb{S}_2$-*telomere*), while an odd path has either both endpoints in $\mathbb{S}_1$ or both endpoints in $\mathbb{S}_2$. A vertex that is not a telomere in $\mathbb{S}_1$ nor in $\mathbb{S}_2$ is said to be *non-telomeric*. In the breakpoint graph a non-telomeric vertex has degree 2. We call *i-cycle* a cycle of length i and *j-path* a path of length j. We also denote by c_i the number of cycles of length i, by p_j the number of paths of length j, by c the total number of cycles and by p_e the total number of even paths. An example is given in Fig. 2(a).

Breakpoint Distance. For canonical genomes $\mathbb{S}_1$ and $\mathbb{S}_2$ the *breakpoint distance*, denoted by d_{BP}, is defined as follows [9]:

$$d_{BP}(\mathbb{S}_1, \mathbb{S}_2) = n_* - \left(|\mathcal{A}(\mathbb{S}_1) \cap \mathcal{A}(\mathbb{S}_2)| + \frac{|\mathcal{T}(\mathbb{S}_1) \cap \mathcal{T}(\mathbb{S}_2)|}{2} \right), \quad \text{where } n_* = |\mathcal{F}_*|.$$

If $\mathbb{S}_1 = \{(1\,\overline{3}\,2)\,[4]\}$ and $\mathbb{S}_2 = \{(1\,2)\,[3\,\overline{4}]\}$, then $\mathcal{A}(\mathbb{S}_1) \cap \mathcal{A}(\mathbb{S}_2) = \{1^t 2^h\}$, $\mathcal{T}(\mathbb{S}_1) \cap \mathcal{T}(\mathbb{S}_2) = \{4^t\}$ and $n_* = 4$, giving $d_{BP}(\mathbb{S}_1, \mathbb{S}_2) = 2.5$. Since a common adjacency corresponds to a 2-cycle and a common telomere corresponds to a 0-path in $BG(\mathbb{S}_1, \mathbb{S}_2)$, the breakpoint distance can be rewritten as

$$d_{BP}(\mathbb{S}_1, \mathbb{S}_2) = n_* - \left(c_2 + \frac{p_0}{2} \right).$$

DCJ Distance. A *double cut and join* (DCJ) is the operation that breaks two adjacencies or telomeres[1] of a genome and rejoins the open ends in a different way [10]. A DCJ models several rearrangements, such as inversion, translocation, fission and fusion. The minimum number of DCJs that transform one genome into the other is their *DCJ distance* d_{DCJ}, that can be derived from $BG(\mathbb{S}_1, \mathbb{S}_2)$ [2]:

$$d_{DCJ}(\mathbb{S}_1, \mathbb{S}_2) = n_* - \left(c + \frac{p_e}{2} \right) = n_* - \left(c_2 + c_4 + \ldots + c_\infty + \frac{p_0 + p_2 + \ldots + p_\infty}{2} \right).$$

If $\mathbb{S}_1 = \{(1\,\overline{3}\,2)\,[4]\}$ and $\mathbb{S}_2 = \{(1\,2)\,[3\,\overline{4}]\}$, then $n_* = 4$, $c = 1$ and $p_e = 2$ (see Fig. 2(a)). Consequently, their DCJ distance is $d_{DCJ}(\mathbb{S}_1, \mathbb{S}_2) = 2$.

The Class of σ_k Distances. Given $BG(\mathbb{S}_1, \mathbb{S}_2)$, for $k \in \{2, 4, 6, \ldots, \infty\}$ we denote by σ_k the cumulative sums $\sigma_k = c_2 + c_4 + \ldots + c_k + \frac{p_0 + p_2 + \ldots + p_{k-2}}{2}$. Then the σ_k distance of $\mathbb{S}_1$ and $\mathbb{S}_2$ is defined to be [5]:

$$d_{\sigma_k}(\mathbb{S}_1, \mathbb{S}_2) = n_* - \sigma_k.$$

It is easy to see that the σ_2 distance equals the breakpoint distance and that the σ_∞ distance equals the DCJ distance.

2.2 Comparing a Singular and a Duplicated Genome

Let $\mathbb{S}$ be a singular and $\mathbb{D}$ be a duplicated genome over the same gene families, that is, $\mathcal{F}(\mathbb{S}) = \mathcal{F}(\mathbb{D})$. The number of genes in $\mathbb{D}$ is twice the number of genes in $\mathbb{S}$ and we need to somehow equalize the contents of these genomes, before searching for common adjacencies and common telomeres of $\mathbb{S}$ and $\mathbb{D}$ or transforming one

---

[1] A broken adjacency has two open ends and a broken telomere has a single one.

genome into the other with DCJ operations. This can be done by *doubling* $\mathbb{S}$, with a rearrangement operation mimicking a *whole genome duplication*: it simply consists of doubling each adjacency and each telomere of $\mathbb{S}$. However, when $\mathbb{S}$ has one or more circular chromosomes, it is not possible to find a unique layout of its chromosomes after the doubling: indeed, each circular chromosome can be doubled into two identical circular chromosomes, or the two copies are concatenated to each other in a single circular chromosome. Therefore, in general the doubling of a genome $\mathbb{S}$ results in a set of doubled genomes denoted by $2\mathbb{S}$. Note that $|2\mathbb{S}| = 2^r$, where r is the number of circular chromosomes in $\mathbb{S}$. For example, if $\mathbb{S} = \{[1\,\overline{3}\,2]\,(4)\}$, then $2\mathbb{S} = \{\mathbb{B}_1, \mathbb{B}_2\}$ with $\mathbb{B}_1 = \{[1\,\overline{3}\,2]\,[1\,\overline{3}\,2]\,(4)\,(4)\}$ and $\mathbb{B}_2 = \{[1\,\overline{3}\,2]\,[1\,\overline{3}\,2]\,(4\,4)\}$ (see Fig. 1). All genomes in $2\mathbb{S}$ have exactly the same multisets of adjacencies and of telomeres, therefore we can use a special notation for these multisets: $\mathcal{A}(2\mathbb{S}) = \mathcal{A}(\mathbb{S}) \cup \mathcal{A}(\mathbb{S})$ and $\mathcal{T}(2\mathbb{S}) = \mathcal{T}(\mathbb{S}) \cup \mathcal{T}(\mathbb{S})$.

Each family in a duplicated genome can be $\binom{a}{b}$-*singularized* by adding the index a to one of its occurrences and the index b to the other. A duplicated genome can be entirely singularized if each of its families is singularized. Let $\mathfrak{S}_b^a(\mathbb{D})$ be the set of all possible genomes obtained by all distinct ways of $\binom{a}{b}$-singularizing the duplicated genome $\mathbb{D}$. Similarly, we denote by $\mathfrak{S}_b^a(2\mathbb{S})$ the set of all possible genomes obtained by all distinct ways of $\binom{a}{b}$-singularizing each doubled genome in the set $2\mathbb{S}$.

The Class of σ_k Double Distances. For $k = 2, 4, 6, \ldots$, each σ_k double distance of a singular genome $\mathbb{S}$ and a duplicated genome $\mathbb{D}$ is defined as follows [3,9]:

$$d_{\sigma_k}^2(\mathbb{S}, \mathbb{D}) = d_{\sigma_k}^2(\mathbb{S}, \check{\mathbb{D}}) = \min_{\mathbb{B} \in \mathfrak{S}_b^a(2\mathbb{S})} \{d_{\sigma_k}(\mathbb{B}, \check{\mathbb{D}})\}, \text{ where } \check{\mathbb{D}} \text{ is any genome in } \mathfrak{S}_b^a(\mathbb{D}).$$

Observe that $d_{\sigma_k}^2(\mathbb{S}, \check{\mathbb{D}}) = d_{\sigma_k}^2(\mathbb{S}, \check{\mathbb{D}}')$ for any $\check{\mathbb{D}}, \check{\mathbb{D}}' \in \mathfrak{S}_b^a(\mathbb{D})$.

σ_2 (Breakpoint) Double Distance. The *breakpoint double distance* of $\mathbb{S}$ and $\mathbb{D}$, denoted by $d_{\mathrm{BP}}^2(\mathbb{S}, \mathbb{D})$, is equivalent to the σ_2 double distance. The solution here can be found with a greedy algorithm [9]: each adjacency or telomere of $\mathbb{D}$ that occurs in $\mathbb{S}$ can be fulfilled. If an adjacency or telomere that occurs twice in $\mathbb{D}$ also occurs in $\mathbb{S}$, it can be fulfilled twice in any genome from $2\mathbb{S}$. Then,

$$d_{\mathrm{BP}}^2(\mathbb{S}, \mathbb{D}) = 2n_* - |\mathcal{A}(2\mathbb{S}) \cap \mathcal{A}(\mathbb{D})| - \frac{|\mathcal{T}(2\mathbb{S}) \cap \mathcal{T}(\mathbb{D})|}{2}, \text{ where } n_* = |\mathcal{F}(\mathbb{S})|.$$

σ_∞ (DCJ) double distance. For the *DCJ double distance* d_{DCJ}^2, that is equivalent to the σ_∞ double distance, the solution space is more complex: computing the DCJ double distance of genomes $\mathbb{S}$ and $\mathbb{D}$ was proven to be NP-hard [9].

3   Equivalence of σ_k Double Distance and σ_k Disambiguation

A nice way of representing the solution space of the σ_k double distance is by using a modified version of the breakpoint graph [3,9].

3.1 Ambiguous Breakpoint Graph

Given a singular genome $\mathbb{S}$ and a duplicated genome $\mathbb{D}$, their *ambiguous break-point graph* $ABG(\mathbb{S}, \check{\mathbb{D}}) = (V, E)$ is a multigraph representing the adjacencies of any element in $\mathbb{S}_b^a(2\mathbb{S})$ and a genome $\check{\mathbb{D}} \in \mathbb{S}_b^a(\mathbb{D})$. The vertex set V comprises, for each family X in $\mathcal{F}(\mathbb{S})$, the two pairs of *paralogous vertices* X_a^h, X_b^h and X_a^t, X_b^t. We can use the notation $\hat{u}$ to refer to the paralogous counterpart of a vertex u. For example, if $u = X_a^h$, then $\hat{u} = X_b^h$.

The edge set E represents the adjacencies. For each adjacency in $\check{\mathbb{D}}$ there exists one $\check{\mathbb{D}}$-edge in E linking its two extremities. The $\mathbb{S}$-edges represent all adjacencies occurring in all genomes from $\mathbb{S}_b^a(2\mathbb{S})$: for each adjacency $\gamma\beta$ of $\mathbb{S}$, we have the *pair of paralogous edges* $\mathcal{E}(\gamma\beta) = \{\gamma_a\beta_a, \gamma_b\beta_b\}$ and its *complementary counterpart* $\widetilde{\mathcal{E}}(\gamma\beta) = \{\gamma_a\beta_b, \gamma_b\beta_a\}$. Note that $\widetilde{\widetilde{\mathcal{E}}}(\gamma\beta) = \mathcal{E}(\gamma\beta)$. The *square* of $\gamma\beta$ is then $\mathcal{Q}(\gamma\beta) = \mathcal{E}(\gamma\beta) \cup \widetilde{\mathcal{E}}(\gamma\beta)$. The $\mathbb{S}$-edges in the ambiguous breakpoint graph are therefore the squares of all adjacencies in $\mathbb{S}$. Let a_* be the number of squares in $ABG(\mathbb{S}, \check{\mathbb{D}})$. Obviously we have $a_* = |\mathcal{A}(\mathbb{S})| = n_* - \kappa(\mathbb{S})$, where $\kappa(\mathbb{S})$ is the number of linear chromosomes in $\mathbb{S}$. Again, we can use the notation $\hat{e}$ to refer to the paralogous counterpart of an $\mathbb{S}$-edge e. For example, if $e = \gamma_a\beta_a$, then $\hat{e} = \gamma_b\beta_b$. An ambiguous breakpoint graph is shown in Fig. 2(b1).

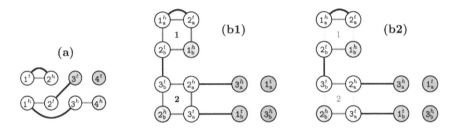

Fig. 2. (a) Breakpoint graph of genomes $\mathbb{S}_1 = \{(1\,2)[3\,\overline{4}]\}$ and $\mathbb{S}_2 = \{(1\,\overline{3}\,2)[4]\}$. Edge types are distinguished by colors: $\mathbb{S}_1$-edges are drawn in blue and $\mathbb{S}_2$-edges are drawn in black. Similarly, vertex types are distinguished by colors: an $\mathbb{S}_1$-telomere is marked in blue, an $\mathbb{S}_2$-telomere is marked in gray, a telomere in both $\mathbb{S}_1$ and $\mathbb{S}_2$ is marked in purple and non-telomeric vertices are white. This graph has one 2-cycle, one 0-path and one 4-path. (b1) Graph $ABG(\mathbb{S}, \check{\mathbb{D}})$ for genomes $\mathbb{S} = \{[1\,2\,3]\}$ and $\check{\mathbb{D}} = \{[1_a\,2_a\,\overline{3}_a\,1_b]\,[\overline{3}_b\,2_b]\}$. Edge types are distinguished by colors: $\check{\mathbb{D}}$-edges are drawn in black and $\mathbb{S}$-edges (squares) are drawn in red. (b2) Induced breakpoint graph $BG(\tau, \check{\mathbb{D}})$ in which all squares are resolved by the solution $\tau = (\{1_a^h 2_a^t, 1_b^h 2_b^t\}, \{2_a^h 3_a^t, 2_b^h 3_b^t\})$, resulting in one 2-cycle, two 0-paths, one 2-path and one 4-path. This is also the breakpoint graph of $\check{\mathbb{D}}$ and $\mathbb{B} = \{[1_a\,2_a\,3_b], [1_b\,2_b\,3_a]\} \in \mathbb{S}_b^a(2\mathbb{S})$. (Color figure online)

The elements of $2\mathbb{S}$ have the same pairs of identical linear chromosomes. Each pair corresponds to four $\mathbb{S}$-*telomeres*, that are not part of any square. The number of $\mathbb{S}$-telomeres is then $4\kappa(\mathbb{S})$. If $\kappa(\mathbb{D})$ is the number of linear chromosomes in $\mathbb{D}$, the number of $\check{\mathbb{D}}$-*telomeres* is $2\kappa(\mathbb{D})$.

3.2   The Class of σ_k Disambiguations

Resolving a square $\mathcal{Q}(\cdot) = \mathcal{E}(\cdot) \cup \widetilde{\mathcal{E}}(\cdot)$ corresponds to *choosing* either $\mathcal{E}(\cdot)$ or $\widetilde{\mathcal{E}}(\cdot)$, while the complementary pair is *masked*. If we number the squares of $ABG(\mathbb{S}, \check{\mathbb{D}})$ from 1 to a_*, a *solution* can be represented by a tuple $\tau = (\mathcal{L}_1, \mathcal{L}_2, \ldots, \mathcal{L}_{a_*})$, where each $\mathcal{L}_i$ contains the pair of paralogous edges (either $\mathcal{E}_i$ or $\widetilde{\mathcal{E}}_i$) that are chosen (kept) in the graph for square $\mathcal{Q}_i$. The graph *induced* by τ is a simple breakpoint graph, which we denote by $BG(\tau, \check{\mathbb{D}})$. Figure 2(b2) shows an example.

Given a solution τ, its k-score is the cumulative sum σ_k with respect to $BG(\tau, \check{\mathbb{D}})$. The problem of finding a solution τ for $ABG(\mathbb{S}, \check{\mathbb{D}})$ so that its k-score is maximized is called σ_k *disambiguation*, that is equivalent to the minimization problem of computing the σ_k double distance of $\mathbb{S}$ and $\mathbb{D}$, therefore the complexities of solving the σ_k disambiguation and the σ_k double distance for any $k \geq 2$ must be the same [9]. As already mentioned, for σ_2 the double distance/disambiguation can be solved in linear time and for σ_∞ the double distance/disambiguation is NP-hard. An optimal solution for the σ_k disambiguation of $ABG(\mathbb{S}, \check{\mathbb{D}})$ gives its *k-score*, denoted by $\sigma_k(ABG(\mathbb{S}, \check{\mathbb{D}}))$. If $k < k'$, any k-cycle contributes to any k'-score, therefore $\sigma_k(ABG(\mathbb{S}, \check{\mathbb{D}})) \leq \sigma_{k'}(ABG(\mathbb{S}, \check{\mathbb{D}}))$.

Approach for Solving the σ_k Disambiguation. Two $\mathbb{S}$-edges in $ABG(\mathbb{S}, \check{\mathbb{D}})$ are *incompatible* when they belong to the same square and are not paralogous. A cycle or a telomere-to-telomere path in $ABG(\mathbb{S}, \check{\mathbb{D}})$ is *valid* when it does not contain any pair of incompatible edges, that is, when it alternates between $\mathbb{S}$-edges and $\check{\mathbb{D}}$-edges. Any valid cycle or path can be called *piece*. Two distinct pieces in $ABG(\mathbb{S}, \check{\mathbb{D}})$ are either *intersecting*, when they share at least one vertex, or *disjoint*. Finally, a *player* is either a valid cycle whose length is at most k or a valid even path whose length is at most $k - 2$. Note that any player is a piece and that any solution τ of $ABG(\mathbb{S}, \check{\mathbb{D}})$ is composed of disjoint pieces.

Given a solution $\tau = (\mathcal{L}_1, \mathcal{L}_2, \ldots, \mathcal{L}_i \ldots, \mathcal{L}_{a_*})$, the *switching* of its i-th element is denoted by $\widetilde{s}(\tau, i)$ and gives $(\mathcal{L}_1, \mathcal{L}_2, \ldots, \widetilde{\mathcal{L}}_i \ldots, \mathcal{L}_{a_*})$. A choice of paralogous edges resolving a given square $\mathcal{Q}_i$ can be *fixed* for any solution, meaning that $\mathcal{Q}_i$ can no longer be switched. In this case, $\mathcal{Q}_i$ is itself said to be *fixed*.

4   First Steps to Solve the σ_k Disambiguation

In this section we give straightforward extensions of results developed in our previous study for circular genomes [3].

4.1   Common Adjacencies and Telomeres Are Conserved

If τ is an optimal solution for the σ_k disambiguation and if a player $C \in BG(\tau, \check{\mathbb{D}})$ is disjoint from any player distinct from C in any other optimal solution, then C must be part of all optimal solutions and is itself said to be *optimal*.

Lemma 1 (extended from [3]**).** *For any σ_k disambiguation, all existing 0-paths and 2-cycles in $ABG(\mathbb{S}, \mathbb{D})$ are optimal.*

This lemma is a generalization of the σ_2 disambiguation and guarantees that all common adjacencies and telomeres are conserved in any σ_k double distance, including the NP-hard σ_∞ case. All 0-paths are isolated vertices, therefore they are selected independently of the choices for resolving the squares. A 2-cycle, in its turn, always includes one $\mathbb{S}$-edge from some square (such as square 1 in Fig. 2(b2)). From now on we assume that squares that have at least one $\mathbb{S}$-edge in a 2-cycle are fixed so that all existing 2-cycles are induced.

4.2   Symmetric Squares Can Be Fixed Arbitrarily

Let v and $\hat{v}$ be a pair of paralogous vertices in $\mathcal{Q}$. The square $\mathcal{Q}$ is said to be *symmetric* when either (i) there is a $\check{\mathbb{D}}$-edge connecting v and $\hat{v}$, or (ii) v and $\hat{v}$ are $\check{\mathbb{D}}$-telomeres, or (iii) v and $\hat{v}$ are directly connected to $\mathbb{S}$-telomeres by $\check{\mathbb{D}}$-edges, as illustrated in Fig. 3. Note that, for any σ_k disambiguation, a symmetric square $\mathcal{Q}$ can be fixed arbitrarily: the two ways of resolving $\mathcal{Q}$ would lead to solutions with the same score. We then assume that $ABG(\mathbb{S}, \check{\mathbb{D}})$ has no symmetric squares.

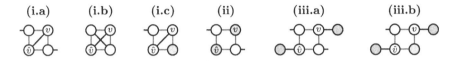

Fig. 3. Possible symmetric squares in the ambiguous breakpoint graph.

4.3   A Linear Time Greedy Algorithm for the σ_4 Disambiguation

Although two valid 4-cycles can intersect with each other, since our graph is free of symmetric squares, two valid 2-paths cannot intersect with each other. Moreover, a 2-path has no $\check{\mathbb{D}}$-edge connecting squares, therefore it cannot intersect with a 4-cycle. For the σ_4 disambiguation it is then clear that, (i) any valid 2-path is always optimal and (ii) a 4-cycle that does intersect with another one is always optimal. We also know that intersecting 4-cycles are always part of co-optimal solutions [3]. An optimal solution can then be obtained greedily: after fixing squares containing edges that are part of 2-cycles, traverse the remainder of the graph and, for each valid 2-path or 4-cycle C that is found, fix the square(s) containing $\mathbb{S}$-edges that are part of C, so that C is induced. When this part is accomplished the remaining squares can be fixed arbitrarily.

4.4   Pruning $ABG(\mathbb{S}, \check{\mathbb{D}})$ for the σ_6 Disambiguation

Players of the σ_6 disambiguation can intersect with each other and not every player is induced by at least one optimal solution. For that reason a more elaborated procedure is required, whose first step is a linear time preprocessing in which from $ABG(\mathbb{S}, \check{\mathbb{D}})$ first all edges are removed that are incompatible with the existing 2-cycles, and then all remaining edges that cannot be part of a player [3]. This results in a $\{6\}$-*pruned* ambiguous breakpoint graph $PG(\mathbb{S}, \check{\mathbb{D}})$.

The edges that are not pruned and are therefore present in $PG(\mathbb{S}, \check{\mathbb{D}})$ are said to be *preserved*. A square that has preserved edges from distinct paralogous pairs is still ambiguous and is called a $\{6\}$-*square*. Otherwise it is resolved and can be fixed according to the preserved edges. Additionally, if none of its edges is preserved, a square is arbitrarily fixed.

The smaller $PG(\mathbb{S}, \check{\mathbb{D}})$ has all relevant parts required for finding an optimal solution of σ_6 disambiguation, therefore the 6-scores of both graphs are the same: $\sigma_6(ABG(\mathbb{S}, \check{\mathbb{D}})) = \sigma_6(PG(\mathbb{S}, \check{\mathbb{D}}))$. A clear advantage here is that the pruned graph might be split into smaller connected components, and it is obvious that the disambiguation problem can be solved independently for each one of them. Each connected component G of $PG(\mathbb{S}, \check{\mathbb{D}})$ is of one of the two types [3]:

1. *Ambiguous*: G includes at least one $\{6\}$-square;
2. *Resolved (trivial)*: G is simply a player.

Let $\mathcal{C}$ and $\mathcal{P}$ be the sets of resolved components, so that $\mathcal{C}$ has all resolved cycles and $\mathcal{P}$ has all resolved paths. Furthermore, let $\mathcal{B}$ be the set of ambiguous components of $PG(\mathbb{S}, \check{\mathbb{D}})$. If we denote by $\sigma_6(G)$ the 6-score of an ambiguous component $G \in \mathcal{B}$, the 6-score of $PG(\mathbb{S}, \check{\mathbb{D}})$ can be computed with the formula:

$$\sigma_6(PG(\mathbb{S}, \check{\mathbb{D}})) = |\mathcal{C}| + \frac{|\mathcal{P}|}{2} + \sum_{G \in \mathcal{B}} \sigma_6(G).$$

For solving the σ_6 disambiguation the only missing part is finding, for each ambiguous component $G \in \mathcal{B}$, an optimal solution τ_G including only the $\{6\}$-squares of G. From now on, by $\mathbb{S}$-edge, $\mathbb{S}$-telomere, $\check{\mathbb{D}}$-edge and $\check{\mathbb{D}}$-telomere, we are referring only to the elements that are preserved in $PG(\mathbb{S}, \check{\mathbb{D}})$.

5   Solving the General σ_6 Disambiguation in Linear Time

Here we present the most relevant contribution of this work: an algorithm to solve the σ_6 disambiguation in linear time, for genomes with linear chromosomes. The proofs omitted here can be found in the extended version of this work [4].

5.1   Intersection Between Players

A player in the σ_6 disambiguation can be either a $\{2,4\}$-*path*, that is a valid 2- or 4-path, or a $\{4,6\}$-*cycle*, that is a valid 4- or 6-cycle.

Let a $\check{\mathbb{D}}\mathbb{S}\check{\mathbb{D}}$-*path* be a subpath of three edges, starting and ending with a $\check{\mathbb{D}}$-edge. This is the largest segment that can be shared by two players: although there is no room to allow distinct $\{2, 4\}$-paths and/or valid 4-cycles to share a $\check{\mathbb{D}}\mathbb{S}\check{\mathbb{D}}$-path in a graph free of symmetric squares, a $\check{\mathbb{D}}\mathbb{S}\check{\mathbb{D}}$-path can be shared by at most two valid 6-cycles. Furthermore, if distinct $\check{\mathbb{D}}\mathbb{S}\check{\mathbb{D}}$-paths intersect at the same

Ď-edge e and each of them occurs in two distinct 6-cycles, then the Ď-edge e occurs in four distinct valid 6-cycles. In Fig. 4 we characterize this exceptional situation[2], which consists in the occurrence of a *triplet*, composed of exactly three connected ambiguous squares in which at most two vertices, necessarily in distinct squares, are pruned out. In a *saturated* triplet, the squares in each pair are connected to each other by two Ď-edges connecting paralogous vertices in both squares; if a single Ď-edge is missing, that is, the corresponding vertices have outer connections, we have an *unsaturated* triplet. This structure and its score can be easily identified, therefore we will assume that our graph is free from triplets. With this condition, Ď-edges can be shared by at most two players:

Proposition 1 (extended from [3]). *Any Ď-edge is part of either one or two (intersecting) players in a graph free of symmetric squares and triplets.*

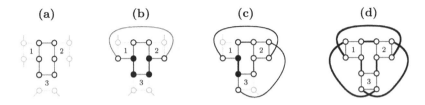

Fig. 4. (a) Resolved component (score = 1): a 6-cycle alternating (black) Ď- and (blue) S-edges, without intersections. (b) Two 6-cycles share one ĎSĎ-path composed of the two black Ď-edges with the blue S-edge in between. (c) Unsaturated triplet with score = 1: every ĎSĎ-path including the same Ď-edge (the thick black one) occurs in two distinct 6-cycles. The thick black Ď-edge occurs in four 6-cycles, all other black edges occur in two 6-cycles. (d) Saturated triplet with score = 2: every ĎSĎ-path occurs in two distinct 6-cycles, every black edge occurs in four 6-cycles. (Color figure online)

As a consequence of Proposition 1, we know that any S-edge of a {6}-square Q is part of exactly one player [3]. Two {6}-squares Q and Q' are *neighbors* when a vertex of Q is connected to a vertex of Q' by a Ď-edge. For the case of circular genomes, all players are cycles. In each component G, since both Ď-edges inciding at the endpoints of an S-edge would induce the same {4,6}-cycle, the choice of an S-edge e of a {6}-square Q (and its paralogous edge $\hat{e}$) would imply a unique way of resolving all neighbors of Q, and, by propagating this to the neighbors of the neighbors and so on, all squares of G would be resolved, resulting in what we called *straight solution* τ_G. The score of G would be given either by τ_G, or by its complementary alternative $\tilde{\tau}_G$, obtained by switching all ambiguous squares of τ_G, or by both in case of co-optimality. Unfortunately, for genomes with linear chromosomes, where paths can intersect in a telomere, the procedure above no longer suffices. For that reason, we need to proceed with a further

---

[2] In our previous paper [3] we overlooked this particular case, that can fortunately be treated in a preprocessing. Otherwise the solution presented there is complete.

characterization of each ambiguous component G of $PG(\mathbb{S}, \check{\mathbb{D}})$, allowing us to split the disambiguation of G into smaller subproblems.

As we will present in the following, the solution for arbitrarily large components can be split into two types of problems, which are analogous to solving the *maximal independent set* of auxiliary subgraphs that are either simple paths or double paths. In both cases, the solutions can be obtained in linear time.

5.2   Intersection Graph of an Ambiguous Component

If two $\{2,4\}$-paths intersect in their $\mathbb{S}$-telomere, this intersection must include the incident $\check{\mathbb{D}}$-edge. Therefore, when we say that an intersection occurs at an $\mathbb{S}$-telomere, this automatically means that the intersection is the $\check{\mathbb{D}}$-edge inciding in an $\mathbb{S}$-telomere. A valid 4-cycle has two $\check{\mathbb{D}}$-edges and a valid 6-cycle has three $\check{\mathbb{D}}$-edges. Besides the one at the $\mathbb{S}$-telomere, a valid 4-path has one $\check{\mathbb{D}}$-edge while a valid 2-path has none - therefore the latter cannot intersect with a $\{4,6\}$-cycle. When we say that 4-paths and/or $\{4,6\}$-cycles intersect with each other in a $\check{\mathbb{D}}$-edge, we refer to an *inner* $\check{\mathbb{D}}$-edge and not one inciding in an $\mathbb{S}$-telomere.

The auxiliary *intersection graph* $\mathcal{I}(G)$ of an ambiguous component G has a vertex with weight $\frac{1}{2}$ for each $\{2,4\}$-path and a vertex with weight 1 for each $\{4,6\}$-cycle of G. Furthermore, if two players intersect, we have an edge between the respective vertices. The intersection graphs of all ambiguous components can be built during the pruning procedure without increasing its linear time complexity. Note that an independent set of maximum weight in $\mathcal{I}(G)$ corresponds to an optimal solution of G. Although in general this problem is NP-hard, in our case the underlying ambiguous component G imposes a regular structure to $\mathcal{I}(G)$, allowing us to find such an independent set in linear time.

5.3   Path-Flows in the Intersection Graph

A *path-flow* in $\mathcal{I}(G)$ is a maximal connected subgraph whose vertices correspond to $\{2,4\}$-paths. A *path-line* of length ℓ in a path-flow is a series of ℓ paths, such that each pair of consecutive paths intersect at a telomere. Assume that the vertices in a path-line are numbered from left to right with integers $1, 2, \ldots, \ell$. A *double-line* consists of two parallel path-lines of the same length ℓ, such that vertices with the same number in both lines intersect in a $\check{\mathbb{D}}$-edge and are therefore connected by an edge. A 2-path has no free $\check{\mathbb{D}}$-edge, therefore a double-line is exclusively composed of 4-paths. If a path-line composes a double-line, it is *saturated*, otherwise it is a *unsaturated*. Since each 4-path of a double-line has a $\check{\mathbb{D}}$-edge intersection with another and each 4-path can have only one $\check{\mathbb{D}}$-edge intersection, no vertex of a double-line can be connected to a cycle in $\mathcal{I}(G)$.

Let us assume that a double-line is always represented with one upper path-line and one lower path-line. A double-line of length ℓ has 2ℓ vertices and exactly two independent sets of maximal weight, each one with ℓ vertices and weight $\frac{\ell}{2}$: one includes the paths with odd numbers in the upper line and the paths with even numbers in the lower line, while the other includes the paths with even numbers in the upper line and the paths with odd numbers in the lower line.

Since a double-line cannot intersect with cycles, it is clear that at least one of these independent sets will be part of a global optimal solution for $\mathcal{I}(G)$. A maximal double-line can be of three different types:

1. *Isolated*: corresponds to the complete graph $\mathcal{I}(G)$. Here, but only if the length ℓ is even, the double line can be cyclic: in this case, in both upper and lower lines, the last vertex intersects at a telomere with the first vertex. Being cyclic or not, any of the two optimal local solutions can be fixed.
2. *Terminal*: a vertex v located at the end of one of the two lines intersects with one unsaturated path-line. At least one of the two optimal local solutions would leave v unselected; we can safely fix this option.
3. *Link*: intersects with unsaturated lines at both ends. The intersections can be:
   (a) *single-sided*: both occur at the ends of the same saturated line, or
   (b) *alternate*: the left intersection occurs at the end of one saturated line and the right intersection occurs at the end of the other.

   Let v' be the outer vertex connected to a vertex v belonging to the link at the right and u' be the outer vertex connected to a vertex u belonging to the link at the left. Let a *balanced link* be alternate of odd length, or single-sided of even length. In contrast, an *unbalanced link* is alternate of even length, or single-sided of odd length. If the link is unbalanced, one of the two local optimal solutions leaves both u and v unselected; we can safely fix this option. If the link is balanced, we cannot fix the solution before-hand, but we can reduce the problem, by removing the connections uu' and vv' and adding the connection $u'v'$. Since both u' and v' must be the ends of unsaturated lines, this procedure simply concatenates these two lines into a single unsaturated path-line. (See Fig. 5.) Finding a maximum independent set of the remaining unsaturated path-lines is a trivial problem that will be solved last; depending on whether one of the vertices u' and v' is selected in the end, we can fix the solution of the original balanced link.

5.4 Cycle-Bubbles in the Intersection Graph

A *cycle-bubble* in $\mathcal{I}(G)$ is a maximal connected subgraph whose vertices correspond to $\{4,6\}$-cycles. Let H be the subgraph of the underlying pruned ambiguous breakpoint graph including all edges that compose the cycles of a cycle-bubble. The optimal solution for H is either the straight solution τ_H or its alternative $\widetilde{\tau}_H$ (algorithm from our previous work [3]). If both τ_H and $\widetilde{\tau}_H$ have the same score, then H is said to be *balanced*, otherwise it is said to be *unbalanced*.

Proposition 2. *Let an ambiguous component G have cycle-bubbles $H_1, ..., H_m$. There is an optimal solution for G including, for each $i = 1, ..., m$: (1) the optimal solution for H_i, if H_i is unbalanced; or (2) either τ_{H_i} or $\widetilde{\tau}_{H_i}$, if H_i is balanced.*

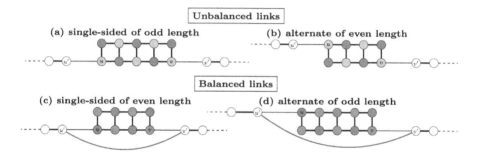

Fig. 5. Double-lines that are balanced and unbalanced links. The yellow solution that in cases (a-b) leaves u and v unselected can be fixed so that an independent set of the adjacent unsaturated path-line(s) can start at v' (and u'). In cases (c-d) either the yellow or the green solution will be fixed later; it will be the one compatible with the solution of the unsaturated-line ending in u' concatenated to the one starting in v'.

Proof. If the whole component G corresponds to one bubble H, the statement is clearly true. Otherwise, we need to examine the intersection with paths. One critical case is a 6-cycle C intersecting three 4-paths, but then there is at least one 2-path to compensate the solution including C (Fig. 6). In general, the best we can get by replacing cycles by paths are co-optimal solutions [4]. □

(a)   **(b)**

Fig. 6. Ambiguous and intersection graphs including a single 6-cycle C (solid edges). Dotted edges are exclusive to paths and dashed gray edges are pruned out. In (a) and (b), C intersects with three 4-paths $P_{11} = u_1..v_1$, $P_{22} = u_2..v_2$ and $P_{33} = u_3..v_3$. In (a), the yellow solution including C also has the three 2-paths $P_{12} = u_1..v_2$, $P_{23} = u_2..v_3$ and $P_{31} = u_3..v_1$, being clearly superior. In (b), the yellow solution still has the 2-path P_{12}, having the same score of the green solution with three 4-paths. (Color figure online)

As a consequence of Proposition 2, if a cycle-bubble is unbalanced, its optimal solution can be fixed so that the unsaturated path-lines around it can be treated separately.

Balanced Cycle-Bubbles Intersecting Path-Flows. If a cycle-bubble H is balanced and intersects with path-flows, then it requires a special treatment. The case in which the intersection involves a single cycle C is also very easy, because certainly either τ_H or $\tilde{\tau}_H$ leaves C unselected and we can safely fix this option. More difficult is when the intersection involves at least two cycles. However, as we will see, here the only case that can be arbitrarily large is easy to handle. Let a cycle-bubble be a *cycle-line* when it consists of a series of valid 6-cycles, such that each pair of consecutive cycles intersect at a $\mathbb{D}$-edge.

Proposition 3. *Cycle-bubbles involving 9 or more cycles must be a cycle-line.*

Proof. A non-linear bubble reaches its "capacity" with 8 cycles, see Fig. 7. □

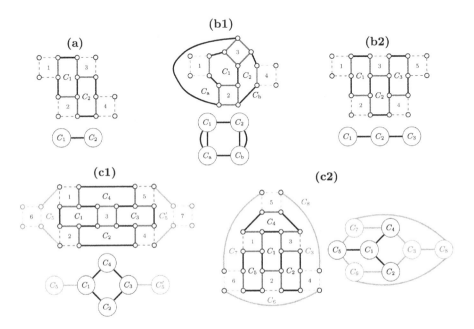

Fig. 7. By increasing the complexity of a bubble we quickly saturate the space for adding cycles to it. Starting with (a) a simple cycle-line of length two, we can either (b1) connect the open vertices of squares 2 and 3, obtaining a cyclic cycle-line of length 4 that cannot be extended, or (b2) extend the line so that it achieves length three. From (b2) we can obtain (c1) a cyclic cycle-line of length 4 that can be extended first by adding cycle C_5 next to C_1 and then either adding C_5' next to C_3 or closing C_6, C_7 and C_8 so that we get (c2). In both cases no further extensions are possible. Note that (c2) can also be obtained by extending a cycle-line of length three and transforming it in a *star* with three branches, that can still be extended by closing C_3, C_6, C_7 and C_8.

Besides having its size limited to 8 cycles, the more complex a non-linear cycles-bubble becomes, the less space it has for paths around it. Therefore, the problem involving these bounded instances could be solved by brute force or complete enumeration. Many of the cases are shown in [4].

Our focus now is the remaining situation of a balanced cycle-line with intersections involving at least two cycles. Recall that cycles can only intersect with unsaturated path-lines. An intersection between a cycle- and a path-line is a *plug connection* when it occurs between vertices that are at the ends of both lines.

Proposition 4. *Cycle-lines of length at least 4 can only have plug connections.*

Proof. Figure 8 shows that a cycle-line of length at least four only has "room" for intersections with path-lines next to its leftmost or rightmost cycles.   □

Fig. 8. A cycle-line of length 4 or larger only allows plug connections.

Balanced cycle-lines with two cycles can have connections to path-lines that are not plugs, but these bounded cases can be solved by complete enumeration or brute force. For arbitrarily large instances, the last missing case is a balanced cycle-line with plug connections at both sides, called a *balanced link*. The procedure here is the same as that for double-lines that are balanced links, where the local solution can only be fixed after fixing those of the outer connections.

5.5   What Remains Is a Set of Independent Unsaturated Path-Lines

If what remains is a single unsaturated path-line of even length, it can even be cyclic. In any case, an optimal solution of each unsaturated path-line can be trivially found: the one selecting all paths with odd numbers must be optimal. Fix this solution and, depending on the connections between the selected vertices of the unsaturated path-line and vertices from balanced cycle-bubbles or balanced double-lines, fix the compatible solutions for the latter ones.

6   Final Remarks and Discussion

This work is an investigation of the complexity of the double distance under a class of problems called σ_k distances, which are between the breakpoint (σ_2) and the DCJ (σ_∞) distance. Extending our previous results that considered only circular genomes, here we presented linear time algorithms for computing the double distance under the σ_4, and under the σ_6 distance, for inputs including linear chromosomes. Our solution relies on the ambiguous breakpoint graph.

The solutions we found so far are greedy with all players being optimal in σ_2, greedy with all players being co-optimal in σ_4 and non-greedy with non-optimal players in σ_6, all of them running in linear time. More specifically for the σ_6

case, after a pre-processing that fixes symmetric squares and triplets, at most two players share an edge. However we can already observe that, as k grows, the number of players sharing a same edge also grows. For that reason, we believe that, if for some $k \geq 8$ the complexity of the σ_k double distance is found to be NP-hard, the complexity is also NP-hard for any $k' > k$. In any case, the natural next step in our research is to study the σ_8 double distance.

Besides the double distance, other combinatorial problems related to genome evolution and ancestral reconstruction, including median and guided halving, have the distance problem as a basic unit. And, analogously to the double distance, these problems can be solved in polynomial time (but differently from the double distance, not greedy and linear) when they are built upon the breakpoint distance, while they are NP-hard when they are built upon the DCJ distance [9]. Therefore, a challenging avenue of research is doing the same exploration for both median and guided halving problems under the class of σ_k distances.

References

1. Bafna, V., Pevzner, P.A.: Genome rearrangements and sorting by reversals. In: Proceedings of FOCS, vol. 1993, pp. 148–157 (1993)
2. Bergeron, A., Mixtacki, J., Stoye, J.: A unifying view of genome rearrangements. In: Bücher, P., Moret, B.M.E. (eds.) WABI 2006. LNCS, vol. 4175, pp. 163–173. Springer, Heidelberg (2006). https://doi.org/10.1007/11851561_16
3. Braga, M.D.V., Brockmann, L.R., Klerx, K., Stoye. J.: A linear time algorithm for an extended version of the breakpoint double distance. In: Proceedings of WABI 2022, vol. 242(13). LIPICs, pp. 1–16 (2022)
4. Braga, M.D.V., Brockmann, L.R., Klerx, K., Stoye, J.: Investigating the complexity of the double distance problems. arXiv: 2303.04205 (2023)
5. Chauve, C.: Personal communication in Dagstuhl Seminar no. 18451 - Genomics, Pattern Avoidance, and Statistical Mechanics (November 2018)
6. El-Mabrouk, N., Sankoff, D.: The reconstruction of doubled genomes. SIAM J. Comput. **32**(3), 754–792 (2003)
7. Hannenhalli, S., Pevzner, P.A.: Transforming men into mice (polynomial algorithm for genomic distance problem). In: Proceedings of FOCS, vol. 1995, pp. 581–592 (1995)
8. Hannenhalli, S., Pevzner, P.A.: Transforming cabbage into turnip: polynomial algorithm for sorting signed permutations by reversals. J. ACM **46**(1), 1–27 (1999)
9. Tannier, E., Zheng, C., Sankoff, D.: Multichromosomal median and halving problems under different genomic distances. BMC Bioinform. **10**, 120 (2009)
10. Yancopoulos, S., Attie, O., Friedberg, R.: Efficient sorting of genomic permutations by translocation, inversion and block interchange. Bioinformatics **21**(16), 3340–3346 (2005)

The Floor Is Lava - Halving Genomes with Viaducts, Piers and Pontoons

Leonard Bohnenkämper[(✉)] 🄳

Faculty of Technology and Center for Biotechnology (CeBiTec), Bielefeld University, Bielefeld, Germany
lbohnenkaemper@techfak.uni-bielefeld.de

Abstract. The Double Cut and Join (DCJ) model is a simple and powerful model for the analysis of large structural rearrangements. After being extended to the DCJ-indel model, capable of handling gains and losses of genetic material, research has shifted in recent years toward enabling it to handle natural genomes, for which no assumption about the distribution of markers has to be made.

Whole Genome Duplications (WGD) are events that double the content and structure of a genome. In some organisms, multiple WGD events have been observed while loss of genetic material is a typical occurrence following a WGD event. Natural genomes are therefore the ideal framework, under which to study this event.

The traditional theoretical framework for studying WGD events is the Genome Halving Problem (GHP). While the GHP is solved for the DCJ model for genomes without losses, there are currently no exact algorithms utilizing the DCJ-indel model.

In this work, we make the first step towards halving natural genomes and present a simple and general view on the DCJ-indel model that we apply to derive an exact polynomial time and space solution for the GHP on genomes with at most two genes per family.

Supplementary material including a generalization to natural genomes can be found at https://doi.org/10.6084/m9.figshare.22269697.

Keywords: Genome Rearrangement · Genome Halving · DCJ-indel

1 Introduction

Whole Genome Duplications (WGD) are evolutionarily far-reaching events in which the content and structure of a genome is doubled. There are lineages with multiple known WGD events [8] and WGDs are typically followed by large-scale loss of genetic content [12].

For genome rearrangement studies, the classic framework for analyzing WGD events, is the *Genome Halving Problem (GHP)*. Broadly speaking, the task formulated by the GHP is, given a present-day genome, to reconstruct a genome with the specific structure that is expected to arise after a WGD. This genome is required to have the minimal distance under some rearrangement model to

K. Jahn and T. Vinař (Eds.): RECOMB-CG 2023, LNBI 13883, pp. 51–67, 2023.
https://doi.org/10.1007/978-3-031-36911-7_4

the given present-day genome. A popular rearrangement model is the *Double-Cut and Join (DCJ)* model [2,14]. This model has been applied to the GHP in 2008 by Mixtacki [10], but as a pure rearrangement model, this solution was not able to account for the losses frequently observed after WGD events. Savard et al. [11] were able to provide a solution to the GHP under DCJ that accounts for lost markers, but not for the loss event itself, meaning they were not able to account for its contribution to the distance. This is likely due to the fact that the DCJ-indel model, which accounts for costs of segmental indel events, was only first formulated around the same time [5].

Another issue touched on briefly here is the reliance of the DCJ-indel model on input genomes with markers occurring at most two times. We call markers occurring more than two times *ambiguous*. Genomes with arbitrary distributions of markers are referred to as *natural*. So far, there are multiple publications studying the distance of natural genomes [4,13], but with multiple rounds of WGDs and other duplication events, natural genomes seem even more suited to study the GHP.

This has been recognized by Avdeyev and Alekseyev, who created an approximate solution for a generalized variant of the GHP [1]. However, as far as we are aware, no exact solutions for the GHP under the DCJ-indel model have been proposed as of yet.

In Sects. 2 and 3, we develop a simple view on the DCJ-indel model that can be derived independently from the ones presented in [5] and [7] . We apply this conceptualization to solve the GHP for genomes without ambiguous, but lost or gained markers in polynomial time in Sect. 4. We note that the GHP is NP-hard for natural genomes under the DCJ-indel model. The hardness proof as well as an ILP for natural genomes based on the formula derived in Sect. 4 are available in the supplement to this work (Sections D,E) [3]. The ILP solving the case of natural genomes will be evaluated in detail in an extended version of this manuscript.

2 Problem Definition

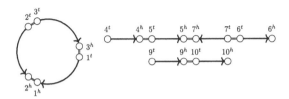

Fig. 1. Genome with one circular and two linear chromosomes. Adjacencies are drawn as double lines.

For the purposes of this work, we view a genome $\mathbb{G}$ as a graph $(X_\mathbb{G}, M_\mathbb{G} \cup A_\mathbb{G})$. $\mathbb{G}$ consists of markers $m := \{m^t, m^h\} \in M_\mathbb{G}$ and their *extremities*, that is their beginning m^t (read "m tail") and an end m^h (read "m head"), which make up the genome's vertices, $X_\mathbb{G} := \{m^t \mid m \in M_\mathbb{G}\} \cup \{m^h \mid m \in M_\mathbb{G}\}$. The structure of a genome is captured by undirected edges $A_\mathbb{G}$, called *adjacencies*. As a shorthand notation, we write ab for an adjacency $\{a, b\}$. Both the set of adjacencies $A_\mathbb{G}$ and the set of markers $M_\mathbb{G}$ form a matching on the extremities $X_\mathbb{G}$.

Each path in $\mathbb{G}$ is simple and alternates between markers and adjacencies. We call a path $p = p_1, p_2, ..., p_{k-1}, p_k$ a *chromosome segment* if it starts and ends with a marker, i.e. $\{p_1, p_2\}, \{p_{k-1}, p_k\} \in M_\mathbb{G}$. If p cannot be extended without repetition, we call p a *chromosome*. If $p_k p_1 \in A_\mathbb{G}$, we refer to p as *circular*. Otherwise (i.e. $\deg(p_1) = \deg(p_k) = 1$), we refer to it as *linear* and call p_1 and p_k *telomeres*. An example of a genome is given in Fig. 1.

Each marker is unique, so in order to find structural similarities, we borrow a concept from biology, namely *homology*. We model homology as an equivalence relation on the markers, i.e. $m \equiv n$ for some $m, n \in M_\mathbb{G}$. We also derive an equivalence relation on extremities with $m^t \equiv n^t \wedge m^h \equiv n^h \iff m \equiv n$, but $m^t \not\equiv n^h \forall m, n$, and a derived equivalence relation on adjacencies, $ab \equiv cd \iff a \equiv c \wedge b \equiv d$ or $a \equiv d \wedge b \equiv c$. The equivalence class $[m]$ of a marker m is also called a *family*. We call a marker m *singular* if it has no homologue but itself in the given problem instance. We call a circular (linear) chromosome consisting only of singular markers a *circular (linear) singleton*.

We now introduce the *Supernatural Graph (SNG)*, a graph structure that will be useful later and meanwhile perfectly illustrates our conceptualization of homology.

Definition 1. *The* Supernatural Graph $\mathcal{SNG}(\mathbb{G}, \equiv)$ *of a genome $\mathbb{G}$ with homology $(\equiv)$ is a graph with vertices $V = X_\mathbb{G}$ and two types of undirected edges $E = E_\gamma \cup E_\xi$, namely adjacency edges $E_\gamma := A_\mathbb{G}$ and extremity edges $E_\xi = \{\{g, h\} \mid g, h \in X_\mathbb{G}, g \neq h, g \equiv h\}$.*

This graph is a variant of the *natural graph* as used in [10] with the difference that adjacencies are modeled as edges instead of vertices and the possibility of more than one homologue per extremity. An example of a SNG can be found in Fig. 2.

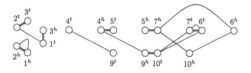

Fig. 2. SNG for genome $\mathbb{A}$ from Fig. 1 resulting from a homology relation $(\equiv_a)$ with the following families: $\{1, 2\}, \{3\}, \{4, 9\}, \{5\}, \{6, 7, 10\}$. Extremities are arranged in the same way as in Fig. 1.

Genomes following a Whole Genome Duplication (WGD) possess a very particular structure due to the nature of this event. Not only is its marker content duplicated, but also its adjacency information is preserved. On simpler homologies, namely if each family contains at most two markers per genome, this is easy to detect. We call such homologies *resolved*. For more complex homologies, it is helpful to find an artificial resolved homology, called a *matching*.

Definition 2. *A matching* ($\overset{*}{\equiv}$) *on a given homology* ($\equiv$) *is a resolved homology for which holds* $m\overset{*}{\equiv}n \implies m \equiv n$ *for any pair of markers* m, n.

We note that when the homology is resolved, at most one extremity edge connects to each vertex in the SNG. Because adjacencies form a matching on the extremities, the resulting SNG consists of only simple cycles and paths. We therefore call such a SNG *simple*. An example is given in Fig. 3. We also note that the simple SNG $\mathcal{SNG}(\mathbb{G}, \overset{*}{\equiv})$ of a genome $\mathbb{G}$ with matching ($\overset{*}{\equiv}$) on the original homology ($\equiv$) contains all adjacency edges of $\mathcal{SNG}(\mathbb{G}, \equiv)$ and a subset of its extremity edges. The simple SNG $\mathcal{SNG}(\mathbb{G}, \overset{*}{\equiv})$ is called a *consistent decomposition* of $\mathcal{SNG}(\mathbb{G}, \equiv)$.

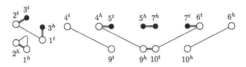

Fig. 3. The simple SNG for genome $\mathbb{A}$ from Fig. 1 resulting from a resolved homology relation ($\overset{*}{\equiv}_a$) is a consistent decomposition of the SNG of Fig. 2 because ($\overset{*}{\equiv}_a$) is a matching on ($\equiv_a$). ($\overset{*}{\equiv}_a$) has the following families: $\{1, 2\}, \{3\}, \{4, 9\}, \{5\}, \{6, 10\}, \{7\}$. Lava vertices (see Sect. 3) filled black.

A genome resulting from a WGD is called *structurally doubled*. To formally define that each marker as well as adjacency information is doubled, we define it via a matching.

Definition 3. *A genome* $\mathbb{G}$ *is called* structurally doubled (SD) *under homology* ($\equiv$) *if and only if there is a matching on* ($\equiv$), *under which each marker and adjacency is in an equivalence class with exactly one other element of* $\mathbb{G}$.

Note that for a resolved homology, this definition corresponds directly with the definition of a *perfectly duplicated* as in [10]. We give an example of a SD genome in Fig. 4.

Over time this specific structure after a WGD is typically erased by rearrangements and losses. Rearrangements in our transformation distance are modeled by the *Double-Cut-And-Join (DCJ)* operation. In principle, a DCJ operation cuts in two places (adjacencies or telomeres) in the genome and reconnects the incident extremities. More formally, we can write as in [2]:

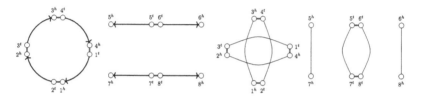

Fig. 4. The genome $\mathbb{B}$ depicted on the left is SD under the resolved homology relation $(\overset{*}{\equiv}_b)$ with families $\{1,3\}, \{2,4\}, \{5,7\}, \{6,8\}$, as illustrated by the SNG on the right. Note that there are also unresolved homology relations, for example $(\equiv_b)$ with families $\{1,3\}, \{2,4,6,8\}, \{5,7\}$, for which $\mathbb{B}$ is SD because $(\overset{*}{\equiv}_b)$ is a matching on $(\equiv_b)$.

Definition 4. *A DCJ operation* transforms up to two adjacencies $ab, cd \in A_{\mathbb{A}}$ *or telomeres* s, t *of genome* $\mathbb{A}$ *in one of the following ways:*
- $ab, cd \to ac, bd$ *or* $ab, cd \to ad, bc$ − $ab \to a, b$
- $ab, s \to as, b$ *or* $ab, s \to bs, a$ − $s, t \to st$

To account for losses or possibly gains following a WGD, we introduce segmental insertions and deletions.

Definition 5. *An* insertion *of length k transforms a genome $\mathbb{A}$ into $\mathbb{A}'$ by adding a chromosome segment $p = p_1, p_2, ..., p_{2k-1}, p_{2k}$ to the genome. It may additionally have either $p_{2k}p_1 \in A_{\mathbb{A}'}$, apply the transformation $ab \to ap_1, p_{2k}b$ for an adjacency ab or the transformation $s \to p_1s$ for a telomere s. A* deletion *of length k removes the chromosome segment $p = p_1, ..., p_{2k}$ and creates the adjacency ab if previously $ap_1, p_{2k}b \in A_{\mathbb{A}}$.*

To reconstruct a SD ancestor to a present day genome, we require this ancestor to be as closely related as possible under our model. We define the GHP first as an abstract problem with an arbitrary set of operations.

Problem 1 (Genome Halving Problem). Given a genome $\mathbb{G}$ and a homology $(\equiv)$, find the minimum length $h(\mathbb{G}, \equiv) = k$ for any legal scenario $\mathbb{G} = \mathbb{G}_0 \overset{o_1}{\to} \mathbb{G}_1..., \overset{o_k}{\to} \mathbb{G}_k$ transforming $\mathbb{G}$ into a structurally doubled genome $\mathbb{G}_k$ with operations $(o_i)_{i=1}^k$. We call $h(\mathbb{G}, \equiv)$ the *halving distance* of $\mathbb{G}$ under $(\equiv)$.

For the DCJ-indel model, the operations used are DCJs (Definition 4) and indels (Definition 5) of arbitrary length. Initially, any sequence of DCJs and indels is legal. However, we will soon see that it makes sense to restrict legal scenarios to avoid an excess of indels.

Problem 2 (DCJ-indel Genome Halving Problem). Given a genome $\mathbb{G}$ and a homology $(\equiv)$, find the minimum length $h_{DCJ}^{id}(\mathbb{G}, \equiv) = k$ for any scenario $\mathbb{G} = \mathbb{G}_0 \overset{o_1}{\to} \mathbb{G}_1..., \overset{o_k}{\to} \mathbb{G}_k$ transforming $\mathbb{G}$ into a structurally doubled genome $\mathbb{G}_k$ with DCJ and indel operations $(o_i)_{i=1}^k$. We call $h_{DCJ}^{id}(\mathbb{G}, \equiv)$ the *DCJ-indel halving distance* of $\mathbb{G}$ under $(\equiv)$.

The original DCJ-indel model allowed indel operations only for singular markers, in that only singular markers could deleted and at most one homologue of a singular marker could be inserted [5]. When restricting legal scenarios in Problem 2 to only include such indel operations, we speak of the *restricted halving distance*.

Note that the restricted halving distance only works properly for a resolved homology, because in an unresolved homology, families with an odd number of markers greater than one cannot be dealt with. However, we can use a matching to circumvent this problem. Since being SD implies a matching ($\dot{\equiv}$) for the reconstructed ancestor $\mathbb{G}_k$, by applying the rearrangement scenario backwards, we know that there must exist the same matching on $\mathbb{G}$, although some markers might not have a homologue. The resulting problem formulation can be used to solve Problem 2 as we show in supplement Section A.1 [3].

Problem 3. Given a genome and a homology ($\equiv$), find a matching ($\dot{\equiv}$) on ($\equiv$), such that the *restricted DCJ-indel halving distance* $\overline{h_{DCJ}^{id}}(\mathbb{G}, \dot{\equiv})$ is minimized.

A typical issue that arises in these types of problems once one introduces segmental insertion and deletion operations is that without restrictions, the shortest way to sort is often by just deleting one genome and inserting another, resulting in an empty matching. We therefore work within an established framework, namely *Maximum Matching* [9].

Definition 6. *A matching ($\dot{\equiv}$) on a homology ($\equiv$) is called a* maximum matching *if for each family under ($\equiv$) there is at most one singular marker under ($\dot{\equiv}$).*

We can then formulate the GHP for the Maximum Matching model as follows:

Problem 4. Given a genome and a homology ($\equiv$), find a valid maximum matching ($\dot{\equiv}$) on ($\equiv$), such that the restricted DCJ-indel halving distance $\overline{h_{DCJ}^{id}}(\mathbb{G}, \dot{\equiv})$ is minimized.

For resolved homologies ($\overset{*}{\equiv}$), we note that a maximum matching must have the same families, that is ($\overset{*}{\equiv}$) = ($\dot{\equiv}$). Thus, for resolved homologies, we do not need to determine a matching. For arbitrary homologies, we find

Proposition 1. *Determining a maximum matching $\dot{\equiv}$ on a natural genome $\mathbb{G}$ with homology ($\equiv$), such that its restricted DCJ-indel halving distance $\overline{h_{DCJ}^{id}}(\mathbb{G}^{\dot{\pm}})$ is minimized, is an NP-hard problem.*

A proof for this proposition can be constructed building on Caprara's alternating cycle decomposition problem [6]. We give it in Section D of the supplementary material to this work [3].

This proof, but also our derivation of the halving distance formula, rely heavily on properties of the SNG. These properties are shared with other data structures, such as the Multi-Relational Diagram in [4]. In order to make our results as reusable as possible, we start by making observations about a generalized version of the SNG:

Definition 7. *A rearrangement graph* $\mathcal{G} = (V, E_\gamma \cup E_\xi, g)$ *is an undirected graph with two types of edges, namely a matching* E_γ *called* adjacency edges, *additional edges* E_ξ *called* extremity edges *and a labeling* g *assigning each vertex* $v \in V$ *to a genome* $g(v)$ *with* $g(v) = g(u)$ *if* $uv \in E_\gamma$.

Similar to the SNG, we refer to rearrangement graphs with at most one extremity edge connecting to each vertex as *simple*. Because then E_γ and E_ξ both form matchings on V, simple rearrangement graphs too consist only of simple cycles and simple paths. We make a few general observations about the DCJ-indel model on simple rearrangement graphs in Sect. 3 before coming back to the concrete case of the halving problem in Sect. 4.

3   The Properties of Simple Rearrangement Graphs

In this section, we examine the general properties of rearrangement graphs which we will later use in Sect. 4 to derive our result on the halving distance. We denote the universe of simple cycles and paths by Ω and the components of a concrete simple rearrangement graph $\mathcal{G}$ by $\Omega(\mathcal{G})$. The universe here means the collection of all possible such components in all possible rearrangement graphs. Oftentimes, we are interested in certain subsets of $\Omega(\mathcal{G})$, such as the set of cycles or paths and the way they react if a DCJ operation is applied to the adjacencies and telomeres of the graph. To that end, we refer to any subset $\mathbb{K} \subset \Omega$ as a *component subset* and define $\mathbb{K}(\mathcal{G}) := \mathbb{K} \cap \Omega(\mathcal{G})$. We refer to $\mathbb{K}(\mathcal{G})$ as an *instance of a component subset*.

We will characterize DCJ and indel operations by their effects on different quantities, such as cardinalities of component subset instances. To that end, we define the following:

Definition 8. *Given a rearrangement graph* $\mathcal{G}$ *before and a rearrangement graph* $\mathcal{G}'$ *after a given sequence of operations* o *as well as a quantity* $q(\mathcal{G})$ *before and the same quantity* $q(\mathcal{G}')$ *after the operation, we define the* difference in q *induced by* o *as*

$$\Delta_o q = q(\mathcal{G}') - q(\mathcal{G}). \tag{1}$$

Before we can characterize the effect of different operations, we need to distinguish between different component subsets by different characteristics, such as the genome of certain vertices or the number of extremity edges. We use a similar distinction as in [7]. We call a component *odd* if the number of extremity edges it contains is odd and *even* otherwise. We denote the component subset of all cycles by $\mathbb{C}$ and those of even and odd cycles by $\mathbb{C}_o$ and $\mathbb{C}_|$ respectively. For paths, we opt for a more fine-grained distinction. We refer to the vertices a path starts or ends in as *endpoints*. Since endpoints have degree 1, they either lack an extremity edge or an adjacency edge. We refer to vertices without an extremity edge as *extremity edge* and denote them by their genome in lower case letters (e.g. a). We refer to the rest of the vertices as *safe vertices* and call those vertices without an incident adjacency edge *telomeres*. We denote telomeres by their genome in upper case letters (e.g. A). Note that a vertex can both be a lava vertex and a telomere. We can then write the component subset of all paths with endpoints α, β as $\mathbb{P}_{\alpha\beta}$. We denote the subset of even paths of this kind by $\mathbb{P}_{\alpha\circ\beta}$ and for odd paths as $\mathbb{P}_{\alpha|\beta}$. For example, the component subset of odd paths ending in a telomere in a genome $\mathbb{A}$ and a lava vertex in a genome $\mathbb{B}$ is denoted by $\mathbb{P}_{A|b}$. As a more intuitive classification, we refer to paths ending in two distinct lava vertices as *pontoons*, to paths ending in two safe telomeres as *viaducts* and to paths ending in both a telomere and a lava vertex as *piers*. As a shorthand, we use the notation $k := |\mathbb{K}(\mathcal{G})|$ for the cardinality of an instance of a component subset $\mathbb{K}$ and K for a generic member $K \in \mathbb{K}(\mathcal{G})$. An example of this notation as well as of the terms we introduced here is given in Fig. 5.

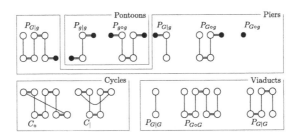

Fig. 5. Classification of components in a simple SNG. Lava vertices filled black.

In order to not create chimeric genomes, we only allow operations if all vertices involved have the same genome label. We can then conceptualize a DCJ operation not as an operation $ab, cd \rightarrow ac, bd$ transforming adjacencies ab, cd of a rearrangement graph $\mathcal{G}$ into ac, bd of $\mathcal{G}'$, but as transforming the components $K_{ab}, K_{cd} \rightarrow K_{ac}, K_{bd}$ containing these adjacencies with $K_{ab}, K_{cd} \in \Omega(\mathcal{G}), K_{ac}, K_{bd} \in \Omega(\mathcal{G}')$. As an immediate result from this notation, we can see that for any component subset $\mathbb{K}$, the instances $\mathbb{K}(\mathcal{G})$ and $\mathbb{K}(\mathcal{G}')$ can differ by at most 4 elements, that is $\mathbb{K}(\mathcal{G}) \setminus \mathbb{K}(\mathcal{G}') \subseteq \{K_{ab}, K_{cd}\}$ and $\mathbb{K}(\mathcal{G}') \setminus \mathbb{K}(\mathcal{G}) \subseteq \{K_{ac}, K_{bd}\}$. We call the components whose adjacencies are transformed *sources* and the components resulting from the operation *resultants*. All DCJ operations (see Definition 4) have at most two adjacencies as input and output and thus at most two sources and resultants, so we conclude:

Observation 1. *For any component subset $\mathbb{K}$ and any DCJ operation d holds*

$$-2 \leq \Delta_d |\mathbb{K}| \leq 2. \tag{2}$$

We notice that this can be generalized for any number of component subsets as long as they are disjoint.

Corollary 1. *For any collection of disjoint component subsets L and any DCJ operation d holds*

$$-2 \leq \Delta_d \sum_{\mathbb{K} \in L} |\mathbb{K}| \leq 2. \tag{3}$$

Cycles form a special case of Observation 1. Notice that a DCJ operation with a cycle C and another component K as sources will always integrate the cycle into K, forming a composite component K'. The only other way to reduce the number of cycles is to linearize it, obtaining a viaduct (see also Fig. 6).

Observation 2. *The only operations reducing the number of cycles are of the form $C, K \to K'$ or $C \to P_{XX}$, with K, K' some component and P_{XX} an even or odd viaduct.*

Fig. 6. The only DCJ operations reducing the number of cycles in a rearrangement graph are of the form $C, K \to K'$ or $C \to P_{XX}$. Squiggled lines represent arbitrary paths in the graph.

We also notice that the cardinality of the more specific component subsets of even and odd cycles can only be reduced in the same manner. All DCJ operations have an inverse. Therefore, the only operations increasing the respective cardinalities must be the "mirror image" of those seen in Fig. 6.

Corollary 2. *For the component subsets of even and odd cycles $\mathbb{C}_\circ, \mathbb{C}_|$ and any DCJ operation d holds $-1 \leq \Delta_d |\mathbb{C}_\circ| \leq 1$ and $-1 \leq \Delta_d |\mathbb{C}_|| \leq 1$.*

For even cycles, things are even more specific. In this case, for $C \to P_{XX}$ the viaduct P_{XX} will always be even. Additionally, for $C, K \to K'$, the components K and K' do not differ in parity or endpoints. We can thus observe:

Observation 3. *If the number of even cycles is changed by DCJ d ($\Delta_d |\mathbb{C}_\circ| \neq 0$), the number of piers, pontoons and odd viaducts does not change, that is, $\Delta_d |\mathbb{P}_{\alpha(*)\beta}| = 0$ for any $\alpha, \beta \in \{X, x, Y, y\}$ and $(*) \in \{\circ, |\}$ with $\alpha \neq \beta$ or $(*) = |$.*

Next, we observe that DCJ operations affect only adjacency edges, so lava vertices in sources are transferred to resultants. Thus, at most one viaduct can be among the resultants of an operation with a lava vertex in its sources (see also Fig. 7).

Observation 4. *If the number of any type of pier or pontoon is reduced by DCJ d ($\exists (*_1) \in \{\circ, |\} : \Delta_d |\mathbb{P}_{W(*_1)z}| < 0 \vee \Delta_d |\mathbb{P}_{w(*_1)z}| < 0$), the number of any type of viaduct can increase by at most 1, that is, $\Delta_d |\mathbb{P}_{X(*_2)Y}| \leq 1$ for any genomes $\mathbb{X}, \mathbb{Y}, \mathbb{W}, \mathbb{Z}, (*_2) \in \{\circ, |\}$.*

Fig. 7. DCJ operations involving at least one pier or pontoon as a source have at least one resultant that is a pier or pontoon.

Pontoons contain two lava vertices, which are distributed to both resultants (see also Fig. 8). Thus, if their number is reduced, the number of viaducts cannot increase.

Observation 5. *If DCJ d reduces the number of any type of pontoons ($\exists (*_1) \in \{\circ, |\} : \Delta_d |\mathbb{P}_{w(*_1)z}| < 0$), the number of any type of viaduct does not increase, that is, $\Delta |\mathbb{P}_{X(*_2)Y}| \leq 0$ for any genomes $\mathbb{X}, \mathbb{Y}, \mathbb{W}, \mathbb{Z}, (*_2) \in \{\circ, |\}$.*

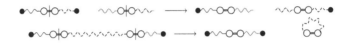

Fig. 8. DCJ operations with a pontoon as a source either result in two components with at least one lava vertex each or a pontoon and a cycle.

Lastly, a DCJ operation cannot change the number of extremity edges in the components it affects. Thus, the total parity of sources and resultants of a DCJ operation is preserved. In order to capture this, we use the notation $(*_i) + (*_j)$ for arbitrary $(*_i), (*_j) \in \{\circ, |\}$ with $(*_i) + (*_j) = \circ$ if $(*_i) = (*_j)$ and $(*_i) + (*_j) = |$ otherwise. We can then write.

Observation 6. *For any DCJ operation d of the form $P_{\alpha(*_1)\beta}, P_{\gamma(*_2)\varsigma} \to P_{\alpha(*_3)\gamma}, P_{\beta(*_4)\varsigma}$ holds $(*_1) + (*_2) = (*_3) + (*_4)$.*

We have so far thoroughly investigated the effects of the DCJ operation. We now briefly touch on indel operations. Since we have seen in Problem 3 that we can reformulate the GHP, such that only singular markers have to be part of indels, we permit only deletions removing lava vertices or insertions creating homologues for lava vertices. First, we study only the deletion of a single marker and the insertion of a single adjacency, which refer to as *uni-indels* collectively. While a uni-indel deletion is always a legal operation under the DCJ-indel model, a uni-indel insertion is not necessarily legal. However, we will soon see that they are useful for describing insertions, so we permit them as an intermediate step of an incomplete insertion. Both types of uni-indels remove two lava vertices from the graph and connect their adjacent vertices into one component. The similarity is visualized in Fig. 9. Since the number of extremity edges is changed by 2 if at all, the total parity is again conserved.

Observation 7. *Uni-indels are of the form*

$$P_{\alpha(*_1)y}, P_{y(*_2)\beta} \rightarrow P_{\alpha(*_3)\beta} \quad or \quad P_{y(*_4)y} \rightarrow C_{(*_5)} \tag{4}$$

for some genomes $\mathbb{X}, \mathbb{Y}, \mathbb{Z}$ *where* $\alpha \in \{x, X\}, \beta \in \{z, Z\}, (*_1) + (*_2) = (*_3),$ *and* $(*_4) = (*_5).$

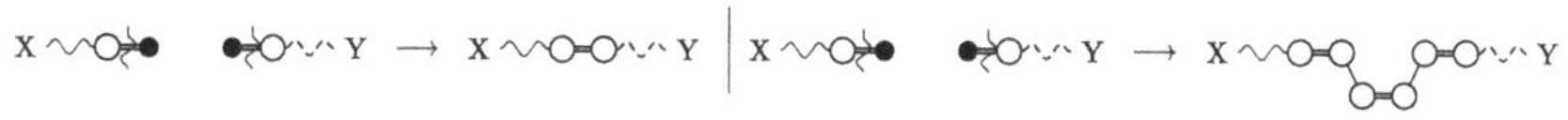

Fig. 9. Uni-indels transform the rearrangement graph in similar ways, joining two paths ending in a lava vertex into a path or cycle (if X, Y are connected). Left: Deletion of a single marker. Right: Insertion of a single adjacency.

We can conceptualize a deletion of l markers as first summarizing the stretch of markers only separated by $l - 1$ pontoons of length 0 and then applying a uni-indel to the "summary" marker.

Observation 8. *For a deletion* δ *of markers* $m_1, ..., m_l$ *there is a uni-indel* u *on a rearrangement graph where* $m_1, ..., m_l$ *are replaced by* $\tilde{m}$ *with* $\Delta_\delta \mathbb{K} = \Delta_u \mathbb{K}$ *for any component subset* $\mathbb{K}$ *that does not contain pontoons of length 0.*

We can conceptualize an insertion of l markers as first inserting the circular chromosome $m_1, ..., m_l$ (i.e. by l uni-indels) and then possibly applying a single DCJ-operation integrating the chromosome into the target adjacency.

Observation 9. *For an insertion* ι *of markers* $m_1, ..., m_l$ *there are uni-indels* $(u_i)_{i=1}^l$ *and a DCJ operation* d *on the same rearrangement graph for which holds* $\Delta_\iota \mathbb{K} = \Delta_{du_1 u_2 ... u_l} \mathbb{K}$ *or* $\Delta_\iota \mathbb{K} = \Delta_{u_1 u_2 ... u_l} \mathbb{K}.$

4   DCJ-Indel Halving for Genomes with Resolved Homology

We now have all ingredients to address the GHP for a resolved homology. Similar to [7], the following can be shown (see supplement Section A.2 [3]).

Proposition 2. *The restricted DCJ-indel halving distance for a genome* $\mathbb{G}$ *with circular singletons* $C_1, .., C_k$ *under resolved homology* $(\overset{*}{\equiv})$ *is*

$$\overline{h_{DCJ}^{id}}(\mathbb{G}, \overset{*}{\equiv}) = \overline{h_{DCJ}^{id}}(\mathbb{G}', \overset{*}{\equiv}) + k \tag{5}$$

where $\mathbb{G}'$ *contains the same chromosomes as* $\mathbb{G}$ *except* $C_1, .., C_k.$

Circular singletons can therefore be dealt with in preprocessing and we need not consider them from now on. We start by establishing a lower bound for the problem without circular singletons.

Proposition 3. *For a genome* $\mathbb{G}$ *with a resolved homology* $(\overset{\star}{\equiv})$ *containing no circular singletons holds*

$$\overline{h_{DCJ}^{id}}(\mathbb{G}, \overset{\star}{\equiv}) \geq n - c_\circ + \left\lceil \frac{p_{g|g} + \max(p_{G\circ g}, p_{G|g}) - p_{G|G}}{2} \right\rceil, \tag{6}$$

where $n = |\{[m] \mid m \in M_{\mathbb{G}}, |[m]| = 2\}|$.

Recall that $p_{g|g}$ is the number of odd pontoons, that is, odd paths ending in two lava vertices, $p_{G\circ g}$ and $p_{G|g}$ are the numbers of even and odd piers respectively, which end in a lava vertex as well as a telomere and $p_{G|G}$ is the number of odd viaducts, which end in two telomeres. For an example, see Fig. 5. To more easily address individual terms, we use the following shorthands,

$$H := n - c_\circ + Q := n - c_\circ + \left\lceil \frac{q}{2} \right\rceil \tag{7}$$

$$:= n - c_\circ + \left\lceil \frac{p_{g|g} + \max(p_{G\circ g}, p_{G|g}) - p_{G|G}}{2} \right\rceil. \tag{8}$$

We start our proof of Proposition 3 by showing the following:

Proposition 4. *For a genome* $\mathbb{G}$ *with a resolved homology* $(\overset{\star}{\equiv})$ *containing no circular singletons holds that* $\mathbb{G}$ *is structurally doubled iff* $n - c_\circ + Q = 0$.

Proof. If there are no singular markers, H reduces to

$$n - c_\circ + \left\lceil \frac{-p_{G|G}}{2} \right\rceil = n - \left(c_\circ + \left\lfloor \frac{p_{G|G}}{2} \right\rfloor\right), \tag{9}$$

which is the DCJ halving formula by Mixtacki and therefore has already been shown to be 0 if and only if $\mathbb{G}$ is SD [10]. If there are singular markers, obviously $\mathbb{G}$ cannot be SD, so the forward direction is trivial. For the backward direction, notice that every extremity of a non-singular marker needs to either contribute to a 2-cycle or 1-path to reduce $n - (c_\circ + \lfloor \frac{p_{G|G}}{2} \rfloor)$ to zero. Therefore, every adjacency and telomere containing an extremity of a non-singular marker must have an equivalent. Thus, extremities without an equivalent can occur only as part of singleton chromosomes. Since our premise is that the genome contains no circular singletons, only linear singletons can remain. These, however, contain even piers $P_{G\circ g}$ at their ends, thus $\max(p_{G\circ g}, p_{G|g}) > 0$ and with that $H > 0$, a contradiction. Therefore, $\mathbb{G}$ must be SD.

We observe that DCJ operations can change H by at most 1.

Proposition 5. *For any DCJ operation* d *holds* $\Delta_d(n - c_\circ + Q) \geq -1$.

Proof. DCJ operations do not affect n. As we have seen in Observation 3, if $c_\circ$ changes, none of the terms in $Q = \left\lceil \frac{p_{g|g} + \max(p_{G\circ g}, p_{G|g}) - p_{G|G}}{2} \right\rceil$ can change at the same time. Since $\Delta_d c_\circ \geq -1$ for a DCJ operation d (see Corollary 2), we only need to concern ourselves with Q. To resolve the maximizations in the formula, we observe that the maximum can change at most as one of its elements.

Observation 10. *For any operation o holds*

$$\Delta_o \max(x, y) \geq \Delta_o x \ or \ \Delta_o \max(x, y) \geq \Delta_o y. \tag{10}$$

Together with Corollary 1, we are able to derive

Corollary 3. *For a given DCJ operation d, there is $x \in \{p_{Gog}, p_{G|g}\}$ with*

$$\Delta_d(p_{g|g} + \max(p_{Gog}, p_{G|g})) \geq \Delta_d(p_{g|g} + x + p_{G|G}) \overset{Cor. \ 1}{\geq} -2. \tag{11}$$

We see that the only way the numerator $q = p_{g|g} + \max(p_{Gog}, p_{G|g}) - p_{G|G}$ could be reduced by more than two is if $p_{G|G}$ is increased and another term is decreased at the same time. Because of Observation 5 we know that this cannot be $p_{g|g}$. Because any DCJ operation with a lava vertex in one of its resultants creates at most one viaduct (see Observation 4), the only operations that could decrease q by more than 2 are of the form $P_{G(*_1)g}, P_{G(*_2)g} \rightarrow P_{G|G}, P_{g(*_3)g}$. From Observation 6, we know that $(*_3) = (*_1) + (*_2) + |$ and thus, either the sources are of different parity, i.e. $P_{G(*_1)g}, P_{G(*_2)g} = P_{Gog}, P_{G|g}$, from which follows $\Delta \max(p_{Gog}, p_{G|g}) \geq -1$ or one resultant is an odd pontoon, i.e. $P_{g(*_3)g} = P_{g|g}$ and therefore $\Delta p_{g|g} = 1$. In both cases, we have $\Delta q \geq -2$ and therefore $\Delta H \geq 1$. Thus, this concludes our proof of Proposition 5.

We are left to examine the effect of indel operations. Regarding Observation 7, we see that a uni-indel either concatenates two piers or pontoons, thereby possibly creating a viaduct or a cycle from an even pontoon. From this follows that a uni-indel u has either $\Delta_u c_o \leq 1$ and $\Delta_u Q = 0$ or $\Delta_u c_o = 0$ and $\Delta_u Q \geq -1$. Since none of the component subsets in H contains (even) pontoons of length 0, we conclude using Observation 8:

Observation 11. *For any deletion δ holds $\Delta_\delta H \geq -1$.*

Conceptualizing insertions as an insertion of a circular chromosome followed by a DCJ (see Observation 9), we see that the insertion of a circular chromosome with k markers increases n by k. On the other hand, the k uni-indels creating its adjacencies decrease $-c_o + Q$ by at most k. The final DCJ has $\Delta H \geq -1$. Therefore,

Observation 12. *For any insertion ι holds*

$$\Delta_\iota H = \Delta_\iota n + \Delta_\iota(-c_o + Q) \geq k - k - 1 = -1. \tag{12}$$

Together, from Observations 11, 12 and Propositions 4, 5 follows Proposition 3.

We give a sorting algorithm that attempts to achieve the lower bound given in Proposition 3 in Algorithm 1. Every step in the algorithm is conceived as a DCJ operation $X, Y \rightarrow W, Z$ transforming X, Y into W, Z. This is not always possible without creating a circular singleton, which can only arise if an even pontoon $Z = P_{gog}$ of length 0 is created. In these cases, we have written the operation as $X, Y \rightarrow W, (Z)^*$ instead. If creating Z would generate a circular singleton, we simply replace the operation by the deletion $X, Y \rightarrow W$. This is possible because if Z would be part of a circular singleton, it means that there is a chromosome segment connecting X and Y. Alternatively, the deletion can also be replaced by an equivalent, but biologically possibly more meaningful insertion (for details see supplement Section B [3]).

Algorithm 1. DCJ-indel Halving

1: **while** $p_{g|g} > 1$ **do** $P_{g|g}, P_{g|g} \rightarrow P_{g \circ g}, (P_{g \circ g})^*$

2: **if** $p_{g|g} = 1$ **then**   ▷ $p_{g|g} \in \{0, 1\}$

3:     **if** $p_{G|g} < p_{G \circ g}$ **then** $P_{G \circ g}, P_{g|g} \rightarrow P_{G|g}, (P_{g \circ g})^*$

4:     **else**

5:        **if** $p_{G \circ g} < p_{G|g}$ **then** $P_{G|g}, P_{g|g} \rightarrow P_{G \circ g}, (P_{g \circ g})^*$

6:        **else** $P_{g|g} \rightarrow C_|, (P_{g \circ g})^*$

7: **while** $p_{G \circ g} > 0 \wedge p_{G|g} > 0$ **do** $P_{G \circ g}, P_{G|g} \rightarrow (P_{g \circ g})^*, P_{G|G}$

8: ▷ At most one of $p_{G|g}$ and $p_{G \circ g}$ is not 0.

9: **while** $P_{G \circ g} > 0$ **do** $P_{G \circ g}, P_{G \circ g} \rightarrow P_{G \circ G}, (P_{g \circ g})^*$ ▷ Both $p_{G|g}$ and $p_{G \circ g}$ are even.

10: **while** $P_{G|g} > 0$ **do** $P_{G|g}, P_{G|g} \rightarrow P_{G \circ G}, (P_{g \circ g})^*$

11: ▷ No piers, only even pontoons remaining

12: **for all** $P_{g \circ g}$ **do**

13:     $P_{g \circ g} \rightarrow (P'_{g \circ g})^*, C_\circ$ ▷ Choose operation, such that $|P'_{g \circ g}| = 0$.

14: Perform DCJ-halving   ▷ All remaining paths are viaducts.

We now regard the assertions written as comments. Most assertions in the algorithm simply follow from the preceding while-conditions. Only the assertion in Line 9 might need some clarification. We can see that this assertion holds because if there is at least one remaining pier, we can create a bijection between piers. This can be done by "skipping" from one lava vertex to the other lava vertex of the same marker, thereby either reaching either another pier or a pontoon where we can repeat the procedure until we finally reach another pier. Since there is only one type of pier in the graph at that point (see Line 8), the number of that type must be even.

Our algorithm reduces H by 1 in almost every step by either having $\Delta q = -2$ or $-\Delta c_\circ = -1$ (see also supplement Section C [3]). The only exception is Step 6. Unfortunately, it is not generally true that the numerator q is odd in this case, so the algorithm only reaches a modified bound. However, we will show that this new bound is correct.

Theorem 1. *For the Maximum Matching halving of a genome $\mathbb{G}$ with resolved homology ($\overset{*}{=}$) without circular singletons holds*

$$\overline{h_{DCJ}^{id}}(\mathbb{G}, \overset{*}{=}) = n - c_\circ + \left\lceil \frac{p_{g|g} + \max(p_{G \circ g}, p_{G|g}) - p_{G|G} + \delta}{2} \right\rceil \tag{13}$$

where $\delta = 1$ if $p_{g|g}$ is odd and $p_{G|g} = p_{G \circ g}$, $\delta = 0$ otherwise.

Proof. Algorithm 1 shows that Theorem 1 is an upper bound on the halving distance. What remains to be seen is whether it is a lower bound as well.

We write $H' := n - c_o + \left\lceil \frac{q+\delta}{2} \right\rceil$. Since δ is non-negative and 0 if the genome is SD, Proposition 4 holds for both H and H'. Because of Proposition 3, the only way H' could be reduced by more than 1 is if $\Delta\delta = -1$. We discuss DCJ operations first. If $\Delta \max(p_{Gog}, p_{G|g}) \geq 1$, after cancellation the terms remaining are $\Delta q \geq \Delta(p_{g|g} - p_{G|G}) \geq -2$ (see Observations 1 and 5). Therefore, if $\Delta H' \leq -2$, we know that the maximization term $\Delta \max(p_{Gog}, p_{G|g})$ is not positive. We now distinguish two cases, in which $\Delta\delta = -1$.

(I) For the first, let us presume that after the operation, $p_{Gog} \neq p_{G|g}$. In conjunction with the fact that a DCJ operation has at most two sources, this means that either $\Delta p_{Gog} < 0$ or $\Delta p_{G|g} < 0$, but not both. Therefore $\Delta \max(p_{Gog}, p_{G|g}) = 0$. Plugging this and $\delta = -1$ in the formula, we arrive at $\Delta q = \Delta p_{g|g} - \Delta p_{G|G} - 1$. Using Observation 5, we then know that $\Delta q \geq \min(\Delta p_{g|g} - 1, -\Delta p_{G|G} - 1)$. As at least one of the sources of the operation must be a pier, we know that $\Delta p_{g|g} \geq -1$ and using Observation 4 that $\Delta - p_{G|G} \geq -1$. Therefore, for this case $\Delta(q + \delta) \geq -2$.

(II) Let us now presume that after the operation still $p_{Gog} = p_{G|g}$. From $\Delta\delta = -1$ then follows that $\Delta p_{g|g} \in \{-1, 1\}$.

(i) Let $\Delta p_{g|g} = -1$. It then follows that $\Delta \max(p_{Gog}, p_{G|g}) \geq 0$, because one source is already "blocked" by the odd pontoon and we would need two piers as sources to reduce the term. Since $\Delta p_{G|G} \leq 0$ (see Observation 5), we have $\Delta(q + \delta) \geq -1 + 0 - 0 - 1 = -2$ in this case.

(ii) For $\Delta p_{g|g} = 1$, we then have $\Delta(q+\delta) = \Delta(\max(p_{Gog}, p_{G|g}) - p_{G|G})$. Because of $p_{Gog} = p_{G|g}$, we have $\Delta \max(p_{Gog}, p_{G|g}) \geq -1$. Using Observation 4, we know that $\Delta(\max(p_{Gog}, p_{G|g}) - p_{G|G}) \geq -2$.

We thus see that in all cases for DCJ operations holds $\Delta(q + \delta) \geq -2$.

For uni-indels, we see using Observation 7 that to reduce δ by 1, they can either fuse two piers of the same parity, fuse a pier with a pontoon or form an odd cycle from an odd pontoon. All of these have $\Delta \max(p_{Gog}, p_{G|g}) \geq 0$, $\Delta p_{G|G} = 0$, $\Delta c_o = 0$ and $\Delta p_{g|g} \geq -1$. Thus, for uni-indels $\Delta H' \geq -1$. Using Observations 8 and 9, we can conclude that indels also have $\Delta H' \geq -1$. Thus, H' is an upper as well as lower bound for the restricted halving distance for a resolved homology.

5   Conclusion and Outlook

After presenting some general statements about a generalization of simple SNGs, we were able to derive a fairly simple formula for the halving distance under the DCJ-indel model for genomes with resolved homology. We note that the problem is NP-hard on arbitrary homologies. However, due to its simplicity, the formula is generalizable to natural genomes using an ILP similar to those in in [13] and [4]. We will present this ILP and an extensive analysis of its performance in an extended version of this work.

Acknowledgements. I thank my Master Thesis supervisors, Jens Stoye and Marília D. V. Braga for their helpful comments in discussions regarding notation, terms and the overall structure of the paper. Thanks also to Tizian Schulz for giving the paper a read and to Diego Rubert for pointing me to the book by Fertin et al. I furthermore thank the anonymous reviewers for their comments, which helped me a lot to increase the readability of the text. Lastly, I thank Daniel Doerr for making me switch from water to lava and for suggesting the catchy first part of this work's title.

References

1. Avdeyev, P., Alekseyev, M.A.: Linearization of ancestral genomes with duplicated genes. In: Proceedings of the 11th ACM International Conference on Bioinformatics, Computational Biology and Health Informatics. BCB '20, Association for Computing Machinery, New York, NY, USA (2020). https://doi.org/10.1145/3388440.3412484
2. Bergeron, A., Mixtacki, J., Stoye, J.: A unifying view of genome rearrangements. In: Bücher, P., Moret, B.M.E. (eds.) Algorithms in Bioinformatics, pp. 163–173. Springer, Berlin Heidelberg, Berlin, Heidelberg (2006)
3. Bohnenkämper, L.: The floor is lava - halving genomes - supplementary material (Mar 2023). https://doi.org/10.6084/m9.figshare.22269697, https://figshare.com/articles/journal_contribution/supplement-14-03_pdf/22269697/1
4. Bohnenkämper, L., Braga, M.D.V., Doerr, D., Stoye, J.: Computing the rearrangement distance of natural genomes. J. Comput. Biol. **28**(4), 410–431 (2021). https://doi.org/10.1089/cmb.2020.0434, pMID: 33393848
5. Braga, M.D.V., Willing, E., Stoye, J.: Double cut and join with insertions and deletions. J. Comput. Biol **18**(9), 1167–1184 (2011). https://doi.org/10.1089/cmb.2011.0118 pMID: 21899423
6. Caprara, A.: Sorting by reversals is difficult. In: Proceedings of The First Annual International Conference On Computational Molecular Biology, pp. 75–83 (1997)
7. Compeau, P.E.: DCJ-indel sorting revisited. Algorithms for molecular biology : AMB **8**(1), 6–6 (Mar 2013). https://doi.org/10.1186/1748-7188-8-6, https://pubmed.ncbi.nlm.nih.gov/23452758, 23452758[pmid]
8. Dehal, P., Boore, J.L.: Two rounds of whole genome duplication in the ancestral vertebrate. PLoS biology **3**(10), e314–e314 (Oct 2005). https://doi.org/10.1371/journal.pbio.0030314, https://pubmed.ncbi.nlm.nih.gov/16128622, 16128622[pmid]
9. Fertin, G., Labarre, A., Rusu, I., Tannier, E., Vialette, S.: Combinatorics of Genome Rearrangements. The MIT Press, Computational Molecular Biology (2009)
10. Mixtacki, J.: Genome halving under DCJ revisited. In: Hu, X., Wang, J. (eds.) Computing and Combinatorics, pp. 276–286. Springer, Berlin Heidelberg, Berlin, Heidelberg (2008)
11. Savard, O.T., Gagnon, Y., Bertrand, D., El-Mabrouk, N.: Genome halving and double distance with losses. J. Comput. Biol. **18**(9), 1185–1199 (2011). https://doi.org/10.1089/cmb.2011.0136pMID: 21899424
12. Scannell, D.R., Byrne, K.P., Gordon, J.L., Wong, S., Wolfe, K.H.: Multiple rounds of speciation associated with reciprocal gene loss in polyploid yeasts. Nature **440**(7082), 341–345 (Mar 2006). https://doi.org/10.1038/nature04562

13. Shao, M., Lin, Y., Moret, B.M.: An exact algorithm to compute the double-cut-and-join distance for genomes with duplicate genes. J. Comput. Biol. **22**(5), 425–435 (2015). https://doi.org/10.1089/cmb.2014.0096 pMID: 25517208
14. Yancopoulos, S., Attie, O., Friedberg, R.: Efficient sorting of genomic permutations by translocation, inversion and block interchange. Bioinformatics 21(16), 3340–3346 (06 2005). https://doi.org/10.1093/bioinformatics/bti535

Two Strikes Against the Phage Recombination Problem

Manuel Lafond[1]([⊠]), Anne Bergeron[2], and Krister M. Swenson[3]

[1] Université de Sherbrooke, Sherbrooke, Canada
manuel.lafond@USherbrooke.ca
[2] Université du Québec à Montréal, Montréal, Canada
[3] CNRS, LIRMM, Université de Montpellier, Montpellier, France

Abstract. The recombination problem is inspired by genome rearrangement events that occur in bacteriophage populations. Its goal is to explain the transformation of one bacteriophage population into another using the minimum number of recombinations. Here we show that the combinatorial problem is NP-Complete, both when the target population contains only one genome of unbounded length, and when the size of the genomes is bounded by a constant. In the first case, the existence of a minimum solution is shown to be equivalent to a 3D-matching problem, and in the second case, to a satisfiability problem. These results imply that the comparison of bacteriophage populations using recombinations may have to rely on heuristics that exploit biological constraints.

1 Introduction

Genetic recombinations or, more generally, the exchange of DNA material between organisms, have been a source of computational problems since the 1865 report of Gregor Mendel on plant hybridization [1]. Recombinations occur in the reproduction of all living organisms, including asexual reproduction, and are fundamental producers of diversity. In this paper, we study the computational complexity of problems related to *modular recombination*, which is a form of exchange pervasive in viruses that infect bacteria, called *phages*.

The biological theory of modular recombination was proposed a few decades ago by Botstein [4], who envisioned " *... viruses as belonging to large interbreeding families, members of which share only a common genome organization consisting of interchangeable genetic elements each of which carries out an essential biological function.*" The common genome organization that Botstein refers to is the preservation of the order of biological functions, called *modules*, along the virus genome, although the actual sequences that carry the function may diverge substantially.

The computational models were slower to emerge, since genomic data about "large interbreeding families" were not commonplace until a few years ago. In 2010 a study of a few dozen sequenced strains of *Staphyloccoccus aureus* was conducted [9], and a scenario of interbreeding was inferred on the population [14].

K. Jahn and T. Vinař (Eds.): RECOMB-CG 2023, LNBI 13883, pp. 68–83, 2023.
https://doi.org/10.1007/978-3-031-36911-7_5

The recent availability of other datasets monitoring phage populations evolving through time [11, 12] or space [7, 13] suggested the problem of computing the minimum number of recombination events that transforms one population of phages into another. In a previous paper [2] we developed a heuristic with approximation bounds based on certain properties of the input and found that, on phages infecting bacteria responsible for cheese fermentation, our heuristic performed well. The question remained, however, as to the computational complexity of the optimization problem.

We answer that question in this article, showing that two basic problems related to the comparison of phage populations are computationally difficult. The first one reduces the problem of finding a perfect 3D-matching to reconstructing a single phage, from a population of phages that represents the triples of the 3D-matching instance, with a minimum number of recombinations. The second one reduces a variant of a classic satisfiability problem to the reconstruction of a population of phages, with only 4 modules that represent variables and clauses, with a minimum number of recombinations.

2   Basic Definitions and Properties

Phage genomes can adopt either a circular or linear shape during their life cycle. Genomic data found in databases are linearized by choosing, as a starting point, one module shared by all members of a family, yielding the following representation of phages.

Given an alphabet $\mathcal{A}$, a phage p with n modules can be represented by $p = p[0..n-1]$ where $p[a] \in \mathcal{A}$. The *recombination* operation at positions a and b between two phages p and q :

$$p = p[0..a-1]|p[a..b-1]|p[b..n-1]$$
$$q = q[0..a-1]|q[a..b-1]|q[b..n-1]$$

yields new phages c and d:

$$c = p[0..a-1]|q[a..b-1]|p[b..n-1]$$
$$d = q[0..a-1]|p[a..b-1]|q[b..n-1].$$

Positions a and b are called the *breakpoints* of the recombination. The recombining phages are called *parents*, and the newly constructed phages, their *children*. This relation allows us, when several recombinations are considered, to refer to *descendants* and *ancestors*, of both phages and positions; each recombination creates two descendants to the two parents, while the each character in each of the children has exactly one ancestral character from the parents. Note that, naturally, ancestor and descendant relationships are transitive through the generations.

A *recombination scenario* S from population $\mathcal{P}$ to population $\mathcal{Q}$ is a sequence of recombinations that constructs all phages of $\mathcal{Q}$ using phages of $\mathcal{P}$ and their descendants.

Note that no phage is discarded in the process, in the sense that $\mathcal{P}$ *grows* until it is a superset of $\mathcal{Q}$.

The problem that we address in this article is the following:

MINIMUM PHAGE POPULATION RECONSTRUCTION (MinPPR)

Input: Populations $\mathcal{P}$ and $\mathcal{Q}$ of equal-length phages, and an integer r.
Question: Does there exist a recombination scenario S from $\mathcal{P}$ to $\mathcal{Q}$ of length at most r?

A *break* in a phage q with respect to the set of phages $\mathcal{P}$ is a position b such that for all parents p in $\mathcal{P}$, $p[b-1..b] \neq q[b-1..b]$. A recombination *heals* a break b of a phage q if it creates a child c such that $c[b-1..b] = q[b-1..b]$. In order to be healed, a break b must be one of the breakpoints of the recombination.

Since a recombination can heal at most two breaks in a single phage, if a phage q has n breaks with respect to the set of phages $\mathcal{P}$, then the minimum number of recombinations to construct q is $\lfloor \frac{n+1}{2} \rfloor$.

A crucial remark is that, even if all the breaks are healed, the reconstruction of a phage q with $n = 2r$ breaks with respect to a set of parents might require more than r recombinations. This is the case, for example, if two parents $p_1 = 10111$ and $p_2 = 11101$ are used to reconstruct $q = 11111$: phage q has no break with respect to the set $\{p_1, p_2\}$, but one recombination is necessary to reconstruct q. This recombination must cut an already healed break in p_1 or p_2, and we say that the break is *reused*.

Definition 1. *In a recombination scenario, a break is said to be* reused *if it is a breakpoint of more than one recombination in the scenario.*

Finally, there is an easy upper bound for the number of recombination necessary to reconstruct a phage:

Proposition 1. *If there exists a scenario that reconstructs a phage q with n modules from a population $\mathcal{P}$, then there exists one of length at most $n-1$.*

Proof. A scenario exists if, for each position b, there exists a phage $p_b \in \mathcal{P}$ such that $p_b[b] = q[b]$, otherwise no recombination can produce the value $q[b]$ at position b. We first recombine p_0 and p_1 using breakpoints 1 and 2, to produce a child that equals q on its first 2 positions, and proceed in a similar way up to position $n-1$.   □

Here we study the decision problem where one asks if $\mathcal{Q}$ can be generated from $\mathcal{P}$ using at most r recombinations, for some given r. Let us first argue that the problem is in NP. A given scenario of r recombinations can be verified in time proportional to r, $|\mathcal{P}|$, and $|\mathcal{Q}|$, but this is not polynomial if r is not polynomial in $|\mathcal{P}|$ and $|\mathcal{Q}|$ (e.g. if r is exponential). However, Proposition 1 gives an upper bound on the number of required recombinations based on the number of modules. Hence, we may assume that r is bounded by a polynomial in $|\mathcal{P}|$ and $|\mathcal{Q}|$ and a scenario can be verified in polynomial time, and thus the problem is in NP.

3   Reconstructing One Target Genome

We first consider the case in which the population Q consists of a single phage of unbounded length. We reduce the 3D-PERFECT-MATCHING problem to it, where we receive a set of triples $T = \{(i_1, j_1, k_1), \ldots, (i_n, j_n, k_n)\} \subseteq [1..m]^3$, where $m \geq 2$ is an integer [10]. The goal is to find a subset $T' \subseteq T$ of size m such that for any two distinct $(i, j, k), (i', j', k') \in T'$, we have $i \neq i', j \neq j'$, and $k \neq k'$. Such a set T' is called a perfect 3D-matching.

Since there is a single phage in Q, the alphabet is the set $\{0, 1\}$, and $Q = 11111 \ldots 1111$ will be the only element of the target population. We consider the following phages, each of length $15m + 2$, that form the input population $\mathcal{P}$. See example in Fig. 1.

1. For each element $(i, j, k) \in T$, we construct a phage P_{ijk} that has three 1's in positions $5i$, $5j + 5m$ and $5k + 10m$, and 0's elsewhere.
2. For each element $(i, j, k) \in T$, we associate three phages, $P_{ij\text{-}}$, $P_{\text{-}jk}$, and $P_{i\text{-}k}$ with two 1's respectively in positions $5i + 1$ and $5j - 1 + 5m$, $5j + 1 + 5m$ and $5k - 1 + 10m$, $5i - 1$ and $5k + 1 + 10m$, and 0's elsewhere.
3. $\hat{P}$ has 0's in every position in which one of the above phages has a 1.

	$i = 1$	$i = 2$	$j = 1$	$j = 2$	$k = 1$	$k = 2$
P_{122}	. . . . 1 . . .	. . . .	 1	. . . .	. . . .	. . . 1 . .
P_{212}	. . . .	. . 1 . . .	. 1	. . . .	. . . .	. . . 1 . .
P_{211}	. . . .	. . 1 . . .	. 1	. . . .	. 1	. . . . . .
P_{222}	. . . .	. . 1 . . .	. . . .	. 1	. . . .	. . . 1 . .
$P_{22\text{-}}$	. . . .	. . . . 1 .	 1	. . . .	. . . .	. . . . . .
$P_{12\text{-}}$	. . . . . 1 . .	. . . . . .	 1	. . . .	. . . .	. . . . . .
$P_{21\text{-}}$	. . . .	. 1 . . 1 .	. . . .	. . . .	. . . .	. . . . . .
$P_{\text{-}22}$	. . . .	. . . . . .	. . . . . . 1 .	. . . .	. 1	. . . . . .
$P_{\text{-}12}$	. . . .	. . . . . .	. 1	. . . .	. 1	. . . . . .
$P_{\text{-}11}$	. . . .	. . . . . .	. 1	. . . . . . 1 .	. . . .	. . . . . .
$P_{2\text{-}2}$	. . . . . . 1 .	. . . . . .	. . . .	. . . .	. . . .	. . . . . 1 .
$P_{1\text{-}2}$	. 1	. . . . . .	. . . .	. . . .	. . . .	. . . . . 1 .
$P_{2\text{-}1}$	. . . . . . 1 .	. . . . . .	. . . .	. . . .	. . . 1 . . .	. . . . . .
$\hat{P}$	1 1 1 . . . 1 1	. . . . 1 1	. . 1 1 . . . 1	. 1 . . . . 1 1	. 1 . 1 . . 1 1	. . . 1 . . . 1
Q	1 1 1 1 1 1 1 1	1 1 1 1 1 1	1 1 1 1 1 1 1 1	1 1 1 1 1 1 1 1	1 1 1 1 1 1 1 1	1 1 1 1 1 1 1 1

Fig. 1: Example input with $T = \{(1, 2, 2), (2, 1, 2), (2, 1, 1), (2, 2, 2)\}$ and $m = 2$. The 1's related to phage P_{222} are in red. Dots are used to represent the value 0, in order to better highlight the relative positions of the 1's. (Color figure online)

We show that T has a 3D-matching if and only if $\mathcal{P}$ can generate Q with at most $6m$ recombinations, implying:

Theorem 1. *The* MINIMUM PHAGE POPULATION RECONSTRUCTION *problem is NP-complete, even when population Q has a single phage.*

We have already established that the problem is in NP. For NP-hardness, we show that there exists a 3D-matching $T' \subseteq T$ if and only if it is possible to reconstruct Q from $\mathcal{P}$ using at most $6m$ recombinations.

3.1   The ($\Rightarrow$) Direction

Suppose that there is a 3D-matching $T' \subseteq T$. For each $(i, j, k) \in T'$, it is possible to apply three recombinations to $P_{ijk}, P_{ij-}, P_{-jk}$ and P_{i-k} to obtain $00\ldots01110\ldots01110\ldots01110$, where the first 111 is centered at column i, the second 111 is centered at column j, the third 111 is centered at column k. The genome $\hat{P}$ has 000 in these three triples of positions, and so with three more recombinations we can bring in these 111 into $\hat{P}$. Use Fig. 2 as an illustration. This costs 6 events. Since T' is a perfect 3D-matching, we can repeat this m times to fill in all the remaining 000's in $\hat{P}$, hence achieving cost $6m$.

3.2   The ($\Leftarrow$) Direction

We next show that if Q can be reconstructed from $\mathcal{P}$ using at most $6m$ recombinations, then T admits a 3D-matching. We first establish several properties that hold in general for scenarios that transform $\mathcal{P}$ into Q, before proving the main result of the section.

Given a scenario S that reconstructs phage Q, we identify the following subsets of parents:

1. S_{ijk} contains phages of the form P_{ijk} that belong to the scenario.
2. S_{xy} contains phages of the form P_{ij-}, P_{-jk} or P_{i-k} that belong to the scenario.

Let $\mathcal{P} = S_{ijk} \cup S_{xy} \cup \{\hat{P}\}$ be the set of parents that initially belong to scenario S. We prove that scenario S reconstructs phage Q in $6m$ recombinations only if the set S_{ijk} corresponds to a perfect matching.

By construction, phage Q has $12m$ breaks with respect to the set of phages $\mathcal{P}$. Since a recombination can heal at most two breaks of a single phage, we need at least $6m$ recombinations to reconstruct Q. In a scenario of length $6m$, no break can be reused.

We distinguish two types of breaks: *red* breaks connect a phage in S_{xy} to a phage in S_{ijk}, and *green* breaks connect a phage in S_{xy} to phage $\hat{P}$, (see Fig. 2). We say that a recombination is *red* when its two breaks are red, and *green* if they are green. There is an equal number of red and green breaks in Q, thus, in a scenario of length $6m$, the number of red recombinations is equal to the number of green recombinations, and is at most $3m$, allowing for eventual red-green recombinations.

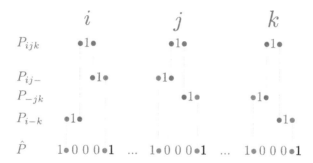

Fig. 2: Red and green breaks. *Red* breaks connect a phage in S_{xy} to a phage in S_{ijk}, and *green* breaks connect a phage in S_{xy} to phage $\hat{P}$. Phage Q has $6m$ red breaks and $6m$ green breaks. (Color figure online)

	$i = 1$	$i = 2$	$j = 1$	$j = 2$	$k = 1$	$k = 2$
P_{122}	. . . . 1 . . . . 1 . . . . 1 . .
P_{211}	. . . . 1 . . . . 1 . . . . 1 . . . .
P_{12-}	. . . . . 1 . . . . 1 . . . .
P_{21-}	. . . . 1 . . 1 . . . .
P_{-11}	. . . . 1 . . . . 1 . . . .
P_{-22}	. . . . 1 . . . . 1 . . .
P_{1-2}	. . . 1 . . . . 1 .
P_{2-1}	. . . . 1 . . . . 1 . . . . . .
$\hat{P}$	1 1 1 . . . 1 1 . . . 1 1 . . . 1 1 . . . 1 1 . . . 1 1 . . . 1
Q	1 1 1 1 1 1 1 1 1 1 1 1 1 1 1 1 1 1 1 1 1 1 1 1 1 1 1 1 1 1 1 1

Fig. 3: A possible output for the example of Fig. 1, with the sets S_{ijk} and S_{xy} used by a recombination scenario of length $6m = 12$. Here $|S_{ijk}| = m$, and $\{(1,2,2),(2,1,1)\}$ is a perfect matching.

The following easy result links properties of a recombination scenario to the existence of a perfect matching:

Lemma 1. *In any scenario that reconstructs Q, $|S_{ijk}| \geq m$. If $|S_{ijk}| = m$, then the set $\{(i,j,k)|P_{ijk} \in S_{ijk}\}$ is a perfect matching.*

Proof. In order to reconstruct Q, all values of i, j and $k \in [1..m]$ must appear at least once in the indices of elements $P_{ijk} \in S_{ijk}$, thus $|S_{ijk}| \geq m$. If $|S_{ijk}| = m$, then all values of i, j and $k \in [1..m]$ appear exactly once implying that the set $\{(i,j,k)|P_{ijk} \in S_{ijk}\}$ corresponds to a perfect matching. See Fig. 3 for an example. $\square$

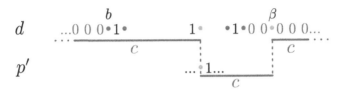

Fig. 4: A recombination between a descendant d of $p \in S_{xy}$ and a phage p'. The recombination creates phage c, in blue, which contains both break b and a 1 from a phage p', where the 1 is located in a column from p's red interval. This recombination must have one breakpoint between b and the other red break of p, and one breakpoint β after the location of the 1 in phage p'. Breakpoint β cannot heal a break because it is between two 0's. (Color figure online)

In order to show that $|S_{ijk}| = m$, we first introduce three lemmas that constrain the order of recombinations contained in a scenario of length $6m$. The first one concerns the *red interval* of a phage in S_{xy}, which contains the 0's adjacent to its red breaks, along with the (circularly) intervening columns that are all 0's. See Fig. 4 for an example of a red interval.

Lemma 2 (red interval). *All descendants of a phage $p \in S_{xy}$ must heal red breaks shared with p before acquiring a 1 in p's red interval.*

Proof. Consider a descendant of p where one of its red breaks b is not yet healed, along with the first recombination producing a child c that contains b and a 1 in p's red interval. If this recombination does not heal b, then it must first heal a break between b and the 1 in p's red interval, thereby creating a phage c containing both the 1, and the break b. This implies that the second break β of this recombination must be in p's red interval, which is a contradiction since the second break cannot be healed in an interval with all 0's. See Fig. 4 for an illustration. □

The second lemma establishes the property that all four breaks spanning two consecutive groups of 1's in $\hat{P}$ will be healed in the same ancestral lineage of Q.

Lemma 3 (sticky breaks). *Consider a break that is healed when phage p is produced by a recombination in a scenario of length $6m$. Any adjacent break must be healed in a descendant of p.*

Proof. Consider the 1 in column x that is adjacent to a single healed break in phage p, so that $p[x-1..x+1] = 110$, or $p[x-1..x+1] = 011$. Consider the 1 that is put in column $x+1$, or $x-1$, while healing the remaining break with column x, producing some child c. Since any break is healed exactly once, the 1's to either side of a healed break must be ancestors of 1's in Q. Therefore, the 1 in column x of both phages p and c must be the ancestor of a 1 in Q, which is only possible if c is a descendant of p. □

Lemma 4 (independent $S_{\mathbf{xy}}$). *Consider phages p_1 and p_2 in $S_{\mathbf{xy}}$. There cannot exist a descendant of both phages with an unhealed red break from both p_1 and p_2.*

Proof Define the green interval of a phage $p \in S_{\mathbf{xy}}$ to be the interval containing the 0's next to the green breaks in p, along with the (circularly) intervening columns that are all 0's. If the green intervals of p_1 and p_2 do not intersect, the red interval lemma gives the result, since there can be no 1 in either red interval before both of the red breaks in a phage are healed.

Suppose their green intervals intersect and, without loss of generality, that the green interval for p_1 starts to the left of the green interval of p_2. Say that there is a descendant containing the left 1 of p_1 and the left 1 of p_2. The red interval lemma implies that a recombination happened directly to the left of the 1 in p_2, which healed that red break. Say that there is a descendant with the left 1 of p_1 and the right 1 of p_2. Due to the red interval lemma, this implies a recombination happened directly to the right of the 1 in p_2, which healed that red break. By symmetry, the other cases are covered by those already listed. □

The next lemma states a desirable property of recombination scenarios of length $6m$, saying that if a phage is in $S_{\mathbf{xy}}$, then its two – unique – siblings are also in $S_{\mathbf{xy}}$. We say that phages p and p' *eventually recombine* in a scenario S if there exists a recombination in S between p, or one of its descendants, and p', or one of its descendants.

Lemma 5. *In a scenario S of length $6m$, if a phage $P_{ijk} \in S_{\mathbf{ijk}}$ eventually recombines using the red breaks of a phage in $S_{\mathbf{xy}}$, then all three phages $P_{ij\text{-}}$, $P_{\text{-}jk}$ and $P_{i\text{-}k}$ eventually recombine with P_{ijk} using both of their red breaks.*

Proof. Consider the first time that a phage P_{ijk} appears in scenario S, and suppose that this recombination involves a phage in $S_{\mathbf{xy}}$, or its descendant. Without loss of generality we may assume this phage is $P_{i\text{-}k}$, due to the circularity of the genomes and symmetry of our construction. We will show that both $P_{ij\text{-}}$ and $P_{\text{-}jk}$ must eventually recombine with P_{ijk} in the scenario, using both of their red breaks.

By the lemma statement, both red breaks of $P_{i\text{-}k}$ were healed in this recombination producing some child c, having only 0's between columns i and k, with the exception of column j. See Fig. 5 for an illustration. Now, the recombination healing the remaining red break adjacent to column i must be in a descendant of c, due to the sticky breaks lemma. Aside from the descendant of c, the other parent p in this recombination must be a descendant of some phage $P_{ij'\text{-}} \in S_{\mathbf{xy}}$, for some j'.

In the following, we show that the only possible companion breakpoint for this recombination occurs when $P_{ij'\text{-}} = P_{ij\text{-}}$. Since every recombination must heal two breaks, there is a companion breakpoint between columns i and j, between columns j and k, or (circularly) between columns k and i.

If the companion breakpoint lies (circularly) between k and i, this implies that p has a 1 in the red interval of $P_{ij'\text{-}}$, which is impossible by Lemma 2.

Say the companion breakpoint lies between columns i and j. In order to heal two breakpoints, there must be a break b between columns i and j in a descendant p' of c. If b is adjacent to j, then it can only be healed using a 1 descending from a phage in $S_{\mathbf{xy}}$. If this phage is not P_{ij-}, then the independent $S_{\mathbf{xy}}$ lemma prohibits a common descendant between this phage and $P_{ij'-}$, a contradiction. Say b is not adjacent to j, but rather in the zone of all 0's in c. Then the existence of b implies that there has been a recombination at the breakpoint adjacent to column j in an ancestor of p'. This leads to the same contradiction as in the previous case.

The same argument applies to a breakpoint occurring between j and k.

Now consider the symmetric case, where a recombination heals the remaining break adjacent to column k in phage c. The same reasoning shows that both red breaks of P_{-jk} are used to recombine with a descendant of P_{ijk}.   □

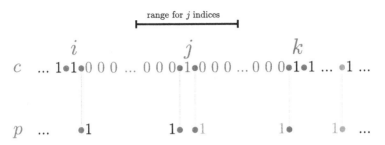

Fig. 5: At the creation of child c, it has only 0's between columns i and k, with the exception of column j. The gray 1's in p are possible breaks that can be healed. (Color figure online)

The previous lemma depends on the assumption that the first recombination with a phage in $S_{\mathbf{ijk}}$ heals the two red breaks of a phage in $S_{\mathbf{xy}}$. The following lemma show that this must be the case.

Lemma 6. *In a scenario S of length $6m$, the first recombination using $P_{ijk} \in S_{\mathbf{ijk}}$ must heal both red breaks of a phage in $S_{\mathbf{xy}}$, be it P_{ij-}, P_{-jk}, or P_{i-k}.*

Proof. Since a phage $P_{ijk} \in S_{\mathbf{ijk}}$ has only red breaks, it must eventually recombine using exactly two red breaks, b_1 and b_2. Suppose that P_{ijk} does not eventually recombine with a single phage of $S_{\mathbf{xy}}$, then b_1 and b_2 are breaks on different phages p_1 and p_2 of $S_{\mathbf{xy}}$. This implies that p_1 and p_2 eventually recombine to produce a child containing both b_1 and b_2. Such a recombination is impossible, due to Lemma 2.   □

Thus we have the result:

Proposition 2. *A scenario S reconstructs Q in $6m$ recombinations only if the set $S_{\mathbf{ijk}}$ is a perfect matching.*

Proof. Lemma 5 shows that for each P_{ijk} in S_{ijk}, the three corresponding $P_{ij\text{-}}$, $P_{\text{-}jk}$ and $P_{i\text{-}k}$ must belong to the scenario. Since there are $3m$ pairs of red breaks, the maximum number of elements of S_{ijk} is m. Lemma 1 gives the result.   □

4   NP-hardness for Genomes of Length 4

In the preceding section, we showed that the MINIMUM PHAGE POPULATION RECONSTRUCTION was hard when the length of the genomes was unbounded. Is it still the case for genome of bounded length? The answer is yes, and we dedicate the remainder of the section to the proof of the following statement:

Theorem 2. *The* MINIMUM PHAGE POPULATION RECONSTRUCTION *problem is NP-complete, even when the genomes of P and Q have length 4.*

We reduce from the BALANCED-4OCC-SAT problem, where we are given a boolean formula ϕ in conjunctive normal form, such that each variable has exactly two positive occurrences in the clauses of ϕ, and exactly two negative occurrences [3].

Consider an instance ϕ of BALANCED-4OCC-SAT with variables $x_1, \ldots, x_n$ and clauses $C_1, \ldots, C_m$. We construct a corresponding instance $(\mathcal{P}, \mathcal{Q}, r)$ of the phage problem. See Fig. 6 for an example with 3 variables and 4 clauses, and Fig. 7 for a more abstract view.

The alphabet for the phages of $\mathcal{P}$ and $\mathcal{Q}$ has, for each variable x_i, a corresponding symbol $i \in [1..n]$, and for each clause C_j, a corresponding symbol c_j. We also add two unique symbols '$-$' and '$\circ$' to the alphabet.

Consider a variable x_i, where $i \in [1..n]$. Let C_g, C_h be the clauses in which x_i occurs positively, and C_r, C_s those in which x_i occurs negatively. Add the following phages to $\mathcal{P}$:

$$X_i = i \circ i \; i$$
$$X_i^+ = c_g \; i \; c_h \; -$$
$$X_i^- = c_r \; i \; c_s \; -$$

and add the following to $\mathcal{Q}$:

$$X_i^* = i \; i \; i \; i$$

Now for each clause C_j, $j \in [1..m]$, add the following to $\mathcal{P}$:

$$D_{j,1} = - \; - \; c_j \, c_j$$
$$D_{j,2} = c_j \; - \; - \; c_j$$

and add the following to $\mathcal{Q}$:

$$D_j^* = c_j \circ c_j c_j$$

We show that ϕ is satisfiable if and only if $\mathcal{Q}$ can be reconstructed from $\mathcal{P}$ with at most $r = n + m$ recombinations.

78   M. Lafond et al.

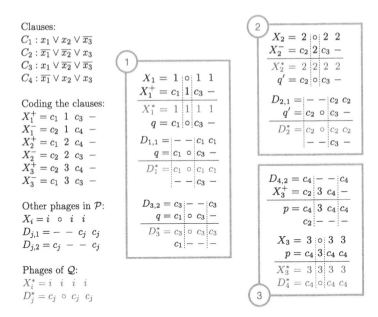

Fig. 6: In this example there are $n = 3$ variables and $m = 4$ clauses, thus $n+m = 7$ phages in $\mathcal{Q}$. One possible recombination scenario of length 7 is depicted. The three recombinations in Group 1 first construct X_1^* using X_1^+ that asserts that clauses 1 and 3 are satisfied when variable x_1 is *true*. The resulting phage q is then used to generate both D_1^* and D_3^*. In Group 2, X_2^* and D_2^* are constructed using X_2^- that asserts that clause 2 is satisfied when when variable x_2 is *false*. Group 3 shows an alternative strategy that first constructs phage $p = c_4\ 3\ c_4\ c_4$, and uses it to simultaneously construct X_3^* and D_4^*.

4.1   The ($\Rightarrow$) Direction

Suppose that ϕ is satisfied by an assignment $\alpha : \{x_1, \ldots, x_n\} \to \{true, false\}$ of the variables. Let us produce $\mathcal{Q}$ from $\mathcal{P}$. For each $i \in [1..n]$, if $\alpha(x_i) = true$, then recombine X_i with X_i^+ by exchanging 2nd positions:

$$
\begin{array}{rcccc}
X_i &=& i & \vdots\ o\ \vdots & i & i \\
X_i^+ &=& c_g & \vdots\ i\ \vdots\ c_h & - \\
\hline
X_i^* &=& i & \vdots\ i\ \vdots & i & i \\
p &=& c_g & \vdots\ o\ \vdots\ c_h & -
\end{array}
$$

This produces children X_i^* and $p = c_g \circ c_h -$, where C_g and C_h are the clauses that are satisfied by setting x_i to *true*. At this point, if D_g^* is not already in $\mathcal{Q}$, then recombine $D_{g,1} = -- c_g c_g$ with $p = c_g \circ c_h -$ by exchanging positions 1 and 2, thereby producing D_g^*:

$$D_{g,1} = \begin{vmatrix} - & - & c_g & c_g \end{vmatrix}$$
$$p = \begin{vmatrix} c_g & \circ & c_h & - \end{vmatrix}$$
$$D_g^* = \begin{vmatrix} c_g & \circ & c_g & c_g \end{vmatrix}$$
$$\begin{vmatrix} - & - & c_h & - \end{vmatrix}$$

For an illustration of the previous two recombinations, see the black edges in Fig. 7. In the same way, if D_h^* is not already in $\mathcal{Q}$, recombine $D_{h,2} = c_h - -c_h$ with $c_j \circ c_h -$ by exchanging positions 2 and 3, which produces D_h^*.

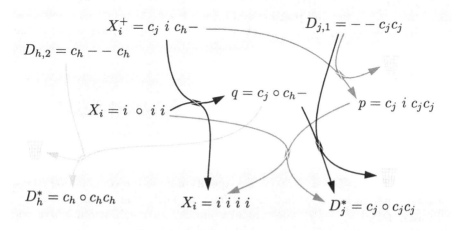

Fig. 7: An illustration of how phages X_i, X_i^+, $D_{j,1}$, and $D_{h,2}$ can recombine to produce X_i^*, D_j^*, and D_h^*. Recombinations are represented by pairs of arrows that meet in the middle at a box with a ⊠ symbol. In black, X_i and X_i^+ first recombine to produce $q = c_j \circ c_h -$, which then recombines with $D_{j,1}$ to produce D_j^* (and some other unused phage). Notice that q can also be used to produce D_h^* in a similar manner. In dark gray, an alternate way to produce X_i^* and D_j^* is depicted. Here, X_i^+ is first recombined with $D_{j,1}$ to produce $p = c_j i c_j c_j$. Phage p is not in $\mathcal{Q}$, but can be recombined with X_i to produce both X_i^* and D_j^* in one operation.

If instead $\alpha(x_i) = false$, recombine X_i with X_i^- by exchanging 2nd positions. This creates X_i^* and $c_r \circ c_s -$, where C_r and C_s are satisfied by setting x_i to $false$. Produce D_r^* and D_s^*, if not already there, as in the previous case.

Since every clause C_j is satisfied by α, for each D_j^*, there will be some X_i in the above procedure that produces D_j^*. Moreover, there are exactly $n + m$ recombinations: one to produce each X_i^*, and one to produce each D_j^*.

4.2   The ($\Leftarrow$) Direction

Suppose that there exists a sequence $S = (R_1, \ldots, R_r)$ of at most $r \leq n + m$ recombinations that reconstructs $\mathcal{Q}$. Let $\mathcal{X} = \{X_1, \ldots, X_n\}$ and $\mathcal{D} = \{D_1, \ldots, D_m\}$ where D_j is either $D_{j,1}$ or $D_{j,2}$, whichever contains the character c_j that is the ancestor of the c_j in the 4th position of D_j^*.

We define a function $f : \mathcal{X} \cup \mathcal{D} \to S$ in the following way: set $f(X_i)$ to the first recombination that creates X_i^*, or an ancestor of X_i^*, with $[1..n]$ in the 2nd position and i in the 4th. Note that the parent used by $f(X_i)$, having i in the 4th position, must also have $\{-, \circ\}$ in the 2nd position, as otherwise there would be a previous recombination with the required properties for being $f(X_i)$.

Set $f(D_j)$ to the first recombination that creates D_j^*, or an ancestor of D_j^*, with $[1..n] \cup \{\circ\}$ in the 2nd position and c_j in the 4th. As previously, note that one of the parents used by $f(D_j)$ contains '$-$' in the 2nd position and c_j in the 4th position.

Proposition 3. *The function f is a bijection.*

Proof. We prove that f is injective, and since there can be at most $n + m$ recombinations in S, we conclude that f is a bijection.

If $i \neq k$, then $f(X_i) \neq f(X_k)$ since equality implies a recombination where at least one parent has an element of $\{-, \circ\}$ in 2nd position, and both children have an element of $[1..n]$.

If $g \neq h$, then $f(D_g) \neq f(D_h)$ since equality implies a recombination where at least one parent has a '$-$' in 2nd position, and both children do not.

Finally, $f(X_i) \neq f(D_j)$ since equality implies a recombination where one parent has a '$-$' in 2nd position, and both children do not.   □

The crucial consequences of Proposition 3 are the three following results:

Proposition 4. *There is exactly one element of $\mathcal{X} \cup \mathcal{D}$ in each recombination of S.*

Proof. Since all phages of $\mathcal{X} \cup \mathcal{D}$ are necessary to produce $\mathcal{Q}$, then each one must be used by at least one recombination. We show that the image of f contains no recombination between two of these phages.

Suppose X_i and X_k recombine. Then this recombination is not $f(X_i)$ or $f(X_k)$ since both parents have a '$\circ$' in 2nd position, implying that none of the children have an element of $[1..n]$.

Suppose D_g and D_h recombine. Then this recombination is not $f(D_g)$ or $f(D_h)$ because both parents have a '$-$' in 2nd position, implying that both children have a '$-$' in 2nd position.

Suppose X_i and D_j are used in a recombination R. Then R is not $f(X_i)$ because one parent has a '$\circ$' in 2nd position, and the other has a '$-$', implying that none of the children have an element of $[1..n]$. So R must be $f(D_j)$, implying that one of the children p is an ancestor of D_j^*, having $\circ$ in 2nd position and c_j in 4th position. In the remaining paragraph we show that there is no recombination from S that makes the c_j in 4th position of p the ancestor of the c_j in 4th position of D_j^*, contradicting the existence of R.

Consider any subsequent recombination R' that uses p as a parent. Each R' cannot be a $f(X_i)$ since p cannot contribute an element of $[1..n]$ in 2nd position, so it must be $f(D_h)$, for $D_h \in \mathcal{D}$ and $h \neq j$. Note that the other parent q, used in R', must have '−' in 2nd position and c_h in 4th position, so the children are then q' with ∘ and c_h in 2nd and 4th positions, and p' with '−' and c_j in 2nd and 4th positions. But, while p' is the child that contains the c_j that could be ancestral to the 4th position of D_j^*, it cannot be used as a parent in a recombination of $f(\mathcal{X} \cup \mathcal{D})$, since it has '−' in 2nd position, and $f(D_j)$ has already been applied. Therefore, p' is not an ancestor of D_j^*. This is true for any such child p' produced by a subsequent recombination, which contradicts that p is an ancestor of D_j^*. This contradict our initial supposition that R is $f(D_j)$.   □

Corollary 1. *The recombination that uses X_i produces X_i^*.*

Proof. Note that X_i is the only phage in $\mathcal{X} \cup \mathcal{D}$ that shares any character with X_i^*. Since both parents of any recombination creating X_i^* must share at least one character with X_i^*, the recombination in S that uses X_i is the only one that can create X_i^*.   □

Corollary 2. *The recombination that uses D_j produces either D_j^* or $c_j i c_j c_j$.*

Proof. By definition D_j is an ancestor of D_j^*, and by Corollary 4 we know that any descendant of D_j must recombine with an element of $\mathcal{X} \cup \mathcal{D}$. Since no other member of $\mathcal{X} \cup \mathcal{D}$ has c_j in positions 1, 3 or 4, a child of D_j must be an ancestor of D_j^* and have that character c_j in those positions. This child must be either D_j^*, $p = c_j i c_j c_j$, or $q = c_j - c_j c_j$.

A subsequent recombination using child $q = c_j - c_j c_j$ does not exist since it would either recombine with an X_k, but not produce X_k^* in contradiction with Corollary 1, or with a D_h, whose 2nd position is also '−'. Therefore, q has no descendant, contradicting that it is an ancestor of D_j^*.   □

We now establish that a recombination scenario S of length $n + m$ implies a valid, and satisfiable truth assignment for ϕ.

For an $X_i \in \mathcal{X}$, we say that X_i *chose* X_i^+ if the only recombination that uses X_i is with an X_i^+ or its descendant, and we say that it chose X_i^- if X_i recombined with X_i^- or its descendant. If X_i chose X_i^+, we set $x_i = true$, and if X_i chose X_i^- we set $x_i = false$. Let us call the resulting assignment α, which we claim is satisfying. We first argue that each x_i is assigned only one value, and then show that each clause is satisfied.

Proposition 5. *A recombination scenario S of length $n + m$ implies a valid, and satisfiable truth assignment for ϕ.*

Proof. We first show that the assignment α is well-defined, *i.e.* each variable x_i is assigned either *true* or *false*, but not both. Since each X_i chooses at least one of X_i^+ or X_i^-, we know that x_i is assigned *true* or *false*. Assume now that x_i is assigned both. Then X_i chose both X_i^+ and X_i^-, meaning that X_i recombines with a phage that descends from both X_i^+ and X_i^-. But the existence of this phage requires a recombination between descendants of both X_i^+ and X_i^-, which contradicts Corollary 4.

We now show that each clause is satisfied by α. By Corollary 2, we know that $D_{j,1}$ and $D_{j,2}$ do not recombine, and only one of the two, that we called D_j, appears in S. Since D_j contributes at most two c_j characters to D_j^*, the third c_j can only be a descendant of an X_i^+ or X_i^-, since they are the only other phages in $\mathcal{P}$ that may contain c_j. By construction, this means that the clause C_j is satisfied by the variable x_i being set to the truth assignment implied by the corresponding X_i^+ or X_i^-. $\qquad\square$

5   Conclusion

The notion of recombination used in this article is the same as the *two-point crossover* function [8] used for characterizing fitness landscapes for the exploration of genotypes. This two-point crossover has since been studied in a general form as a *k-point crossover* [5]. While, to the best of our knowledge, there is no work directly linking the area of fitness landscape exploration to the minimization problem discussed in this article, we hope that these related areas can be fused in the future.

In this paper, we showed that the MINIMUM PHAGE POPULATION RECONSTRUCTION is NP-Complete in two extreme cases: bounded length, and a single target phage. Although negative, such results may come as a relief, since we can turn our focus to algorithms that, by accounting for biological constraints, could provide drastically reduced search spaces for parsimonious solutions. For example, the use of other measures of evolution, or information about community structure [6] might play a significant role in reducing the complexity of the problem: after all, phages recombine all the time, and thrive doing so.

Acknowledgments. ML was partially supported by Canada NSERC Grant number 2019-05817. AB was partially supported by Canada NSERC Grant number 05729–2014. KMS was partially supported by the grant ANR-20-CE48-0001 from the French National Research Agency (ANR).

References

1. Abbott, S., Fairbanks, D.J.: Experiments on Plant Hybrids by Gregor Mendel. Genetics **204**(2), 407–422 (10 2016). https://doi.org/10.1534/genetics.116.195198
2. Bergeron, A., Meurs, M.J., Valiquette-Labonté, R., Swenson, K.M.: On the comparison of bacteriophage populations. In: Jin, L., Durand, D. (eds.) Comparative Genomics, pp. 3–20. Springer International Publishing, Cham (2022)
3. Berman, P., Karpinski, M., Scott, A.: Approximation hardness of short symmetric instances of max-3sat. In: Electronic Colloquium on Computational Complexity (ECCC). vol. TR03-049 (2004)

4. Botstein, D.: A theory of modular evolution for bacteriophages. Ann. N. Y. Acad. Sci. **354**(1), 484–491 (1980). https://doi.org/10.1111/j.1749-6632.1980.tb27987.x
5. Changat, M., et al.: Transit sets of-point crossover operators. AKCE Int. J. Graphs Comb. **17**(1), 519–533 (2020)
6. Chevallereau, A., Pons, B.J., van Houte, S., Westra, E.R.: Interactions between bacterial and phage communities in natural environments. Nat. Rev. Microbiol. **20**(1), 49–62 (2022)
7. Chmielewska-Jeznach, M., Bardowski, J.K., Szczepankowska, A.K.: Molecular, physiological and phylogenetic traits of lactococcus 936-type phages from distinct dairy environments. Sci. Rep. **8**(1), 12540 (2018). https://doi.org/10.1038/s41598-018-30371-3
8. Gitchoff, P., Wagner, G.P.: Recombination induced hypergraphs: a new approach to mutation-recombination isomorphism. Complexity **2**(1), 37–43 (1996)
9. Kahánková, J., Pantuček, R., Goerke, C., Ružičková, V., Holochová, P., Doškar, J.: Multilocus pcr typing strategy for differentiation of Staphylococcus aureus siphoviruses reflecting their modular genome structure. Environ. Microbiol. **12**(9), 2527–2538 (2010)
10. Karp, R.: Reducibility among combinatorial problems. Complexity of Computer Computations, pp. 85–103 (1972)
11. Kupczok, A., Neve, H., Huang, K.D., Hoeppner, Marc P Heller, K.J., Franz, C.M.A.P., Dagan, T.: Rates of Mutation and Recombination in Siphoviridae Phage Genome Evolution over Three Decades. Mol. Biol. Evol. **35**(5), 1147–1159 (02 2018). https://doi.org/10.1093/molbev/msy027
12. Lavelle, K., et al.: A decade of streptococcus thermophilus phage evolution in an irish dairy plant. Appl. Environ. Microbiol. **84**(10) (2018). https://doi.org/10.1128/AEM.02855-17
13. Murphy, J., et al.: Comparative genomics and functional analysis of the 936 group of lactococcal siphoviridae phages. Scientific Reports 6, 21345 EP - (02 2016)
14. Swenson, K.M., Guertin, P., Deschênes, H., Bergeron, A.: Reconstructing the modular recombination history of staphylococcus aureus phages. In: BMC bioinformatics. vol. 14, p. S17. Springer (2013). https://doi.org/10.1186/1471-2105-14-S15-S17

Physical Mapping of Two Nested Fixed Inversions in the X Chromosome of the Malaria Mosquito *Anopheles messeae*

Evgenia S. Soboleva[1] , Kirill M. Kirilenko[1], Valentina S. Fedorova[1] ,
Alina A. Kokhanenko[1], Gleb N. Artemov[1] , and Igor V. Sharakhov[1,2(✉)]

[1] Laboratory of Ecology, Genetics and Environmental Protection, Tomsk State University,
36 Lenin Avenue, Tomsk 634050, Russia
igor@vt.edu
[2] Department of Entomology, The Fralin Life Sciences Institute, Virginia Polytechnic Institute
and State University, 360 West Campus Drive, Blacksburg, VA 24061, USA

Abstract. Chromosomal inversions play an important role in genome evolution, speciation and adaptation of organisms to diverse environments. Mapping and characterization of inversion breakpoints can be useful for describing mechanisms of rearrangements and identification of genes involved in diversification of species. Mosquito species of the Maculipennis Subgroup include dominant malaria vectors and nonvectors in Eurasia, but breakpoint regions of inversions fixed between species have not been mapped to the genomes. Here, we use the physical genome mapping approach to identify breakpoint regions of the X chromosome inversions fixed between *Anopheles atroparvus* and the most widely spread sibling species *An. messeae*. We mapped breakpoint regions of two large nested fixed inversions (~13 Mb and ~ 10 Mb in size) using fluorescence *in situ* hybridization of 53 gene markers with polytene chromosomes of *An. messeae*. The DNA probes were designed based on gene sequences of the annotated *An. atroparvus* genome. The two inversions resulted in five syntenic blocks, of which only two syntenic blocks (encompassing at least 179 annotated genes in the *An. atroparvus* genome) changed their position and orientation in the X chromosome. Analysis of the *An. atroparvus* genome revealed enrichment of DNA transposons in sequences homologous to three of four breakpoint regions suggesting the presence of "hot spots" for rearrangements in mosquito genomes. Our study demonstrated that the physical genome mapping approach can be successfully applied to identification of inversion breakpoint regions in insect species with polytene chromosomes.

Keywords: Chromosome Inversions · X Chromosome · Breakpoint Regions · Synteny Blocks · Malaria Mosquitoes · *Anopheles*

Supplementary Information The online version contains supplementary material available at https://doi.org/10.1007/978-3-031-36911-7_6.

K. Jahn and T. Vinař (Eds.): RECOMB-CG 2023, LNBI 13883, pp. 84–99, 2023.
https://doi.org/10.1007/978-3-031-36911-7_6

1 Introduction

Chromosomal inversions are drivers of genome evolution, adaptation and speciation of diploid organisms (Kirkpatrick and Barton 2006). A reverse order of the genetic material in the chromosome due to an inversion causes suppression of recombination in a heterozygous organism during meiosis and accumulation of divergent alleles. As a result, polymorphic inversions play a role in diversification of population by providing ecological, behavioral, and physiological adaptations to the changing environments. Polymorphic inversions in malaria mosquitoes have been shown to be associated with epidemiologically important traits (Francisco, Ayala and Coluzzi 2005; D. Ayala, Ullastres, and González 2014). Fixed chromosomal inversions contribute to species divergence and are used by researchers for reconstructing species phylogenies as an independent approach to a molecular phylogeny. For example, chromosomal phylogeny has identified ancestral and derived karyotypes in malaria mosquitoes of the Anopheles gambiae complex (Kamali et al. 2012; I. V. Sharakhov 2013; Anselmetti et al. 2018; Fontaine et al. 2015). A recent combination of the whole-genome and inversion-based approaches reconstructed the phylogeny of mosquito species of the *Maculipennis* Subgroup (Yurchenko et al., 2022)

The *Maculipennis* Subgroup consists of malaria mosquito species distributed in northern Eurasia. According to the recent multi-gene phylogeny reconstruction, supported by molecular cytogenetic analysis, the ancestral species of these mosquitoes migrated from North America to Eurasia through Bering Land Bridge about 20 million years ago and later divided into three clades: the Southern Eurasia clade, the European clade, and the Northern Eurasian clade) (Yurchenko et al., 2022). *Anopheles messeae* is a malaria mosquito species that belongs to the Northern Eurasian clade of the *Maculipennis* Subgroup. Although malaria was eliminated from the area of the *An. messeae* distribution more than half of a century ago, this dominant malaria vector is still considered a potential public health threat (Sinka et al. 2010). *Anopheles daciae* is a closely related sibling species that was discriminated from to *An. messeae* based on nucleotide sequence substitutions in the Internal Transcribed Spacer 2 (ITS2) of the ribosomal DNA (rDNA) (Nicolescu et al. 2004). The distribution of *An. messeae* extends from the Ireland Island in the West to to the Amur river region in the East and from Scandinavia and Yakutia in the North to Iran and Northern China in the South (Gornostaeva and Danilov 2002; Djadid et al. 2007; S. Zhang et al. 2017). *Anopheles messeae* along with other members of the Northern Eurasian clade, *An. daciae* and *An. maculipennis*, survive harsh winters by entering a complete diapause. The ability of *An. messeae* to occupy diverse ecological zones may be explained by well-developed chromosomal inversion polymorphism. There are five wide-spread highly-polymorphic inversions located on four chromosome arms: X1, X2 on chromosome X; 2R1 on chromosome 2; 3R1 and 3L1 on different arms of chromosome 3 (Stegniĭ, Kabanova, and Novikov 1976; Naumenko et al. 2020). Also, a number of fixed chromosomal inversions differentiate genomes of the *Maculipennis* Subgroup species. Approximate locations of breakpoints of fixed inversions in the *Maculipennis* Subgroup have been identified by reading the banding patterns of polytene chromosomes, which are giant interphase chromosomes developed via multiple rounds of DNA replications. Polytene chromosomes in Anopheles are found in salivary glands, Malpighian tubules, midgut and fat body of larvae as well as in Malpighian tubules and ovarian nurse cells of adults. Cytogenetic studies identified fixed autosomal inversions and demonstrated shared X chromosome arrangements among *An. atroparvus*, *An. labranchiae* and *An. maculipennis* (Stegniy 1991, 1982). Although fixed rearrangements on the X chromosome between *An. atroparvus* and *An. messeae* have been observed, their precise number and location could not be determined using traditional cytogenetics due to the divergent chromosomal morphology between the species. Precise mapping and characterization of breakpoints in mosquito genomes can be useful for describing mechanisms of rearrangements and identification of genes involved in species differentiation. The genome of *An. atroparvus* (a species from the European clade) is the available reference assembly (Neafsey et al. 2015); (G. Artemov et al. 2018) that can be used for comparative physical mapping with other members of the Maculipennis Subgroup.

The aim of this study was to perform physical genome mapping and molecular characterization of the X chromosomal inversions fixed between the malaria mosquito species *An. messeae* and *An. atroparvus*. We identified and characterized conserved synteny blocks and breakpoint regions of X chromosome inversions fixed between these two species using the genome of *An. atroparvus*.

2   Material and Methods

2.1   Collection of Mosquitoes and Species Identification

Anopheles fourth instar were collected from a natural pond in Togur (Tomsk region, Russia, 58°22′12.7″N 82°51′07.3″E). Larvae were fixed in Carnoy's solution (96% ethanol: glacial acetic acid, 3:1) and stored at −20 °C. Identification of the *An. messeae* species was carried out based on the length and the sequence of internal transcribed sequence 2 (ITS2) with the PCR-RFLP protocol (Artemov et al. 2021).

2.2   Chromosome Preparation and Karyotyping

Salivary glands of fourth instar *Anopheles* larvae were isolated in Carnoy's solution, kept in a drop of 45% acetic acid for 10 min, covered by a coverslip with a filter paper and moderately squashed as previously described (Artemov et al. 2021). For mapping of fixed inversions, the X1 karyotype of males and X11 karyotype of females *An. messeae* were used. Obtained chromosome squashes were compared with the *An. messeae* standard chromosome map that has the X1 arrangement (Fig. S1) (Artemov et al. 2021).

2.3   Marker Genes Selection, DNA Probes Development and Synthesis

In this study we introduce the iterative mapping method that consists of the marker selection, physical orthologous gene mapping, and breakpoint regions identification (see Fig. S2). First, we took 17 markers, which were localized at the 1-Mb distance from each other on the X chromosome of *An. atroparvus* in the previous study (Yurchenko et al., 2022). Second, if one or several markers from this mapping changed localization in comparison with the *An. atroparvus* reference, the new 3–4 gene markers located between juxtaposed intact synteny blocks and translocated (within inversion) markers were selected and mapped to the *An. messeae* X chromosome. Then, the procedure of marker selection and mapping was repeated until no genes in the genome distance between the intact and translocated markers remained. Thus, the nucleotide sequences of *An. atroparvus* genes from the AatrE3 genome assembly (Ye et al. 2012; G. Artemov et al. 2018a) were used for the development of specific primers with the Primer-BLAST online software (Ye et al. 2012). Expected PCR products of 0.5–1 kb in length were checked with the BLAST tool of the VectorBase (Tesler 2002; Amos et al. 2022) against the *An. atroparvus* AatrE3 genome to ensure no similarity with non-target sequences. PCR was conducted according to the standard protocol with the presence of *An. atroparvus* genomic DNA as a template (Ye et al. 2012; G. Artemov et al. 2018b).

2.4 DNA Labeling and Fluorescence *in Situ* Hybridization (FISH)

Gene-specific DNA probes were labeled in the reaction with the Klenow fragment in the presence of heptamer primers with random nucleotide sequences and one of the modified nucleotides TAMRA-5-dUTP or Biotin-11-dUTP (Biosan, Novosibirsk, Russia) (Gleb N. Artemov et al. 2018). Obtained DNA probes were precipitated in ethanol, dried and dissolved in a hybridization mix (2 × SSC, 10% SDS, 50% deionized formamide, 1% Tween 20). FISH was performed according to the standard protocol (Sharakhova et al. 2019). After hybridization, preparations with the Biotin-11-dUTP probes underwent washing in the Block buffer (3% BSA, 2 × SSC, 0.01% Tween 20) at 37 °C for 15 min and the probes were detected using Avidin-FITC (Sigma Aldrich, USA) diluted 1:300 in the Block buffer during 1.5–2 h. After the final washing step in 2 × SSC, 0.01% Tween 20, the slides were dried for 30 s and the antifade solution with DAPI (Abcam, USA) was added under the coverslips.

2.5 Microscopy and Physical Mapping

Chromosome preparations were analyzed with a fluorescence microscope Axio Imager Z1 (Carl Zeiss, Germany). Digital microphotographs of chromosome spreads with fluorescent probe signals were captured with a CCD camera AxioCam MRm and Axio-Vision 4.7.1 software (Carl Zeiss, Germany). The same software as well as the Adobe Photoshop software were used for picture editing. Physical mapping of *An. atroparvus* orthologues on polytene chromosomes of *An. messeae* and *An. daciae* was conducted using the standard cytogenetic map for *An. messeae* with the X1 arrangement (Fig. S1) (Gleb N. Artemov et al. 2021).

2.6 Computational Pair-Wise Analysis of Rearrangements

The Genome Rearrangements In Man and Mouse (GRIMM) program (Tesler 2002) was used to calculate inversion distances in a pair-wise analysis between *An. atroparvus* and *An. messeae*. The GRIMM software uses the Hannenhalli and Pevzner algorithms for computing the minimum number of rearrangement events and for finding optimal scenarios for transforming one genome into another. The signed option for synteny block orientation was used.

2.7   Transposable Element (TE) and Simple Repeat Annotation

Using the Extensive *de novo* TE Annotator (EDTA) pipeline (Ou et al. 2019), a custom library of TEs was developed. Additionally, the RepeatModeler (http://www.repeatmas ker.org/RepeatModeler/) was run to make the library complete. This library was used to annotate TEs and simple tandem repeats in the *An. atroparvus* AatrE3 assembly using RepeatMasker (v. 4.1.2) (Flynn et al. 2020). The rmsk2bed tool from BEDOPS (v. 2.4.41) was used to convert RepeatMasker.out files to convenient.bed files (Neph et al. 2012). The resulting.bed files contained information about the beginning and the end of each repetitive DNA element and information about its type. For further analysis, all types of repetitive DNA were divided according to the RepeatMasker annotation into four groups: 1) TEs class I: Retrotransposons; 2) TEs class II: DNA transposons; 3) Unknown TEs; 4) Simple repeats (Table 1).

Table 1.   Groups of TEs based on the summary file produced by EDTA.

Repetitive DNA family	Repeat name
TEs I: Retrotransposons	LTRs: Gypsy, Copia, Unknown; LINEs
TEs II: DNA TEs	TIRs: CACTA, Mutator, PIF_Harbinger, Tc1_Mariner, hAT; helitron
Unknown TEs	Repeat_region
Simple repeats	Low_complexity, Simple_repeat

The density of repetitive DNA elements from various groups was counted in each breakpoint region (BR). Depending on the categories into which they were divided, repetitive DNA elements in each BR were visualized. In 50-kb genomic regions immediately upstream and downstream of each BR in the *An. atroparvus* genome including the BR itself, bar plots were depicted to show how many repetitive DNA elements of each group are present. Genes were taken from the.gff3 AatrE3 annotation file retrieved from VectorBase (Giraldo-Calderón et al. 2015) and the density of genes in each BR was displayed as described above.

2.8   Statistical Analyses

A statistical analysis was done to determine whether the difference in the density of repetitive DNA elements and genes in BRs is significant when compared to other regions of the X chromosome. The Poisson distribution was chosen to explain our data because the density of repetitive DNA elements or genes is a discrete value. The length of each BR was measured (ranging from 7 kb to 12 kb), and this length served as the chromosome's overall bin size for the calculation of λ (the mean density of the repetitive DNA elements). Based on the obtained λ and x (the density of repetitive DNA elements in a single BR) for each BR, the value of P(x) was calculated. Moreover, coding regions were removed based on the coordinates obtained from the annotation.gff file of the AatrE3 genome assembly. Subsequently, the densities of repetitive DNA elements were recalculated using λ as the mean density in non-coding regions and x as the density in a single BR without coding regions. The resulting value of P(x) was utilized to assess the statistical significance of the difference in repetitive DNA density between BRs and non-coding regions on the X chromosome. The density of genes in each synteny block (SB) with a bin size of 100 kb (optimal for gene density calculation with appropriate detailing) was also estimated and a one-way ANOVA test was run to determine if the density of genes in SBs varies significantly. We define a SB as a region that contains no less than two genes in the same order independently of their orientation or collinearity. For a statistical analysis, the scipy package (v. 1.10) of the Python programming language was used (Virtanen et al. 2020).

3   Results

3.1   Physical Mapping of Inversion Breakpoint Regions

For physical mapping of breakpoint regions of the fixed inversions, the locations of the 53 genes in the X chromosome of *An. messeae* were determined by FISH (Table S1). The genome of *An. atroparvus* was used as a reference for selection of gene markers. Iterative physical mapping was conducted in two steps. In the first step, 17 gene markers (previously used in (Yurchenko et al., 2022) were mapped at the distance about 1 Mb from each other to the X1 chromosome arrangement of *An. messeae* to define large scale chromosome rearrangements. With exception of the three markers (AATE17741, AATE005236, and AATE010434), the gene orders were collinear between *An. messeae* and *An. atroparvus*. We assumed that these three genes are involved in the fixed chromosome rearrangements. In the second step, additional 36 gene markers were mapped to more precisely identified breakpoint regions (BRs) of fixed rearrangements. The selection of the new markers was conducted based on the previous *in situ* hybridization results with the 17 markers (Yurchenko et al., 2022). The range of the BRs was becoming shorter after each round of hybridization until no mappable markers were identified within BRs (Fig. 1A).

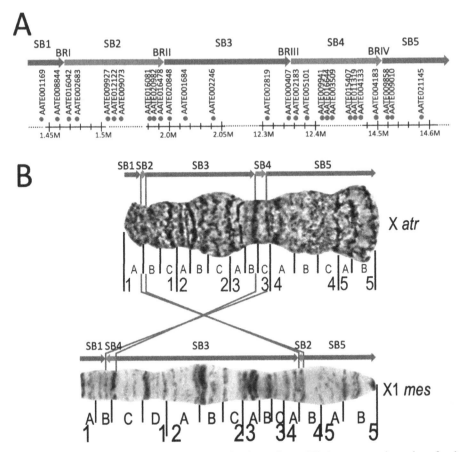

Fig. 1. Physical maps of gene markers and breakpoint regions of X chromosome inversions fixed between *An. atroparvus* and *An. messeae*. A – Genome positions of *An. atroparvus* genes, which were used for identification of rearrangements breakpoints in *An. messeae*. B – The X chromosome position and direction of synteny blocks. Blue and red arrows represent collinear and inverted regions, respectively. Blue and red dots indicate genes located in collinear and inverted regions, respectively. BRI-IV - breakpoint regions, SB1-5 - synteny blocks. The X chromosome maps of *An. atroparvus* (*X atr*) and *An. messeae* (*X1 mes*) are adapted from Artemov et al. 2015 and Artemov et al. 2021, respectively.

According to the annotation of *An. atroparvus* AatrE3 genome assembly, BRI contains one gene (AATE011784) BRIII has two genes (AATE001433 and AATE020008). The sizes of BRs in the *An. atroparvus* genome assembly ranged from 7.4 (BRI) to 12.2 kb (BRIII) (Fig. 2, Table 2).

As a result, we determined four BRs (BRI-IV) of rearrangements fixed between *An. atroparvus* and *An. messeae* (Fig. 1).

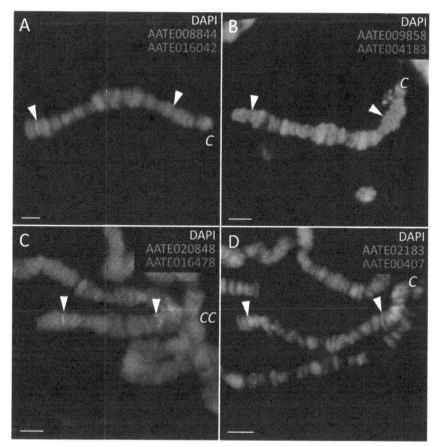

Fig. 2. FISH of DNA-probes developed based on the *An. atroparvus* genes that flank BRs of fixed rearrangements in the *An. messeae* X chromosome. (A) Markers for BRI. (B) Markers for BRII. (C) Markers for BRIII. (D) Markers for BRIV. The positions of the marker genes on X chromosomes are indicated by arrows. *C* – the centromere end of the X chromosome, *CC* – chromocenter. Scale bars – 10 μm.

3.2   Reconstruction of the X Chromosome Rearrangements Using Synteny Blocks

The four BRs divided the X chromosomes of *An. atroparvus* and *An. messeae* into five synteny blocks (SB1-5), which have different sizes (Table 3). The SBs had no significant differences in gene densities, p-value of ANOVA was 0.65, F-value was 0.61. The mean gene density in the X chromosome was 7.31 ± 3.68 per 100 kb with a maximum value of 20 genes per 100 kb.

Table 2. The coordinates of the BR-flanking gene markers in the *An. atroparvus* genome assembly.

BR	"Left" marker coordinates			
	Start	Stop	Gene	Exon*
I	1 458 040	1 458 895	AATE008844	7
II	1 992 544	1 992 840	AATE016478	1
III	12 347 972	12 348 540	AATE000407	4–5
IV	14 508 845	14 509 445	AATE004183	10
BR	"Right" marker coordinates			
	Start	Stop	Gene	Exon*
I	1 466 290	1 466 970	AATE016042	2–3
II	2 000 315	2 000 825	AATE020848	1–2
III	12 360 692	12 362 480	AATE002183	8
IV	14 520 988	14 521 746	AATE009858	1

Table 3. The genomic lengths, gene counts, and gene densities in the five synteny blocks in the X chromosome of the *An. atroparvus* AatrE3 genome.

SB	Length of SB, bp	Gene counts in SB	Mean gene density in SB (per 100 kb)
1	1 458 895	122	8.36
2	526 550	42	7.97
3	10 348 225	722	6.98
4	2 148 753	137	6.38
5	3 273 690	199	6.6
Total	17 756 113	1222	7.26

The SB1, SB3 and SB5 have the same order and orientation in *An. atroparvus* and *An. messeae*, whereas SB2 and SB4 have different order and orientation in the two species (Fig. 1B). Using the GRIMM software (Tesler, 2002), we reconstructed the rearrangement events that can cause the observed differences in SB order and orientation. The following orders and orientations of SBs were used as an input for the GRIMM analysis.

> *An. atroparvus* X
1 2 3 4 5
> *An. messeae* X1
1 − 4 3 − 2 5

The program identified two inversion events that transform the standard X chromosome arrangement of *An. atroparvus* into the X1 arrangement of *An. messeae* (Fig. 3).

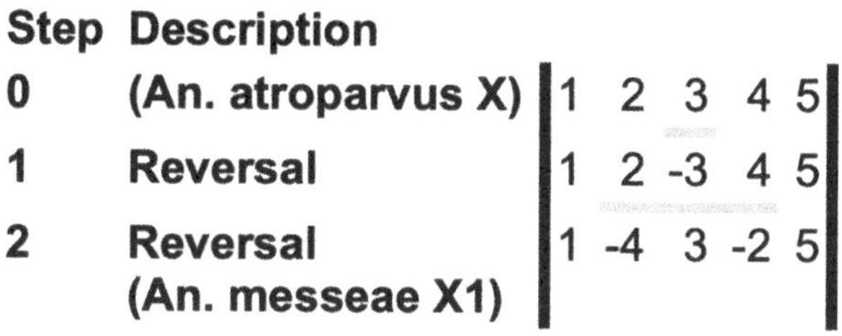

Step Description

Step	Description						
0	(An. atroparvus X)	1	2	3	4	5	
1	Reversal	1	2	-3	4	5	
2	Reversal (An. messeae X1)	1	-4	3	-2	5	

Fig. 3. The result of the GRIMM reconstruction of the fixed rearrangements between the X chromosome of *An. atroparvus* and the X1 chromosome of *An. messeae*. The order and orientation of SBs are indicated by numbers. The yellow lines show inversions (reversals).

The reconstructed inversions differ in length and exploit different pairs of BRs. The larger inversion (~13 Mb) involved SB2, SB3 and SB4 and used BRI and BRIV, whereas the smaller inversion (~10 Mb) involved only SB3 and used BRII and BRIII. The smaller inversion is nested inside the larger inversion. There are two possible sequences of the rearrangement events. The result of these rearrangements is a "swapping" of the SB2 and SB4 position and a change in their orientation. These two SBs make up 16% of the euchromatic part of the X chromosome and 10% of the entire X chromosome of *An. atroparvus*. At least 179 annotated genes located in the SB2 and SB4 changed their position and orientation between the X chromosomes of *An. atroparvus* and *An. messeae*.

3.3   The Genomic Content in the Neighborhoods of the Breakpoint Regions

Since the genome sequence of *An. messeae* is not yet available, we analyzed the genomic content in the BRs using the genome sequence of *An. atroparvus*. We analyzed these BRs for the presence of transposable elements (TEs) and tandem repeats, which are known to cause chromosomal rearrangements (Cáceres et al. 1999; Lonnig and Saedler 2002; Lobachev et al., 2007; Coulibaly et al. 2007). Most TEs in dipteran insects belong to two major classes that are transposed by using an RNA transposition intermediate (Class I) and the "cut and paste" mechanism (Class II) (Wicker et al. 2007). The density of the class I TEs was low in the X chromosome of *An. atroparvus* and only a single retroelement was found near one of the BRs (BRI). However, TEs of Class II were more abundant in the X chromosome and the density of DNA transposons was significantly higher in the BRII, BRIII, and BRIV in comparison with the average density for the non-coding part of X chromosome (p-values: 8.6×10^{-9}, 6.7×10^{-7}, 8.6×10^{-4} respectively) as well as for the whole chromosome for the X chromosome (p-values: 2.7×10^{-7}, 2.3×10^{-5}, 5.2×10^{-3} respectively) (Fig. 4).

BRI was characterized by a high density of simple repeats in comparison with BRII, BRIII and BRIV (Fig. S3). All the BRs regions had different TE content and simple repeats content (Table 4). The BRI contained no TEs, but the other three BRs had

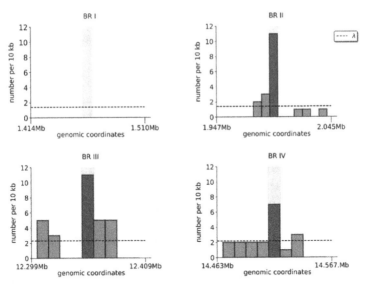

Fig. 4. Density of DNA transposons (Class II TEs) in the BRs and approximately 100-kb genomic neighborhoods of the BRs in the X chromosome of *An. atroparvus*. Light pink bars show the position of BRs within genomic neighborhoods. Red and blue bars show the density of TEs in BRs and genomic neighborhoods, respectively. Genomic neighborhoods are defined as 50-kb genomic regions located immediately upstream and downstream of each BR. λ is a mean density of DNA transposons over the entire chromosome with bin size equal to corresponding BR size. (Color figure online)

at least one copy of CACTA TIR transposon DNA/DTC. Multiple copies of DNA PIF Harbinger MITE/DTH were found in BRII and BRIII, whereas BRIV had four Helitrons. Tandem repeats were found in every BPs as 1–6 nucleotide repeats (Table 4). Analysis of gene density in the BRs showed that all four BRs are located in the genomic regions with similar gene density as in the entire X chromosome (Fig. S4).

Table 4. Repetitive DNA content and coverage in breakpoint regions according to the Repeat-Masker annotation.

BR	Simple repeats	Transposable elements	Total coverage of repetitive DNA in BR
I	$(CAG)_8$, $(CAG)_{10}$, $(AC)_{25}$, $(GA)_{11}$, $(CTC)_9$, $(TTC)_{13}$, $(CCAGC)_5$, $(TATTTA)_8$, A-rich$_{93}$, A-rich$_{39}$, A-rich$_{42}$, A-rich$_{36}$	None	7.4%

(continued)

Table 4. (*continued*)

BR	Simple repeats	Transposable elements	Total coverage of repetitive DNA in BR
II	$(TTC)_{12}$	CACTA TIR transposon DNA/DTC (1), DNA hAT DNA/DTA (1), DNA PIF Harbinger MITE/DTH (8), Helitron (1), Unknown (1)	18.6%
III	$(CGATGC)_9$, $(GCCACC)_7$, $(TGC)_{30}$, $(TAT)_9$, $(CT)_{14}$, $(GGATT)_6$, $(C)_{27}$	CACTA TIR transposon DNA/DTC (3), DNA PIF Harbinger MITE/DTH (2), hAT TIR transposon DNA MITE/DTA (2), hAT TIR transposon DNA/DTA (1), DNA/DTT (1), Helitron (2)	14.4%
IV	$(C)_{24}$	CACTA TIR transposon DNA/DTC (2), Mutator TIR transposon (1), Helitron (4), Unknown (2)	23.9%

BR – breakpoint region. The number in the brackets indicates the number of TEs

4 Discussion

In this study we used the approach of iterative physical mapping of BRs using FISH with DNA probes designed based on gene sequences. We demonstrated that this approach can be successfully applied to insect species with polytene chromosomes, such as *An. messeae*. Increasing the density of the gene markers with each iteration mapping step makes it possible to precisely identify BRs of fixed rearrangements between a species with a reference genome and a target species. This approach considers only multiple linked markers in order to minimize misidentification of BRs due to single gene transpositions.

Evolutionary genomic studies in *Anopheles* mosquito species demonstrated that fixed inversions accumulated about 3 times faster on the X chromosome than on autosomes (Xia et al. 2010; Jiang et al. 2014; Neafsey et al. 2015). It has been shown that the fixation rate of underdominant and advantageous partially or fully recessive rearrangements should be higher for X chromosomes (due to the hemizygosity of males) than for autosomes (Charlesworth, Coyne, and Barton 1987). According to our reconstruction, the X chromosome of *An. messeae* underwent two simultaneous or consecutive inversion events, while only one autosomal inversion is fixed between *An. atroparvus* and *An. messeae* (Stegniy 1991, 1982). The order of these events could not be determined without analysis of the BRs in the *An. messeae* genome. The X1 arrangement is fixed in *An. messeae* species, whereas the X1 arrangement is polymorphic in the sister species *An. daciae* (Naumenko et al., 2020, Artemov et al., 2021). Our previous high-resolution cytogenetic mapping of polytene chromosomes with the X0 and X1 arrangements in *An.*

daciae (Artemov et al., 2021) demonstrated that the BRs of the polymorphic inversion X1 do not coincide with the BRs of the fixed inversions in *An. messeae.* Therefore, the arrangement X0 originated independently in *An. daciae* based on the ancestral X1 arrangement shared by both species.

Inversions are generated by two major mechanisms, ectopic recombination and staggered breaks (Villoutreix et al. 2021; Ling and Cordaux 2010; Ranz et al. 2007). The staggered breaks mechanism requires no specific sequence for realization. In contrast, the ectopic recombination mechanism occurs during recombination between two homologous opposite-oriented genomic sequences (often TEs) on the same chromosome. A recent genomic analysis of multiple *Anopheles* species demonstrated that repetitive elements comprise 4–20% of the assembled genomes and the X chromosome had a higher repeat content than autosomes (Lukyanchikova et al. 2022). As there is no sequenced genome available for *An. messeae*, we analyzed BRs using the *An. atroparvus* genome. Enrichment of TEs and other repetitive sequences at BRs in *An. atroparvus* may suggest the presence of "hot spots" for rearrangements in the ancestral X chromosome. We found that BR neighborhoods contain a variety of TEs (Fig. 4). Some BRs contain TEs belonging to the same families (Table 3), which could be a prerequisite to inversion origination by the ectopic recombination mechanism. Abundance of TEs within inversion BRs is typical for species of the *An. gambiae* complex (Sharakhov et al. 2006) and *An. stephensi* (Thakare et al. 2022). We found DNA transposons but not retrotransposons in the BRs of *An. atroparvus*. This is in contrast with *An. gambiae* in which LTR-Gypsy elements were identified in the 2La inversion BRs (Tubio et al. 2011).

Assuming a similar gene content between *An. messeae* and *An. atroparvus,* the large inversion captured 901 genes or 75% of the chromosome, and the small inversion captured 722 genes or 59% of the chromosome. The high percentage of the genes captured by the inversions in *Anopheles* agrees with the larger sizes of X chromosomal inversions in comparison with autosomal inversions demonstrated for *Drosophila* (Cheng and Kirkpatrick 2019; Connallon and Olito 2022). It has been shown that intermediate-to-large size inversions are maintained as balanced polymorphisms via associative overdominance for a long time and play a role in local adaptation of species (Connallon and Olito 2022). Therefore, it is reasonable to assume that the large X chromosomal inversions before fixation had adaptive values in the ancestral populations of *An. messeae.* Chromosomal inversions are known to suppress genetic recombination in the rearranged region, decreasing gene flow between inverted and non-inverted variants (Kirkpatrick and Barton 2006; Kirkpatrick 2010). As a result, different combinations of gene alleles, called supergenes, are associated with each arrangement (Schwander, Libbrecht, and Keller 2014). One obvious phenotype that distinguishes *An. messeae* from *An. atroparvus* is the ability of females of the former species to develop complete diapause in winter by accumulating large lipid reserves. In contrast, *An. atroparvus* females do not develop large fat bodies but continue taking occasional blood meals during the incomplete diapause in winter (Jetten and Takken, 1994). When the annotated genomes of multiple species of the Maculipennis Subgroup are available, it will be possible to investigate the association of alleles with the ability to develop the complete diapause. Here, we looked at predicted functions of 179 genes belonging to the SB2 and SB4 that changed their chromosome positions in *An. messeae* in comparison with *An. atroparvus* to see if these genes could be involved in the DIAPAUSE-associated pathways. We compared these genes with known genes associated with overwintering diapause in *Culex pipiens pallens* that overwinters at the imago stage as *Anopheles* (Zhang et al. 2019). Five orthologous genes were found in SB2 and SB4. These genes are involved in drug metabolism by cytochrome P450 (KEGG 00982) and glutathione metabolism (KEGG 00480), such as elongation factor 1-gamma (AATE015772), alpha 1,3-glucosidase (AATE015407), and gamma-glutamyltranspeptidase (AATE005125), as well as in butanoate metabolism pathway (KEGG 00650), such as 17-beta-hydroxysteroid dehydrogenase type 6 precursor (AATE017741) and uridine phosphorylase (AATE011973). These rearranged genes could potentially be associated with adaptation to the harsh winter climate and the X chromosomal inversions could facilitate linkage of favorable alleles in *An. messeae*.

Funding. Collection of mosquito samples and mapping experiments were supported by the Russian Science Foundation grant № 21-14-00182. Bioinformatics and statistical analyses were supported by the Tomsk State University Development Programme (Priority-2030).

Availability of Data and Materials. The data and materials used in this study data is archived in a publicly accessible repository (doi.org/https://doi.org/10.5281/zenodo.7749003).

References

Amos, B., et al.: VEuPathDB: The eukaryotic pathogen, vector and host bioinformatics resource center. Nucleic Acids Res. **50**(D1), D898-911 (2022)

Anselmetti, Y., Duchemin, W., Tannier, E., Chauve, C., Bérard, S.: Phylogenetic signal from rearrangements in 18 Anopheles species by joint scaffolding extant and ancestral genomes. BMC Genomics **19**(Suppl 2), 96 (2018)

Artemov, G.N., Bondarenko, S.M., Naumenko, A.N., Stegniy, V.N., Sharakhova, M.V., Sharakhov, I.V.: Partial-arm translocations in evolution of malaria mosquitoes revealed by high-coverage physical mapping of the Anopheles atroparvus genome. BMC Genomics **19**(1), 278 (2018)

Artemov, G.N., et al.: New cytogenetic photomap and molecular diagnostics for the cryptic species of the malaria mosquitoes and from Eurasia. Insects **12**(9) (2021). https://doi.org/10.3390/insects12090835

Artemov, G., Stegniy, V., Sharakhova, M., Sharakhov, I.: The development of cytogenetic maps for malaria mosquitoes. Insects **9**(3), 121 (2018)

Artemov, G.N., et al.: A Standard photomap of ovarian nurse cell chromosomes in the European malaria vector Anopheles atroparvus. Med. Vet. Entomol. **29**(3), 230–237 (2015)

Ayala, D., Ullastres, A., González, J.: Adaptation through chromosomal inversions in Anopheles. Front. Genet. **5**(May), 129 (2014)

Ayala, D., Francisco, J., Coluzzi, M.: Chromosome speciation: humans, drosophila, and mosquitoes. Proc. Nat. Acad. Sci. United States America **102**(Suppl 1), 6535–6542 (2005)

Cáceres, M., Ranz, J.M., Barbadilla, A., Long, M., Ruiz, A.: Generation of a widespread drosophila inversion by a transposable element. Science **285**(5426), 415–418 (1999)

Charlesworth, B., Coyne, J.A., Barton, N.H.: The relative rates of evolution of sex chromosomes and autosomes. Am. Nat. (1987). https://doi.org/10.1086/284701

Cheng, C., Kirkpatrick, M.: Inversions are bigger on the X chromosome. Mol. Ecol. **28**(6), 1238–1245 (2019)

Connallon, T., Olito, C.: Natural selection and the distribution of chromosomal inversion lengths. Mol. Ecol. **31**(13), 3627–3641 (2022)

Coulibaly, M.B., et al.: Segmental duplication implicated in the genesis of inversion 2Rj of Anopheles gambiae. PLoS ONE **2**(9), e849 (2007)

Djadid, N.D., Gholizadeh, S., Tafsiri, E., Romi, R., Gordeev, M., Zakeri, S.: Molecular identification of palearctic members of Anopheles maculipennis in Northern Iran. Malar. J. **6**(January), 6 (2007)

Flynn, J.M., et al.: RepeatModeler2 for automated genomic discovery of transposable element families. Proc. Natl. Acad. Sci. U.S.A. **117**(17), 9451–9457 (2020)

Fontaine, M.C., et al.: Extensive introgression in a malaria vector species complex revealed by phylogenomics. Science (2015). https://doi.org/10.1126/science.1258524

Giraldo-Calderón, G.I., et al.: VectorBase: an updated bioinformatics resource for invertebrate vectors and other organisms related with human diseases. Nucleic Acids Res. **43**(Database issue), D707–D713 (2015)

Gornostaeva, R.M., Danilov, A.V.: On ranges of the malaria mosquitoes (Diptera: Culicidae: Anopheles) of the maculipennis complex on the territory of Russia. Parazitologiia **36**(1), 33–47 (2002)

Jetten, T.H., Takken, W.: Anophelism without malaria in Europe: a review of the ecology and distribution of the genus Anopheles. Wageningen Agric. Univ. Papers **94**(5), 1–69 (1994)

Jiang, X., et al.: Genome analysis of a major urban malaria vector mosquito, Anopheles stephensi. Genome Biol. (2014). https://doi.org/10.1186/s13059-014-0459-2

Kamali, M., Xia, A., Zhijian, T., Sharakhov, I.V.: A new chromosomal phylogeny supports the repeated origin of vectorial capacity in malaria mosquitoes of the Anopheles gambiae complex. PLoS Pathog. **8**(10), e1002960 (2012)

Kirkpatrick, M.: How and why chromosome inversions evolve. PLoS Biol. **8**(9) (2010). https://doi.org/10.1371/journal.pbio.1000501

Kirkpatrick, M., Barton, N.: Chromosome inversions, local adaptation and speciation. Genetics **173**(1), 419–434 (2006)

Ling, A., Cordaux, R.: Insertion sequence inversions mediated by ectopic recombination between terminal inverted repeats. PLoS ONE **5**(12), e15654 (2010)

Lobachev, K.S., Rattray, A., Narayanan, V.: Hairpin- and cruciform-mediated chromosome breakage: causes and consequences in eukaryotic cells. Front. Biosci. (2007). https://doi.org/10.2741/2381

Lonnig, W.-E., Saedler, H.: Chromosome rearrangements and transposable elements. Annu. Rev. Genet. **36**(June), 389–410 (2002)

Lukyanchikova, V., et al.: Anopheles mosquitoes reveal new principles of 3D genome organization in insects. Nat. Commun. **13**(1), 1960 (2022)

Naumenko, A.N., et al.: Chromosome and genome divergence between the cryptic Eurasian malaria vector-species Anopheles messeae and Anopheles daciae. Genes **11**(2) (2020). https://doi.org/10.3390/genes11020165

Neafsey, D.E., et al.: Mosquito genomics. highly evolvable malaria vectors: the genomes of 16 Anopheles mosquitoes. Science **347**(6217), 1258522 (2015)

Neph, S., et al.: BEDOPS: high-performance genomic feature operations. Bioinformatics **28**(14), 1919–1920 (2012)

Nicolescu, G., Linton, Y.-M., Vladimirescu, A., Howard, T.M., Harbach, R.E.: Mosquitoes of the *Anopheles maculipennis* group (Diptera: Culicidae) in Romania, with the discovery and formal recognition of a new species based on molecular and morphological evidence. Bull. Entomol. Res. (2004). https://doi.org/10.1079/ber2004330

Ou, S., et al.: Benchmarking transposable element annotation methods for creation of a streamlined, comprehensive pipeline. Genome Biol. **20**(1), 275 (2019)

Ranz, J.M., et al.: Principles of genome evolution in the drosophila melanogaster species group. PLoS Biol. (2007). https://doi.org/10.1371/journal.pbio.0050152

Schwander, T., Libbrecht, R., Keller, L.: Supergenes and complex phenotypes. Current Biol. CB **24**(7), R288–R294 (2014)

Sharakhova, M.V., Artemov, G.N., Timoshevskiy, V.A., Sharakhov, I.V.: Physical genome mapping using fluorescence in situ hybridization with mosquito chromosomes. Insect Genomics, 177–194 (2019)

Sharakhov, I.V., et al.: Breakpoint structure reveals the unique origin of an interspecific chromosomal inversion (2La) in the Anopheles gambiae complex. Proc. Natl. Acad. Sci. U.S.A. **103**(16), 6258–6262 (2006)

Sharakhov, I.V.: Chromosome phylogenies of malaria mosquitoes. Tsitologiia **55**(4), 238–240 (2013)

Sinka, M.E., et al.: The dominant Anopheles vectors of human malaria in Africa, Europe and the middle east: occurrence data, distribution maps and bionomic précis. Parasit. Vectors **3**(December), 117 (2010)

Stegniĭ, V.N., Kabanova, V.M., Novikov, I.: Study of the karyotype of the malaria mosquito. Tsitologiia **18**(6), 760–766 (1976)

Stegniy, V.N.: Population Genetics and Evolution of Malaria Mosquitoes, p. 137. Tomsk State University Publisher, Tomsk, Russia (1991)

Stegniy, V.N.: Genetic adaptation and speciation in sibling species of the Eurasian maculipennis complex. In: Steiner, W.W.M., Tabachnick, W.J., Rai, K.S., Narang, S. (eds.) Recent Developments in the Genetics of Insect Disease Vectors, pp. 454–464. Stipes, Champaign, 111 (1982)

Tesler, G.: GRIMM: genome rearrangements web server. Bioinformatics **18**(3), 492–493 (2002)

Thakare, A., et al.: The genome trilogy of Anopheles stephensi, an urban malaria vector, reveals structure of a locus associated with adaptation to environmental heterogeneity. Sci. Rep. **12**(1), 3610 (2022)

Tubio, J.M.C., et al.: Evolutionary dynamics of the Ty3/gypsy LTR retrotransposons in the genome of Anopheles gambiae. PloS One **6**(1), e16328 (2011)

Villoutreix, R., Ayala, D., Joron, M., Gompert, Z., Feder, J.L., Nosil, P.: Inversion breakpoints and the evolution of supergenes. Mol. Ecol. **30**(12), 2738–2755 (2021)

Virtanen, P., et al.: Author correction: SciPy 1.0: fundamental algorithms for scientific computing in Python. Nat. Methods **17**(3), 352 (2020)

Wicker, T., et al.: A unified classification system for eukaryotic transposable elements. Nat. Rev. Genet. **8**(12), 973–982 (2007)

Xia, A., et al.: Genome landscape and evolutionary plasticity of chromosomes in malaria mosquitoes. PLoS ONE (2010). https://doi.org/10.1371/journal.pone.0010592

Ye, J., Coulouris, G., Zaretskaya, I., Cutcutache, I., Rozen, S., Madden, T.L.: Primer-BLAST: a tool to design target-specific primers for polymerase chain reaction. BMC Bioinf. **13**(June), 134 (2012)

Yurchenko, A.A., et al.: Phylogenomics revealed migration routes and adaptive radiation timing of holarctic malaria vectors of the maculipennis group (2022). https://doi.org/10.1101/2022.08.10.503503

Zhang, C., et al.: Understanding the regulation of overwintering diapause molecular mechanisms in culex pipiens pallens through comparative proteomics. Sci. Rep. **9**(1), 1–12 (2019)

Zhang, S., et al.: Anopheles vectors in mainland china while approaching malaria elimination. Trends Parasitol. **33**(11), 889–900 (2017)

Gene Order Phylogeny via Ancestral Genome Reconstruction Under Dollo

Qiaoji Xu and David Sankoff[✉]

Department of Mathematics and Statistics, University of Ottawa,
150 Louis Pasteur Pvt., Ottawa, ON K1N 6N5, Canada
sankoff@uottawa.ca

Abstract. We present a proof of principle for a new kind of step-wise algorithm for unrooted binary gene-order phylogenies. This method incorporates a simple look-ahead inspired by Dollo's law, while simultaneously reconstructing each ancestor (sometimes referred to as hypothetical taxonomic units "HTU"). We first present a generic version of the algorithm illustrating a necessary consequence of Dollo characters. In a concrete application we use generalized oriented gene adjacencies and maximum weight matching (MWM) to reconstruct fragments of monoploid ancestral genomes as HTUs. This is applied to three flowering plant orders while estimating phylogenies for these orders in the process. We discuss how to improve on the extensive computing times that would be necessary for this method to handle larger trees.

Keywords: large phylogeny problem · inferring gene-order HTUs · Dollo's law · gene proximities · maximum weight matching · plant orders

1 Introduction

In formal phylogenetics, Dollo models postulate that a character can only appear once in the course of evolution, although it may disappear from several descendent lineages. This idea has a number of combinatorial consequences, and suggests an algorithmic inference of phylogeny that differs significantly from standard approaches. In this paper, we present a proof of principle for such an algorithm in the context of unrooted binary trees.

Our approach is basically hierarchical, agglomerative, combining pairs of input taxa or already constructed subtrees but, crucially, does make use, via Dollo, of a limited amount of information from outside these pairs.

The **input** to the generic version of our algorithm consists of n taxa, sometimes referred to as observed taxonomic units - "OTUs", corresponding to the terminal vertices of the phylogeny to be constructed, each of which represented by a set of distinct characters, where the sets may overlap with each other to varying degrees. The strategy is to construct $n - 2$ **output** sets, each containing some of the same characters, corresponding to the non-terminal vertices,

© The Author(s) 2023
K. Jahn and T. Vinař (Eds.): RECOMB-CG 2023, LNBI 13883, pp. 100–111, 2023.
https://doi.org/10.1007/978-3-031-36911-7_7

or HTUs, of a binary tree constructed at the same time, such that the Dollo condition, or at least some of its important consequences, is maintained.

In contrast to the generic version of the algorithm, for any specific formulation of a Dollo model, the algorithm must be modified so that the output sets conform to the requirements of the problem at hand. These modifications generally mean that Dollo is no longer a sufficient condition, but remains a necessary condition for a character to be included in an output set.

In the main part of this paper, the elements of an input set are all the proximities between oriented genes in one genome. The elements of each output set are chosen from a very large number of proximities, namely those satisfying the Dollo condition, but narrowed down to include only those consistent with an optimal linear ordering, i.e. fragments of chromosome. We have previously invoked these concepts in studying the "small phylogeny problem", where the tree topology is given (e.g., [1–5]), but here we concentrate on the "large phylogeny problem", where we actually construct this phylogeny.

We apply our method to three plant orders to compare the results with known phylogenies, with almost total agreement.

Finally we discuss possible improvements in efficiency and extensions beyond unrooted binary branching phylogenies.

2   Dollo's Law in the Context of Unrooted Binary Trees

The idea that a character is gained only one time and can never be regained if it is lost is realized in an unrooted tree by the property that the set of vertices containing the character are connected. This is a necessary and sufficient condition, valid both for terminal vertices (or degree 1) and internal (ancestral) ones (degree 3 in an unrooted binary tree).

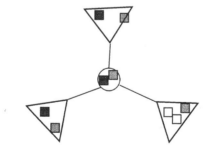

Fig. 1. Necessary condition for characters to appear at internal vertex of binary branching tree. Light shaded character (small square) appears in all three trees (triangles) subtended by the internal vertex (circle). Dark shaded character appears in only two of the trees. Unshaded character appears in only one subtree so does not affect internal vertex. The shaded characters are "phylogenetically validated" with respect to the internal vertex. The unshaded one is not validated.

Formally, the connectedness condition can be satisfied by a set of non-terminal nodes of a tree. For phylogenetic inference, however, we require that a character be present in the input set of at least two terminal vertices, i.e., visible to the observer. Moreover, if it were only in one terminal set, it could not be necessary presence for any of the output sets at the non-terminal vertices.

In an unrooted binary branching tree, each non-terminal vertex subtends three subtrees, as in Fig. 1. For a data set to be used in constructing a phylogenetic tree and the output sets, clearly each character must be present in at least two of the three subtrees, as illustrated in Fig. 1. More precisely, each character must be present at least in one terminal vertex set in at least two of the three subtrees. We call these characters "phylogenetically validated".

3   Generic Algorithm

1. input n sets of characters
2. $i = 1$
3. all n sets are "eligible" vertices
4. while $i \leq n - 3$
   (a) *for each pair (G, H) of the $m = n - i + 1$ eligible vertices calculate a potential ancestral vertex A containing all phylogenetically validated characters (in both G and H, or in one of G or H plus any other eligible vertex).
   (b) pick the pair (G', H') with maximum total number of the characters in their potential ancestor A.
   (c) ancestor vertex A becomes eligible, and the other two, G' and H', become ineligible
   (d) two edges of the tree are defined: AG' and AH'
   (e) $i = i + 1$
5. Now $i = n - 3$, so that there are three eligible vertices G', H', K'. Define three edges of the tree: AG', AH' and AK'
6. calculate ancestral vertex A containing all phylogenetically validated characters (in any two or all three of G', H' and K')
7. output all $2n - 3$ tree edges and all $n - 2$ selected output (ancestral, or HTU) sets.

The asterisked step 4a may be interpreted in at least two different ways. In requiring that a character be present in a subtree, we may mean that character must be present in some input set associated with a terminal vertex of that subtree, or we may require something stronger, that the character be present in the eligible (ancestral) vertex of that subtree.

In constructing a potential ancestor A of two input sets G and H, we define a **Dollo** character to be one in either G or H, but not both, as well as in some eligible vertex other than G or H.

Our sketch of the generic algorithm is meant to illustrate how the Dollo principle allows a kind of look-ahead in the hierarchical construction of the

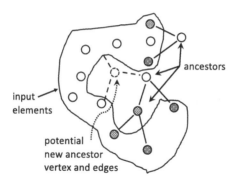

Fig. 2. The enclosed area contains input vertices. Outside the enclosed area are ancestors. Five of the input vertices have already been joined to an ancestor. These are shaded, as is one ancestor that itself been joined to an earlier ancestor. Eligible vertices, either input or ancestral, are not shaded. The dotted lines show an attempt to join an input vertex to an ancestor, with one dotted line potentially linking a Dollo character.

phylogeny (Fig. 2). Without some modification, however, the algorithm can result in some counter-intuitive results.

Consider the input sets (1, 2), (3, 4), (1, 3), (2, 4), where all four characters are also in other sets. The largest potential ancestor (1, 2, 3, 4) is constructed by pairing (1, 2) with (3, 4), or pairing (1, 3) with (2, 4). Note that this ancestor is based entirely on Dollo characters.

The grouping of (1, 2) with (3, 4) in this construction is not intuitively satisfying from the phylogenetic viewpoint since these two input sets have nothing in common. This suggests down-weighting the Dollo characters. For example, if we assign weight $1/2$ to Dollo characters compared to weight 2 to a character in both G and H, the potential ancestor of (1, 2) and (3, 4) only has weight 2, while the ancestor of (1, 2) and (1, 3) has weight 3. It is the latter that will be chosen by the algorithm if we replace "maximum number" by "maximum weight".

A principled way of assigning weights might be $2 + \alpha q/m$ for characters in both G and H plus any other eligible vertex, and $1 + \beta q/m$ for a character in one of G or H plus any other eligible vertex, where q is the number of other eligible vertices out of m containing the character, and $2\beta < \alpha < 1$.

4   Genomics Case: Generalized Adjacencies of Genes

4.1   Algorithm: Ancestor via Maximum Weight Matching (MWM)

In this version of the algorithm, the content of the sets corresponding to the given extant genomes are all the generalized gene adjacencies in the genome. This includes pairs of adjacent genes, taking their orientations or DNA strandedness into account, but also includes pairs of gene that are not immediate neighbours, but that are separated by at most 7 other genes in the gene order on a chromosome.

The ancestor genomes also contain adjacencies, not the simple union of the contents of the two daughter genomes, but only the best set of consistent adjacencies, namely the output of a MWM from among all the adjacencies in the two daughters plus certain adjacencies from other "eligible" genomes. The matching criterion ensures that all the adjacencies in a set associated with an ancestor are compatible with a linear ordering, as in a chromosome, although we are not concerned with actually building chromosomes here.

The definition of orthologous genes in the various input genomes, the construction of the sets of adjacency, and the use of MWM are taken from the first steps of the RACCROCHE method [1–5], which is concerned with actual chromosome reconstruction of the ancestors genomes in a small phylogeny context. This is not our concern here, which is the branching structure of a phylogeny, namely the large phylogeny problem.

The MWM analysis was carried out using Joris van Rantwijk's implementation [6] of Galil's algorithm for maximum matching in general (not just bipartite) graphs [7], based on Edmonds "blossom" method including the search for augmenting paths and his "primal-dual" method for finding a matching of maximum weight [8].

1. input n extant genomes
2. $i = 1$
3. all n genomes are "eligible"
4. while $i \leq n - 3$
    (a) for all pairs (G, H) of the $m - i + 1$ eligible vertices find potential ancestral genome A as the Maximum Weight Matching of phylogenetically validated generalized adjacencies (in both G and H - weight 2, or in one of G or H plus any other eligible vertex - weight 2, or in both G and H plus any other eligible vertex - weight 3).
    (b) pick the pair (G', H') with the highest Maximum Weight Matching score.
    (c) ancestor genome A becomes eligible, and the other two, G' and H' become ineligible.
    (d) two edges of the tree are defined: AG' and AH'
    (e) $i = i + 1$
5. Now $i = n - 3$, so that there are three eligible vertices G', H', K'. Define three edges of the tree : AG', AH' and AK'
6. calculate ancestral vertex A containing all phylogenetically validated adjacencies (in any two or all three of G', H' and K')

A down-weighting scheme for Dollo characters may also be adopted here, although we do not consider that further here.

The use of MWM is what distinguishes our method from Dollo parsimony and other adjacency methods. MWM ensures that the adjacencies that appear in a vertex set are consistent, that they can all be part of a chromosome, whereas all other phylogenetic methods based on adjacencies do not care that some sets of adjacencies are not consistent with a chromosome-based genome. These methods may base their results on a gene that is adjacent to three or four genes, while MWM will only retain adjacencies that allow a gene to be adjacent to only two other genes.

5   Application to Phylogenies of Three Plant Orders

Detailed references to all the genomes mentioned here are given in reference [5], including access codes for the CDS files of the genomes we use in the CoGe platform [9,10].

The phylogenies that serve as validation of our constructs are by and large uncontroversial, based on up-to-date sources, mainly [11–13].

5.1   Asterales

From the family Asteraceae, we used the published genomes of safflower (*Carthamus tinctorius*) and artichoke (*Cynara cardunculu*) from the subfamily Carduoideae, lettuce (*Lactuca sativa*) and dandelion (*Taraxacum mongolicum*) from the subfamily Cichorioideae, and *Mikania micrantha* and *Stevia rebaudian* from the subfamily Asteroideae.

Figure 3 and Table 1 show that our method partitioned the six genomes correctly into three groups.

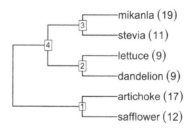

Fig. 3. Partial phylogeny of the order Asterales (family Asteraceae) as correctly reconstructed by our algorithm, with haploid numbers of chromosomes. Labels on interior nodes indicate the algorithm steps at which they were created (cf Table 1).

Table 1. Steps in searching for Asterales ancestors

potential sisters	MWM for potential anc 1	MWM for pot. anc 2	MWM for potential. anc 3	anc 4
Safflower Artichoke	**1844325**			
Lettuce Dandelion	1767015	**1711720**		
Stevia Mikania	1677330	1648609	**1537870**	
anc3_anc2_anc1				**1322071**
Lettuce Safflower	1752682			
Lettuce Stevia	1663257	1646505		
Lettuce Artichoke	1814764			
Lettuce Mikania	1588424	1576347		

(continued)

Table 1. (*continued*)

potential sisters	MWM for potential anc 1	MWM for pot. anc 2	MWM for potential. anc 3	anc 4
Safflower Stevia	1631338			
Safflower Dandelion	1683778			
Safflower Mikania	1565038			
Stevia Dandelion	1590905	1581354		
Stevia Artichoke	1685538			
Dandelion Artichoke	1741613			
Dandelion Mikania	1525367	1518558		
Artichoke Mikania	1612480			
anc1_Lettuce		1644223		
anc1_Stevia		1494957	1467125	
anc1_Dandelion		1573684		
anc1_Mikania		1414983	1399206	
anc2_Stevia			1468319	
anc2_Mikania			1395159	
anc2_anc1			1374650	

5.2   Fagales

From the order Fagales, we used the published genomes of oak (*Quercus robur*) and beech (*Fagus sylvatica*) from the family Fagaceae, birch (*Betula platyphylla*) and hazelnut (*Corylus mandshurica*) from the family Betulaceae, and walnut (*Juglans regia*), pecan (*Carya illinoinensis*) and hickory (*Carya cathayensis*) from the family Juglandaceae.

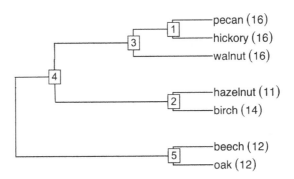

Fig. 4. Partial phylogeny of the order Fagales, as correctly reconstructed by our algorithm, with haploid numbers of chromosomes. Labels on interior nodes indicate the algorithm steps at which they were created (cf Table 2).

Table 2. Steps in searching for Fagales ancestors

potential sisters	MWM for potential anc 1	MWM for pot. anc 2	MWM for pot. anc 3	MWM for pot.anc 4	ancestor 5
Hickory Pecan	**2280458**				
Birch Hazelnut	2133010	**2130194**			
anc1_Walnut		1938190	**1929418**		
anc3_anc2				**1744607**	
anc4_Oak_Beech					**1502714**
Walnut Birch	1780904	1777540			
Walnut Oak	1432598	1414818	1399566		
Walnut Hickory	2138055				
Walnut Pecan	2048931				
Walnut Beech	1662931	1654141	1623485		
Walnut Hazelnut	2026641	2013614			
Birch Oak	1474374	1481635			
Birch Hickory	1980129				
Birch Pecan	1892718				
Birch Beech	1735741	1744305			
Oak Hickory	1612423				
Oak Pecan	1534218				
Oak Beech	1390113	1394735	1366291	1332320	
Oak Hazelnut	1675695	1677963			
Hickory Beech	1851957				
Hickory Hazelnut	2244383				
Pecan Beech	1772551				
Pecan Hazelnut	2148460				
Beech Hazelnut	1951082	1952202			
anc1_Oak		1443626	1460361		
anc1_Beech		1704015	1711078		
anc1_Hazelnut		2108715			
anc1_Birch		1844003			
anc2_Walnut			1817364		
anc2_Oak			1473713	1487666	
anc2_Beech			1743668	1737292	
anc2_anc1			1849853		
anc3_Oak				1410577	
anc3_Beech				1650427	

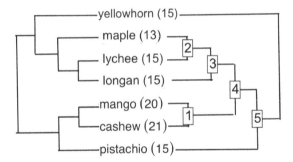

Fig. 5. Partial phylogeny of the order Sapindaless with haploid numbers of chromosomes. Left is the known phylogeny and right is the reconstructed one. Labels on interior nodes at right indicate the algorithm steps at which they were created.

Figure 4 and Table 2 show that our method partitioned the seven genomes correctly into three families.

5.3   Sapindales

While the results on the Asterales and Fagales are heartening, we cannot expect the method to always perform as well. This is illustrated by an analysis of the order Sapindales. From this order, we used the published genomes of cashew (*Anacardium occidentale*), mango (*Mangifera indica*) and pistachio (*Pistacia vera*) from the family Anacardiaceae and maple (*Acer catalpifolium*), longan (*Dimocarpus longan*), lychee (*Litchi chinensis*) and yellowhorn (*Xanthoceras sorbifoli*) from the family Sapindaceae.

As seen in Fig. 5, except for the incorrect placement of yellowthorn, the method separates the two families as expected. Were this genome to be removed, the reconstructed tree would be identical to the known tree, aside from a permutation of the species within the Sapindaceae. Indeed, Table 3 shows that at the fourth step in the algorithm, yellowthorn was almost assigned to join the other Sapindaceae.

6   Discussion and Conclusions

The algorithmic reconstruction of ancient gene orders, and associated phylogenies, has a history of over three decades (e.g., [14–17]). The present work differs from all these in that it is situated in a paradigm [1–5] where only the strictly monoploid ancestors in a phylogeny are reconstructed, or strictly linear chromosomal fragments, whether or not such genomes actually existed or are just inherent in the basic chromosomal organization within the possibly re-occurring polyploid history of the group.

Although our approach builds a hierarchy by combining pairs of OTUs or already constructed subtrees, it is unusual that it does make use, via Dollo, of a

Table 3. Steps in searching for Sapindales ancestors

potential sisters	MWM for potential anc 1	MWM for pot. anc 2	MWM for pot. anc 3	MWM for pot.anc 4	ancestor 5
Cashew Mango	**2114258**				
Lychee Maple	1909424	**1915717**			
anc2_Longan			**1881830**		
anc3_anc1				**1812017**	
anc4_Yellowhorn_Pistachio					**177012**
Lychee Longan	1839623	1847180			
Lychee Yellowhorn	1792086	1718022			
Lychee Cashew	1715727				
Lychee Pistachio	1881466	1885124			
Lychee Mango	1899616				
Longan Maple	1741981	1750100			
Longan Yellowhorn	1775908	1568607	1561594		
Longan Cashew	1566047				
Longan Pistachio	1707939	1713967	1699756		
Longan Mango	1729214				
Maple Yellowhorn	1892491	1652072			
Maple Cashew	1652307				
Maple Pistachio	1821498	1820282			
Maple Mango	1847530				
Yellowhorn Cashew	1840431				
Yellowhorn Pistachio	1966295	1689715	1694495	1511822	
Yellowhorn Mango	1945493				
Cashew Pistachio	1702324				
Pistachio Mango	1920295				
anc1_Lychee		1795812			
anc1_Longan		1621378	1634641		
anc1_Maple		1734318			
anc1_Yellowhorn		1651337	1678486	1636597	
anc1_Pistachio		1814383	1830387	1656675	
anc2_Yellowhorn			1696079		
anc2_Pistachio			1803140		
anc2_anc1			1739109		
anc3_Yellowhorn				1802826	
anc3_Pistachio				1754820	

limited amount of information from outside these pairs. This is in effect a very partial look-ahead.

The goal of this paper was to present evidence that our method can construct accurate or plausible phylogenies, based entirely on comparative gene order. We were not preoccupied with questions of computing time. Indeed, since our reconstruction is based on maximum weight matching (MWM) software on very large graphs, it is bound to be computationally expensive. Moreover, since the current experimental version uses MWM to exhaustively evaluate all possibilities separately at each step in building a hierarchy, only a moderate number of OTUs can be input. There are, however, many possibilities to improving the efficiency, by constraining the search space, by branch and bound techniques, by saving a certain number of partial solutions in parallel, and other techniques.

References

1. Xu, Q., Jin, L., Zheng, C., Leebens Mack, J.H., Sankoff, D.: RACCROCHE: ancestral flowering plant chromosomes and gene orders based on generalized adjacencies and chromosomal gene co-occurrences. In: Jha, S.K., Măndoiu, I., Rajasekaran, S., Skums, P., Zelikovsky, A. (eds.) ICCABS 2020. LNCS, vol. 12686, pp. 97–115. Springer, Cham (2021). https://doi.org/10.1007/978-3-030-79290-9_9
2. Xu, Q., Jin, L., Leebens-Mack, J.H., Sankoff, D.: Validation of automated chromosome recovery in the reconstruction of ancestral gene order. Algorithms **14**(6), 160 (2021)
3. Chanderbali, A.S., Jin, L., Xu, Q., et al.: Buxus and Tetracentron genomes help resolve eudicot genome history. Nat. Communun. **13**, 643 (2022). https://doi.org/10.1038/s41467-022-28312-w
4. Xu, Q., et al.: Ancestral flowering plant chromosomes and gene orders based on generalized adjacencies and chromosomal gene co-occurrences. J. Comput. Biol. **28**(11), 1156–79 (2021)
5. Xu, Q., Jin, L., Zheng, C., Zhang, X., Leebens-Mack, J., Sankoff, D.: From comparative gene content and gene order to ancestral contigs, chromosomes and karyotypes. Sci. Rep. **13**, 6095 (2023)
6. van Rantwijk, J.: Maximum Weighted Matching (2008). http://jorisvr.nl/article/maximum-matching
7. Galil, Z.: Efficient algorithms for finding maximum matching in graphs. ACM Comput. Surv. **18**, 23–38 (1986)
8. Edmonds, J.: Paths, trees, and flowers. Can. J. Math. **17**, 449–67 (1965)
9. Lyons, E., Freeling, M.: How to usefully compare homologous plant genes and chromosomes as DNA sequences. Plant J. **53**, 661–673 (2008)
10. Lyons, E., et al.: Finding and comparing syntenic regions among Arabidopsis and the outgroups papaya, poplar and grape: CoGe with rosids. Plant Physiol. **148**, 1772–1781 (2008)
11. Published Plant Genomes. Usadel lab, Forschungszentrum Jülich, Heinrich Heine University., Düsseldorf (2022). https://www.plabipd.de/
12. Stevens, P.F.: Angiosperm Phylogeny Website. Version 14 (2017). http://www.mobot.org/MOBOT/research/APweb/
13. Chase, M.W., et al.: An update of the Angiosperm Phylogeny Group classification for the orders and families of flowering plants: APG IV. Bot. J. Linn. Soc. **181**, 1–20 (2016)
14. Sankoff, D., Leduc, G., Antoine, N., Paquin, B., Lang, B.F., Cedergren, R.: Gene order comparisons for phylogenetic inference: evolution of the mitochondrial genome. Proc. Natl. Acad. Sci. **15**, 6575–9 (1992)
15. Moret, B.M., Warnow, T.: Advances in phylogeny reconstruction from gene order and content data. Methods Enzymol. **395**, 673–700 (2005)

16. Hu, F., Lin, Y., Tang, J.: MLGO: phylogeny reconstruction and ancestral inference from gene-order data. BMC Bioinf. **15**, 1–6 (2014)
17. Perrin, A., Varré, J.S., Blanquart, S., Ouangraoua, A.: ProCARs: progressive reconstruction of ancestral gene orders BMC genomics **16** S5:S6 (2015)

Prior Density Learning in Variational Bayesian Phylogenetic Parameters Inference

Amine M. Remita[1]([⊠])[iD], Golrokh Vitae[2][iD], and Abdoulaye Baniré Diallo[1,2][iD]

[1] Department of Computer Science, Université du Québec à Montréal,
Montréal, Canada
`remita.amine@courrier.uqam.ca`, `diallo.abdoulaye@uqam.ca`
[2] CERMO-FC Center, Department of Biological Sciences,
Université du Québec à Montréal, Montréal, Canada

Abstract. The advances in variational inference are providing promising paths in Bayesian estimation problems. These advances make variational phylogenetic inference an alternative approach to Markov Chain Monte Carlo methods for approximating the phylogenetic posterior. However, one of the main drawbacks of such approaches is modelling the prior through fixed distributions, which could bias the posterior approximation if they are distant from the current data distribution. In this paper, we propose an approach and an implementation framework to relax the rigidity of the prior densities by learning their parameters using a gradient-based method and a neural network-based parameterization. We applied this approach for branch lengths and evolutionary parameters estimation under several Markov chain substitution models. The results of performed simulations show that the approach is powerful in estimating branch lengths and evolutionary model parameters. They also show that a flexible prior model could provide better results than a predefined prior model. Finally, the results highlight that using neural networks improves the initialization of the optimization of the prior density parameters.

Keywords: Variational Bayesian phylogenetics · Variational inference · Prior learning · Markov chain substitution models · Gradient ascent · Neural networks

1 Introduction

The Bayesian phylogenetic community is exploring faster and more scalable alternatives to the Markov chain Monte Carlo (MCMC) approach to approximate a high dimensional Bayesian posterior [10]. The search for other substitutes is motivated by the falling computational costs, increasing challenges in

Supported by NSERC, FRQNT, Genome Canada and The Digital Research Alliance of Canada.

large-scale data analysis, advances in inference algorithms and implementation of efficient computational frameworks. Some of the alternatives, reviewed in [10], are adaptive MCMC, Hamiltonian Monte Carlo, sequential Monte Carlo and variational inference (VI). Until recently, few studies were interested in applying classical variational inference approaches in probabilistic phylogenetic models [6,19,40]. However, VI started to gain some attraction from the phylogenetic community taking advantage of advances that made this approach more scalable, generic and accurate [46], such as stochastic and black box VI algorithms [15,34], latent-variable reparametrization [25,36], and probabilistic programming [5]. These advancements allowed designing powerful and fast variational-based algorithms to infer complex phylogenetic models [7,12,49] and analyze large-scale phylodynamic data [22]. Few studies have evaluated variational phylogenetic models using simple settings and scenarios to acquire some understanding of the variational inference approach. For example, some analyses assumed a fixed tree topology to estimate continuous parameters (such as branch lengths and evolutionary model parameters) and approximate their posterior [12] or marginal likelihood [13]. Zhang and Matsen [45,48] developed and tested a fully variational phylogenetic method that jointly infers unrooted tree topologies and their branch lengths under the JC69 substitution model [21].

Bayesian methods incorporate the practitioner's prior knowledge about the likelihood parameters through the prior distributions. Defining an appropriate and realistic prior is a difficult task, especially in the case of small data regimes, similar sequences or parameters with complex correlations [17,18,30]. It is important to note that the variational phylogenetic methods assign fixed prior distributions with default hyperparameters to the likelihood parameters, which is a similar practice in MCMC methods [30]. For example, i.i.d. exponential priors with a mean of 0.1 are usually placed on the branch lengths [12,13,48]. However, such a choice could bias the posterior approximation and induce high posterior probabilities in cases where the data are weak or the actual parameter values do not fall within the range specified by the priors [8,17,30,44]. The effect of branch length priors on Bayesian inference using MCMC has been widely investigated [27,44], and explanations and remedies have been proposed [31,35,50]. Therefore, it is crucial to study the impact of the prior (mis)specification on the convergence and the quality of the variational approximation of the posterior and to propose solutions to overcome this problem.

Here, we show that variational phylogenetic inference can also suffer from misspecified priors on branch lengths and less severely on sequence evolutionary parameters. We adopt a different strategy from MCMC to improve the variational posterior approximation by leveraging the structure of the variational objective function and making the prior densities more flexible. To do that, we propose a variational approach to relax the rigidity of the prior densities by jointly learning the prior and variational parameters using a gradient-based method and a neural network-based parameterization. Moreover, we implemented a variational Bayesian phylogenetic framework (nnTreeVB) to evaluate its performance, consistency and behaviour in estimating the branch lengths and evolutionary model parameters using simulated datasets.

2   Background

2.1   Notation

Suppose $\mathbf{X}$, the observed data, is an alignment of M nucleotide sequences with length N, where $\mathbf{X} \in \mathcal{A}^{M \times N}$ and $\mathcal{A} = \{A, G, C, T\}$. We assume that each site (column X_n) in the alignment evolves independently following an alignment-amortized, unrooted and binary tree τ. τ is associated with a vector $\mathbf{b}$ of unobserved $2M - 3$ branch lengths. Each branch length represents the number of hidden substitutions that happen over time from a parent node to a child node. To estimate the branch length vector, we assume that the process of evolution (substitutions) follows a continuous-time Markov chain model parameterized by $\psi = \{\rho, \pi\}$, where ρ is the set of relative substitution rates and π are the relative frequencies of the four nucleotides. We restrict the sum of each set of parameters, ρ and π, to one. We use time-reversible Markov chain models, assuming the number of changes from one nucleotide to another is the same in both ways. Several substitution models could be defined depending on the constraints placed on their parameters ψ. JC69 is the basic model with equal substitution rates and uniform relative frequencies [21], so $\psi = \emptyset$. The general time-reversible (GTR) model sets free all the parameters ψ [38,41]. In this study, we are interested in estimating the optimal branch length vector $\mathbf{b}$ and the model parameters ψ given the alignment $\mathbf{X}$ and a fixed tree topology τ.

2.2   Bayesian Phylogenetic Parameter Inference

The Bayesian approach to the phylogenetic parameter estimation is based on the evaluation of the joint conditional density $p(\mathbf{b}, \psi \mid \mathbf{X}, \tau)$, which we call it the phylogenetic parameter posterior. Using Bayes' theorem, the joint posterior is computed as

$$p(\mathbf{b}, \psi \mid \mathbf{X}, \tau) = \frac{p(\mathbf{X}, \mathbf{b}, \psi \mid \tau)}{p(\mathbf{X} \mid \tau)} = \frac{p(\mathbf{X} \mid \mathbf{b}, \psi, \tau)\, p(\mathbf{b}, \psi)}{\iint p(\mathbf{X} \mid \mathbf{b}, \psi, \tau)\, p(\mathbf{b}, \psi)\, \mathrm{d}\mathbf{b}\, \mathrm{d}\psi}.$$

The Bayes' equation exposes the tree likelihood $p(\mathbf{X} \mid \mathbf{b}, \psi, \tau)$, the joint prior density $p(\mathbf{b}, \psi)$ and the model evidence $p(\mathbf{X} \mid \tau)$. The tree likelihood is the conditional probability of the observed data given the phylogenetic parameters and the tree topology. It is efficiently computed using Felsenstein's pruning algorithm [9,33]. The algorithm assumes the independence between the alignment sites, thus $p(\mathbf{X} \mid \mathbf{b}, \psi, \tau) = \prod_{n=1}^{N} p(\mathbf{X}_n \mid \mathbf{b}, \psi, \tau)$. The rescaling algorithm [2,43] can be used to avoid cases of underflow often occurring in large trees. Usually, we model the joint prior density using independent distributions with fixed hyperparameters for each phylogenetic parameter. Ergo, it factorizes into $p(\mathbf{b}, \psi) = p(\mathbf{b})\, p(\psi) = \prod_{m=1}^{2M-3} p(\mathbf{b}_m)\, p(\rho)\, p(\pi)$. Finally, the model evidence is the marginal probability of the data given the tree topology that integrates over the possible values of the phylogenetic parameters. The integrals over high dimensional variables make the evidence intractable, hindering the computation of the phylogenetic parameter posterior density.

2.3 Variational Inference and Parameterization

A strategy to approximate the intractable posterior density $p(\mathbf{b}, \boldsymbol{\psi} \mid \mathbf{X}, \tau)$ is leveraging the variational inference (VI) approach [3,20], which redefines the inference as an optimization problem. First, VI defines a family of tractable (simple to compute) densities q parameterized by ϕ. Second, it finds among these densities the closest member q_ϕ^* to the posterior density minimizing their Kullback-Leibler divergence $\mathrm{KL}(q_\phi(\mathbf{b}, \boldsymbol{\psi}) \parallel p(\mathbf{b}, \boldsymbol{\psi} \mid \mathbf{X}, \tau))$ by tuning the variational parameters ϕ. However, computing the KL divergence is intractable because it involves the true posterior. Therefore, the VI approach maximizes another objective function, the evidence lower bound (ELBO), which is equivalent to minimizing the KL divergence and does not require the computation of the intractable evidence. The equation of the ELBO is

$$\mathrm{ELBO}(\phi, \mathbf{X}, \tau) = \mathbb{E}_{q_\phi(\mathbf{b}, \boldsymbol{\psi})} \left[\log \left(\frac{p(\mathbf{X} \mid \mathbf{b}, \boldsymbol{\psi}, \tau) \, p(\mathbf{b}, \boldsymbol{\psi})}{q_\phi(\mathbf{b}, \boldsymbol{\psi})} \right) \right] \leq \log p(\mathbf{X} \mid \tau),$$

which can also be written as

$$\mathrm{ELBO}(\phi, \mathbf{X}, \tau) = \mathbb{E}_{q_\phi(\mathbf{b}, \boldsymbol{\psi})} \left[\log p(\mathbf{X} \mid \mathbf{b}, \boldsymbol{\psi}, \tau) \right] - \mathrm{KL} \left(q_\phi(\mathbf{b}, \boldsymbol{\psi}) \parallel p(\mathbf{b}, \boldsymbol{\psi}) \right). \quad (1)$$

Since the tree likelihood function is non-exponential, the expectations are computed using Monte Carlo sampling of the joint approximate posterior (joint variational density) $q_\phi(\mathbf{b}, \boldsymbol{\psi})$.

Following previous variational phylogenetics studies [12,13], we use a Gaussian mean-field variational distribution to model the joint variational density. In this class of variational distributions, each phylogenetic parameter i is independent and follows a Gaussian (normal) distribution defined by its distinct variational parameters $\phi_i = \{\mu_i, \sigma_i\}$. Hence, the joint variational density factorizes into $q_\phi(\mathbf{b}, \boldsymbol{\psi}) = \prod_{m=1}^{2M-3} q_{\phi_{\mathbf{b}_m}}(\mathbf{b}_m) \, q_{\phi_\rho}(\boldsymbol{\rho}) \, q_{\phi_\pi}(\boldsymbol{\pi})$. We apply invertible transformations on the variational densities and adjust their probabilities (using tractable Jacobian determinants) to accommodate the constraints on the phylogenetic parameters (branch lengths must be non-negative, and relative rates and frequencies, each must have a sum of one). Finally, we use a stochastic gradient ascent algorithm [24] and reparameterization gradients [25,36] to optimize the variational parameters ϕ.

3 New Approach

3.1 Gradient-Based Learning Prior Density

We notice from Eq. 1 that maximizing the ELBO induces a regularized maximization of the tree likelihood. The regularization of the likelihood is maintained by the minimization of KL divergence term, which encourages the joint variational $q_\phi(\mathbf{b}, \boldsymbol{\psi})$ and the joint prior $p(\mathbf{b}, \boldsymbol{\psi})$ densities to be close [25,26]. Recall that the optimization is performed with respect to the variational parameters ϕ, and the parameters of the prior distributions are fixed initially. Therefore, minimizing the KL divergence in Eq. 1 squeezes and drives the approximate posterior density towards the joint prior density. The KL divergence could dominate and lead to underfitting the variational model if the data is weak and inconsistent [16,28]. We seek to counterbalance the regularization effect by relaxing the inflexibility of the joint prior distribution. To achieve this, we implement adaptable parameters θ instead of using fixed parameters for the prior densities. The prior parameters θ will be learned (updated) jointly with the variational parameters ϕ during the optimization of the ELBO using gradient ascent. Though the ELBO is now maximized with respect to ϕ and θ, learning θ does not need a reparameterization gradient because the expectation remains computed using Monte Carlo sampling of the joint variational density:

$$\text{ELBO}(\phi, \theta, \mathbf{X}, \tau) = \mathbb{E}_{q_\phi(\mathbf{b}, \psi)} \left[\log p(\mathbf{X} \mid \mathbf{b}, \psi, \tau) \right] - \text{KL} \left(q_\phi(\mathbf{b}, \psi) \parallel p_\theta(\mathbf{b}, \psi) \right). \quad (2)$$

As previously mentioned, we apply independent prior densities on each phylogenetic parameter, including the branch lengths and the substitution model parameters. We assign on the branch lengths independent exponential distributions with rates λ_b. The JC69 model has no free parameters. Thus, the set of prior parameters of this model is $\theta_{JC69} = \{\lambda_b\}$. In the case of the GTR model, we apply Dirichlet distributions on relative substitution rates and relative frequencies with concentrations α_ρ and α_π, respectively, so $\theta_{GTR} = \{\lambda_b, \alpha_\rho, \alpha_\pi\}$.

3.2   Neural Network-Based Prior Parameterization

The new prior parameters θ are initialized independently using fixed values or sampled values from a predefined distribution (e.g. uniform or normal). To add more flexibility to the prior density, we use differentiable feed-forward neural networks to generate the prior parameter values instead of relying on a direct gradient-based update. A neural network (NeuralNet) is constituted of a stack of L layers of neurons, where each neuron is defined by adaptable weight vector $\mathbf{w}$ and bias a. A layer l in a neural network with a vector input ζ_l generates $g_l(\mathbf{W}_l \zeta_l^\mathsf{T} + a_l)$, where g_l is the identity function or a nonlinear real-valued activation function (e.g. ReLU, Softplus and Softmax). Therefore, the prior parameters to be learned θ are constituted of the set of weights and biases of the neural networks instead of the parameters of the distributions.

The rate vector of the branch length prior densities is produced by a neural network with an input of a uniformly-sampled random noise ζ_{λ_b}:

$$\lambda_b = \left[\lambda_{b_1}, \lambda_{b_2}, \dots, \lambda_{b_{2M-3}} \right]^\mathsf{T} = \text{NeuralNet}_{\theta_{\lambda_b}}(\zeta_{\lambda_b}).$$

We use independent neural networks to produce the vectors of concentrations for the relative substitution rates and the relative frequencies of the GTR model:

$$\alpha_\rho = [\alpha_{\rho_1}, \alpha_{\rho_2}, \dots, \alpha_{\rho_6}]^\mathsf{T} = \text{NeuralNet}_{\theta_{\alpha_\rho}}(\zeta_{\alpha_\rho}),$$

$$\alpha_\pi = [\alpha_{\pi_1}, \alpha_{\pi_2}, \alpha_{\pi_3}, \alpha_{\pi_4}]^\mathsf{T} = \text{NeuralNet}_{\theta_{\alpha_\pi}}(\zeta_{\alpha_\pi}),$$

where ζ_{α_ρ} and ζ_{α_π} are independent random noises.

4   Experimental Study

We conceived nnTreeVB, a variational Bayesian phylogenetic framework, to evaluate and compare the consistency, effectiveness and behaviour of the proposed variational phylogenetic models implemented with different prior density schemes. The framework allows us to assess the variational models in estimating one phylogenetic parameter at a time. Thus, `nnTreeVB` simulates multiple datasets varying the prior distribution of the considered phylogenetic parameter and drawing the remaining parameters from their default priors. It fits the variational models implemented either with fixed or adaptable prior distributions on these datasets. The fixed priors using default hyperparameters can match one of the priors used in data generation. The following sections describe the procedure used in the data simulation and the settings of the variatinoal inference models.

4.1   Dataset Simulation Procedure

A dataset comprises of a tree (associated with its branch length vector) and a sequence alignment simulated using prior distributions applied over their parameters. Given a number of taxa M, we build each tree topology by sampling from a uniform distribution to have an equal prior probability. The branch lengths are sampled independently from an exponential distribution with a predefined rate λ. The rate is inversely proportional to the expected number of substitutions along a branch ($\lambda = 1/\mathbb{E}[\mathbf{b}_m]$). Afterwards, we select a substitution model and draw its parameters (if it has any) from their prior distributions. We assumed Dirichlet distribution for the substitution rates and relative frequencies and gamma distribution for positive parameters like the transition/transversion rate ratio κ in K80 [23] and HKY85 [14] models.

Eventually, using the tree and the substitution model and given a sequence length N, we simulate the evolution of a sequence alignment based on a site-wise homogeneity strategy [37]. This strategy evolves sequences from a root sequence (sampled from the distribution of relative frequencies) with the same substitution model over lineages and with the same set of branch lengths for alignment sites. We simulated datasets with 16 and 64 taxa and 1000 and 5000 base pairs (bp) to investigate the impact of the data regime in terms of the tree size and the sequence alignment length, respectively. Finally, we produced 100 replicates of each dataset characterized by a prior setting and a data size condition.

4.2   Variational Inference Settings

Using the `nnTreeVB` framework, we implemented three variational phylogenetic (VP) models that differ in their prior distribution computation and initialization. The first model is the baseline, which uses fixed prior distributions (VP-FPD) with default hyperparameters. The two other models follow the proposed approach described in Sect. 3, which defines adaptable prior parameters to be learned jointly along the variational parameters while optimizing the ELBO. Initializing the prior parameters with i.i.d. samples from a uniform distribution corresponds to the simple VP model with learnable prior distributions (VP-LPD). The last model (VP-NPD) implements feed-forward NeuralNets to generate the prior parameters from a uniform sampled input. Based on preliminary hyperparameters fine-tuning results, we used one hidden layer of size 32 and a ReLU activation, $\max(0, x)$, for each NeuralNet. To ensure the positiveness of a parameter (like rates and concentrations), we apply a softplus function, $\log(1 + e^x)$, on the output layer.

nnTreeVB implements automatic differentiation via PyTorch library [32] in order to estimate the gradient of the ELBO and optimize it. We evaluate the gradient using the Monte Carlo integration by sampling a single data point from the approximate density at each iteration. The variational parameters and adaptable prior parameters are updated using the Adam algorithm [24] with a learning rate value of 0.1 for simple parameters and 0.01 for NeuralNet-generated parameters. The training of each VP model is performed over 2000 iterations and replicated ten times to fit a given dataset. In the end, we estimate each phylogenetic parameter by sampling 1000 data points from the approximate density of each model replicate.

4.3   Performance Metrics

We investigate the convergence and the performance of the variational phylogenetic models through the two components of the ELBO (as in Eqs. 1 and 2): the log likelihood (LogL) and the KL divergence between the approximate and the prior densities (KL-qprior). Moreover, to assess the accuracy of estimating a set of phylogenetic parameters (branch lengths, substitution rates or relative frequencies), we compute the scaled Euclidean distance (Dist) and the Pearson correlation coefficient (Corr) between the estimated vector and the actual one used in the simulation of the sequence alignment. The scaled Euclidean distance is a transformed distance to be in the range $[0, 1]$ calculated as $\text{Dist} = \frac{d}{1+d}$, where d is the Euclidean distance. We also report the tree length (TL) for the evaluation of estimating the branch lengths, which is the sum of their vector. Finally, we used the nonparametric Kruskal-Wallis H test [29] to evaluate for each performance metric the differences between the results of the models.

5   Results

In this section, we present the results of simulation-based experiments we performed to evaluate the performance of the two proposed VP models that learn the prior density parameters, VP-LPD and VP-NPD, and compare them to VP-FPD, which is the default model and baseline used in variational phylogenetic inference. We demonstrate that when a phylogenetic parameter deviates from the default prior distribution, VP-L/NPD models are more efficient in approximating the posterior density. Moreover, we show how using neural network prior parameterization, implemented in VP-NPD, improves the accuracy of the substitution model parameters estimation.

We employed the VP models to estimate the branch lengths and substitution model parameters from datasets simulated with different sequence alignment lengths (1000 and 5000 bp) and tree sizes (16 and 64 taxa). However, along the section, we highlight the results from more challenging datasets with shorter sequences (1000 bp) and larger trees (64 taxa). For each analysis, we fit ten times the three VP models on 100 dataset replicates, sample 1000 estimates from their approximate densities and report their performance metrics' averages and standard deviations.

5.1   Branch Lengths Estimation Performance

First, we assessed the accuracy of branch lengths estimation using datasets generated with distinct *a priori* means over the branch lengths. Table 1 shows the results of VP models tested on datasets with branch lengths drawn with i.i.d. exponential prior means (rates) of 0.001 (1000), 0.01 (100) and 0.1 (10). We used the JC69 substitution model in sequence alignment simulation and variational inference models to avoid any effect of evolutionary model parameters on estimating the branch lengths. For inference, each VP model applies i.i.d. exponential priors on branch lengths. The VP model with fixed prior density, VP-FPD, places an i.i.d. prior with mean 0.1, which is usually used by default in variational phylogenetic software [12,48] for branch lengths and corresponds to the prior used to simulate the third dataset in the current experiment. We notice in Table 1, in the case of branch lengths simulated with a prior mean of 0.001, which is severely divergent from the default prior mean of 0.1, that the VP-FPD model overestimates the total tree length (TL) up to 2 times its actual value. Interestingly, VP-L/NPD models approximate better the branch lengths with smaller average Euclidean distances ($p = 0$, Kruskal-Wallis test), and estimate the TL average values with ratios less than 1.14 to the real TL. However, when the branch lengths follow distributions close to the default prior, VP-FPD has better branch length estimations ($p = 0$, Kruskal-Wallis test on Euclidean distances). Nonetheless, VP-L/NPD models have fairly similar results, and their estimations are better

Table 1. Performance of branch lengths (BL) estimation on datasets simulated with different prior means. The datasets have 64 sequences of length 1000 bp and are simulated with JC69 substitution model. BL values are drawn from an exponential distribution. VP models implement a JC69 model and apply an i.i.d. exponential prior on BL. VP-FPD applies a prior with mean 0.1.

	BL 0.001		BL 0.01		BL 0.1	
	LogL	KL-qprior	LogL	KL-qprior	LogL	KL-qprior
Real data	−2484.7627(95.45)		−9019.0723(571.12)		−43329.2734(2280.71)	
VP-FPD	−2518.1560(93.18)	414.8906(3.89)	−9025.1680(570.85)	304.8004(5.69)	−43344.3680(2281.89)	246.7808(5.41)
VP-LPD	**−2451.9842(95.00)**	29.2950(1.59)	**−9006.1580(571.88)**	85.6934(3.81)	**−43340.0960(2281.90)**	180.7204(3.85)
VP-NPD	−2454.8502(95.16)	**27.5081(3.67)**	−9007.7260(571.90)	85.3090(3.92)	−43340.7120(2281.94)	**180.6803(3.74)**

	TL	Dist	Corr	TL	Dist	Corr	TL	Dist	Corr
Real data	0.1239(0.01)			1.2390(0.11)			12.3896(1.09)		
VP-FPD	0.2528(0.02)	0.0240(0.00)	0.4636(0.10)	1.3463(0.12)	0.0506(0.01)	0.8966(0.03)	**12.4283(1.10)**	**0.1773(0.02)**	0.9806(0.00)
VP-LPD	**0.1407(0.02)**	0.0152(0.00)	**0.5783(0.09)**	**1.2356(0.12)**	0.0479(0.01)	**0.9058(0.02)**	12.4716(1.12)	0.1790(0.02)	**0.9810(0.00)**
VP-NPD	0.1418(0.02)	**0.0151(0.00)**	0.5732(0.09)	1.2360(0.12)	**0.0479(0.01)**	0.9055(0.02)	12.4633(1.13)	0.1791(0.02)	**0.9810(0.00)**

correlated with the real values ($p = 0$, Kruskal-Wallis test). Regardless of dataset peculiarities, VP-LPD approximates better the log likelihood (LogL) compared to the other models. Although all models approximate, up to a point, the actual LogL of the datasets, VP-L/NPD models have smaller KL divergence between the approximate density and prior density (KL-qprior).

Then, we analyzed the consistency of the VP models and the effect of the dataset size on the estimation of branch lengths. Table 2 shows the results of the estimations on datasets simulated with sequence lengths of 1000 and 5000 bp, number of taxa of 16 and 64, and branch lengths sampled from i.i.d exponential priors of mean 0.001. For all VP models, the average distances and correlations between the estimate and the actual branch lengths improve with longer sequence alignments but not always with larger trees. Moreover, the models more accurately estimate the total TL with bigger datasets. The improvement is clearly noticed with the VP-FPD model, where its ratio of the estimate and actual TL decreases from 2.11 with the 1000/16 sequence alignment to 1.19 with the 5000/64 sequence alignment. The TL ratios of the VP-L/NPD models decrease from 1.2 to 1.02 with the same sequence alignments, respectively. In terms of these performance metrics, VP-L/NPD models have better estimates compared to those of VP-FPD, regardless of the dataset size ($p \leq 2.51\mathrm{E}{-}13$, Kruskal-Wallis tests). Also, for all dataset sizes, the VP-LPD model optimizes the LogL better, and VP-L/NPD models have smaller KL-qprior divergences than those of the VP-FPD model.

Table 2. Performance of branch lengths (BL) estimation on datasets simulated with different sizes. The datasets are simulated with JC69 substitution model. BL values are drawn from an exponential distribution with mean **0.001**. VP models implement a JC69 model and apply an i.i.d. exponential prior on BL. VP-FPD applies a prior with mean 0.1.

N	M	Model	LogL	KL-qprior	TL	Dist	Corr
1000	16	Real data	−1670.4626(52.86)		0.0296(0.01)		
		VP-FPD	−1678.7043(51.34)	94.7330(2.13)	0.0625(0.01)	0.0117(0.00)	0.4852(0.18)
		VP-LPD	**−1664.4830(52.45)**	7.1037(0.82)	**0.0362(0.01)**	0.0074(0.00)	**0.5925(0.16)**
		VP-NPD	−1664.9436(52.45)	**6.4874(0.77)**	**0.0362(0.01)**	**0.0073(0.00)**	0.5882(0.16)
	64	Real data	−2484.7627(95.45)		0.1239(0.01)		
		VP-FPD	−2518.1560(93.18)	414.8906(3.89)	0.2528(0.02)	0.0240(0.00)	0.4636(0.10)
		VP-LPD	**−2451.9842(95.00)**	29.2950(1.59)	**0.1407(0.02)**	0.0152(0.00)	**0.5783(0.09)**
		VP-NPD	−2454.8502(95.16)	**27.5081(3.67)**	0.1418(0.02)	**0.0151(0.00)**	0.5732(0.09)
5000	16	Real data	−8196.2695(193.34)		0.0296(0.01)		
		VP-FPD	−8196.1490(192.97)	123.7540(2.51)	0.0361(0.01)	0.0039(0.00)	0.7871(0.09)
		VP-LPD	**−8189.3350(193.73)**	16.8418(1.09)	**0.0310(0.01)**	0.0035(0.00)	0.8042(0.08)
		VP-NPD	−8189.5360(193.71)	**16.5364(1.12)**	**0.0310(0.01)**	0.0035(0.00)	0.8041(0.09)
	64	Real data	−12074.5166(440.76)		0.1239(0.01)		
		VP-FPD	−12089.3860(439.94)	536.9764(5.30)	0.1482(0.01)	0.0076(0.00)	0.8231(0.04)
		VP-LPD	**−12059.1640(441.47)**	70.4068(2.46)	**0.1261(0.01)**	0.0067(0.00)	**0.8404(0.04)**
		VP-NPD	−12061.0200(441.25)	71.1167(7.69)	0.1267(0.01)	**0.0067(0.00)**	0.8393(0.04)

Table 3. Performance of branch lengths (BL) estimation on datasets simulated with external (ext) and internal (int) BL having different prior means. The datasets have 64 sequences of length 1000 bp and are simulated with JC69 substitution model. BL values are drawn from an exponential distribution. VP models implement a JC69 model and apply an i.i.d. exponential prior on BL. VP-FPD applies a prior with mean 0.1.

	BL (ext0.005, int0.1)			BL (ext0.1, int0.005)		
	LogL	KL-qprior	TL	LogL	KL-qprior	TL
Real data	−24300.4082(1692.34)		6.2972(0.73)	−26387.3984(1863.89)		6.7119(0.76)
VP-FPD	−24313.7400(1693.06)	289.6527(5.46)	6.3608(0.73)	−26400.9060(1864.33)	295.4958(5.73)	6.7690(0.76)
VP-LPD	**−24296.1740(1693.27)**	119.6626(3.69)	6.3325(0.74)	**−26383.8200(1864.33)**	128.7518(4.00)	**6.7296(0.77)**
VP-NPD	−24297.7720(1693.24)	**119.4662(3.78)**	**6.3296(0.75)**	−26385.5320(1864.40)	**128.4644(3.98)**	6.7327(0.78)

	All branches		Externals		Internals		All branches		Externals		Internals	
	Dist	Corr	Dist	Corr	Dist	Corr	Dist	Corr	Dist	Corr	Dist	Corr
VP-FPD	**0.1279(0.02)**	0.9871(0.00)	0.0330(0.01)	0.7536(0.09)	**0.1246(0.02)**	0.9818(0.01)	0.1213(0.02)	0.9895(0.00)	**0.1180(0.02)**	0.9854(0.00)	0.0318(0.01)	0.7413(0.09)
VP-LPD	0.1281(0.02)	**0.9875(0.00)**	**0.0281(0.01)**	**0.7981(0.07)**	0.1257(0.02)	**0.9821(0.01)**	**0.1207(0.02)**	**0.9898(0.00)**	0.1183(0.02)	**0.9857(0.00)**	0.0271(0.00)	**0.7870(0.08)**
VP-NPD	0.1280(0.02)	**0.9875(0.00)**	**0.0281(0.01)**	0.7963(0.07)	0.1257(0.02)	**0.9821(0.01)**	0.1209(0.02)	**0.9898(0.00)**	0.1185(0.02)	**0.9857(0.00)**	0.0271(0.00)	0.7847(0.08)

Next, we investigated how the VP models perform on datasets whose trees have external and internal branch lengths drawn from distributions with different means. We entertained two scenarios to build such trees (see Table 3). In the first one, we simulated trees with short external branches and longer internal branches (using prior means of 0.005 and 0.1, respectively). The second scenario simulates trees with long external branches and shorter internal branches (using prior means of 0.1 and 0.005, respectively). Besides this branch length distribution detail, we used the same settings for dataset simulation and model inference as in the previous experiment. Overall, the estimation results shown in Table 3 are relatively similar for the three VP models. However, we note that VP-LPD model optimizes better the LogL, and VP-L/NPD models have smaller KL-qprior divergences, as found in the previous experiment. In the first scenario, branch length estimations of the VP-FPD model have slightly smaller average distances with actual values ($p = 2.63E−05$, Kruskal-Wallis test). The external branches (with means of 0.005) are estimated better with VP-L/NPD models, and the internal branches (with means of 0.1) are estimated better with the VP-FPD model. The second scenario has an opposite outcome compared to the first one. The VP-LPD model estimates branch lengths with smaller average distances to the actual values ($p = 4.19E−189$, Kruskal-Wallis test). The external branches (with means of 0.1) are estimated better with VP-FPD model, and the internal branches (with means of 0.005) are estimated better with VP-L/NPD models. Thus, VP-L/NPD models achieve better estimations of either external or internal branch lengths drawn from distributions different from the default prior.

5.2   Substitution Model Parameters Estimation Performance

After that, we evaluated the performance of the VP models in approximating the posterior densities of the substitution model parameters. Usually, the model parameters are global and inferred for the whole sequence alignment. Here, we are interested in estimating the parameters of the GTR model, parameterized by a set of six substitution rates and another of four relative frequencies. We assume

Table 4. Performance of substitution rates estimation on datasets simulated with different prior means on the transition/transversion rate ratio κ. The datasets have 64 sequences of length 1000 bp and are simulated with HKY85 substitution model. κ values are drawn from gamma distribution. Relative frequencies are drawn from Dirichlet distribution with concentrations of 10. BL values are drawn from an exponential distribution of mean 0.1. VP models implement a GTR model and apply an i.i.d. exponential prior on BL and Dirichlet priors on rates and relative frequencies. VP-FPD applies a Dirichlet prior with concentration 1.

	κ 0.25		κ 1		κ 4	
	LogL	KL-qprior	LogL	KL-qprior	LogL	KL-qprior
Real data	−41403.2227(2158.37)		−42732.2461(2179.98)		−40114.5391(2028.43)	
VP-FPD	−41424.0040(2159.88)	270.0602(5.28)	−42753.3000(2181.57)	270.5125(5.40)	−40134.8240(2029.26)	267.1696(5.47)
VP-LPD	**−41418.5160(2159.88)**	184.4728(4.15)	**−42747.5040(2181.22)**	185.6610(4.08)	**−40128.5920(2029.27)**	181.7221(3.87)
VP-NPD	−41418.7160(2160.08)	**182.2155(4.10)**	−42747.5360(2181.73)	**183.1652(4.08)**	−40128.8320(2029.29)	**179.3205(3.70)**

	Dist	Corr	Dist	Corr	Dist	Corr
VP-FPD	0.0185(0.01)	0.9950(0.00)	0.0190(0.01)	0.5226(0.40)	0.0182(0.01)	0.9980(0.00)
VP-LPD	**0.0175(0.01)**	**0.9955(0.00)**	0.0181(0.01)	0.5360(0.40)	0.0176(0.01)	**0.9981(0.00)**
VP-NPD	**0.0175(0.01)**	0.9954(0.00)	**0.0178(0.01)**	**0.5407(0.39)**	**0.0173(0.01)**	**0.9981(0.00)**

that each set of parameters sums to one, so we can use Dirichlet distributions as priors over them. To simplify the evaluation scenarios for the estimations of the six substitution rates, we used the HKY85 model [14] to simulate the sequence alignments. HKY85 is a special case of the GTR model that defines a ratio of transition and transversion rates named κ and a set of relative frequencies. Thus, varying the ratio κ will change the values of two class rates at once. For the other simulation settings, we sampled branch length values from i.i.d. exponential priors of mean 0.1, κ values from gamma distributions (default mean equals 1), and the relative frequencies from Dirichlet distributions (default concentrations equal 10 for simulation and 1 for inference). We implemented the VP models with a GTR model. We applied i.i.d. exponential priors on branch lengths and Dirichlet priors on substitution rates and relative frequencies. The VP-FPD and VP-L/NPD models implement default and adaptable prior parameters, respectively, for all phylogenetic parameters.

In Table 4, we present the performance of substitution rates estimation on three datasets simulated with ratios κ drawn with prior means of 0.25, 1 and 4. We noted that the VP-L/NPD models estimate the substitution rates with smaller average distances and better average correlations with actual rates ($p = 0$, Kruskal-Wallis tests) more accurately than those of the VP-FPD model, even, surprisingly, for the dataset simulated with default prior means. Moreover, they optimize better the LogL of the three datasets, and their KL-qprior divergences are smaller compared to the default model. Regarding evaluating the estimation of the relative frequencies, we simulated sequence alignments with different nucleotide content prior distributions. Table 5 highlights the results of the VP models applied to three datasets characterized by an AT-rich, equally distributed and GC-rich nucleotide content, respectively. As for the estimation of substitution rates, VP-L/NPD models perform well in estimating the relative frequencies and the LogL of the data. Further, the VP-NPD model has the best estimations for the three datasets compared to the others ($p = 0$, Kruskal-Wallis tests).

Table 5. Performance of relative frequencies estimation on datasets simulated with different prior means. The datasets have 64 sequences of length 1000 bp and are simulated with HKY85 substitution model. Relative frequencies values are drawn from Dirichlet distribution with different nucleotide concentrations. For instance, **Dir(10AT)** means nucleotides **A** and **T** have concentrations of 10 and the other two concentrations are 1. **Dir(10)** means all concentrations equal to 10. κ values are drawn from gamma distribution with mean 1. BL values are drawn from an exponential distribution of mean 0.1. **VP** models implement a GTR model and apply an i.i.d. exponential prior on BL and Dirichlet priors on rates and relative frequencies. **VP-FPD** applies a Dirichlet prior with concentration 1.

	Dir(10AT)		Dir(10)		Dir(10GC)	
	LogL	KL-qprior	LogL	KL-qprior	LogL	KL-qprior
Real data	−34683.3945(3061.60)		−42732.2461(2179.98)		−34382.1719(3514.33)	
VP-FPD	−34703.5800(3063.90)	255.2194(8.09)	−42753.2040(2181.45)	270.5957(5.53)	−34401.6440(3515.04)	254.0669(8.25)
VP-LPD	−34696.9040(3064.24)	172.3171(5.89)	−42747.5880(2181.26)	185.5565(4.02)	−34394.8080(3514.94)	172.0875(6.25)
VP-NPD	**−34696.7960(3064.05)**	**170.4964(5.66)**	**−42747.5880(2181.26)**	**183.1414(4.00)**	**−34394.8080(3514.94)**	**170.3758(6.09)**
	Dist	Corr	Dist	Corr	Dist	Corr
VP-FPD	0.0130(0.01)	0.9994(0.00)	0.0147(0.01)	0.9899(0.02)	0.0123(0.01)	0.9995(0.00)
VP-LPD	0.0127(0.01)	0.9994(0.00)	0.0143(0.01)	0.9907(0.02)	0.0120(0.01)	0.9995(0.00)
VP-NPD	**0.0123(0.01)**	**0.9995(0.00)**	**0.0140(0.01)**	**0.9911(0.02)**	**0.0119(0.01)**	0.9995(0.00)

5.3 Convergence Analysis

Finally, we studied the convergence progression of the VP models when fitted on different datasets generated with three prior means (0.001, 0.01 and 0.1) for branch lengths. Figure 1 illustrates the performance results during the training step using the models' LogL and the KL-qprior divergences. We also report the scaled Euclidean distances of the branch length estimates with their actual values. The figure shows the averages and standard deviations of the metrics, which were calculated over 100 data times ten fit replicates.

First, in all the datasets, the LogL of the three VP models converges to the actual average LogL within the first 400 iterations. However, the LogL convergence is slower in datasets with shorter branches than in those with longer ones.

Second, The convergence trend of the KL-qprior divergences is most particular, as shown in Fig. 1. We noticed that the KL-qprior divergences of the two types of VP models have different progression trajectories during their training across all the datasets. On the one hand, optimizing the VP-FPD models starts with large KL-qprior values (>1000). Then, the KL-qprior values drop sharply (≈100) around the fortieth iteration as the LogL estimations start to converge. After the drop, they return to increase to higher values gradually. On the other hand, the optimization of the VP-L/NPD models starts with small KL-qprior values (<100). They increase slowly as the LogL values start to converge until the fortieth iteration (<200). After that, they return to decrease gradually for datasets with shorter branches, but they continue to increase for datasets with longer branches.

Last, for each dataset, the branch length Euclidean distances of the three models have a similar convergence progression. As with the LogL, the convergence of branch length distances to small values is slower in datasets with shorter branches than in those with longer branches. This trend is more noticeable with

(a) BL simulated with a prior mean 0.001

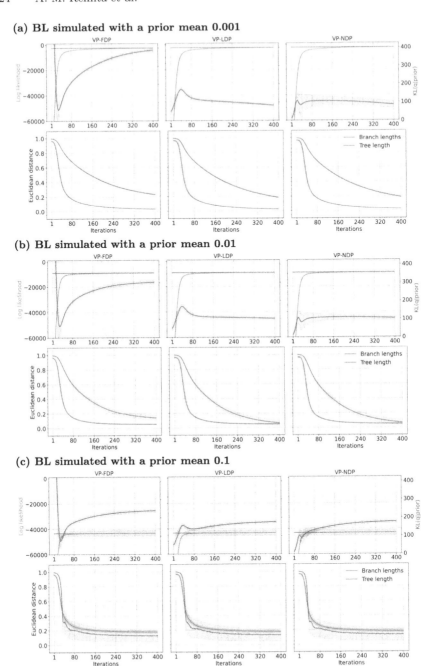

(b) BL simulated with a prior mean 0.01

(c) BL simulated with a prior mean 0.1

Fig. 1. Convergence and performance of the VP models for branch length (BL) estimation on datasets simulated with different prior means. The datasets have 64 sequences of length 1000 bp and are simulated with JC69 substitution model. BL values are drawn from an exponential distribution. VP models implement a JC69 model and apply an i.i.d. exponential prior on BL. VP-FPD applies a prior with mean 0.1. The first 400 training iterations are shown.

the total TL. However, when we compare the convergence of the LogL estimations and the branch length distances, we find that the LogL estimations converge faster than the distances. This finding suggests that the LogL has multiple local and nearby maxima that can be reached with distinct branch length estimates that are not close to the actual values.

Table 6. Running times of the VP models for branch length (BL) estimation on datasets simulated with different prior means. The times are reported in seconds. The datasets have 64 sequences of length 1000 bp and are simulated with JC69 substitution model. BL values are drawn from an exponential distribution. VP models implement a JC69 model and apply an i.i.d. exponential prior on BL. VP-FPD applies a prior with mean 0.1.

	BL 0.001	BL 0.01	BL 0.1
VP-FPD	**83.1292(2.37)**	96.5395(17.69)	102.8235(23.20)
VP-LPD	84.4600(1.66)	**93.1648(12.06)**	101.0446(19.15)
VP-NPD	83.1529(2.01)	96.6440(1.46)	**97.4801(7.55)**

In addition, we measured the empirical running times required to fit the VP models during 2000 iterations (see Table 6). For all the datasets, the running times of the three models are close and of the same order of magnitude. However, the VP-L/NPD models need slightly less time to fit datasets with more divergent sequences. Regardless of the models, the running times increase largely with the increase of the number of taxa and moderately with the increase of the alignment length (results not shown).

6   Discussion

Recent applications of variational Bayesian phylogeny have assigned default prior distributions to the likelihood parameters [12,13,48], which can lead to biased and excessively high posterior probabilities [17,30]. Here, we demonstrated that variational phylogenetic (VP) models using misspecified prior densities are prone to bias when the data are weak. For example, we showed that a VP model estimates twice the total length of the tree when using default and independent exponential priors on branch lengths with relatively similar and short sequences. This finding is supported by several MCMC-based studies that have analyzed and explained the effect of branch length priors on the posterior resulting in very long trees [4,8,27,44]. However, we found that the estimation of the substitution model parameters using VP models with default priors is less biased, even when the actual parameters are far from the range of the default priors. It was reported that sequence evolutionary parameters are relatively insensitive to the prior choice when estimated using the whole sequence alignment and less complex models [1].

In this paper, we introduced a variational phylogenetic approach that provides flexibility to the prior densities, making it insensitive to inappropriate prior specifications. Using a gradient ascent strategy, the approach implements adaptable parameters for the prior distributions that will be jointly learned from the data with the variational approximate parameters. The prior learning in variational inference (VI) is connected to the Empirical Bayes (EB), where EB and VI estimate the prior parameters by maximizing the marginal likelihood and the evidence lower bound, respectively [11,39]. We implemented two VP models with adaptable prior densities, VP-LPD and VP-NPD, using random uniform-sampled and neural networks-generated initializations, respectively. We showed that regardless of the type of initialization, the models perform better in estimating phylogenetic parameters than VP models with fixed priors that differ from the actual values of the parameters. Moreover, the VP-NPD model improved the accuracy of the estimation of the sequence evolutionary parameters. However, its accuracy decreases in estimating the vector of branch lengths. Furthermore, an advantage of using a neural network-based prior parameterization is reducing the burning step (in the first fortieth iterations) and speeding up the model convergence (Fig. 1).

This work is the first step for implementing and evaluating the VP models using adaptable prior densities with different parameterization strategies. Nevertheless, it has some limitations that can be addressed in future work. The current design of prior models does not capture the correlations between the branch lengths nor between phylogenetic parameters. More sophisticated prior models could be evaluated using suitable simulation-based scenarios, such as compound Dirichlet priors for branch lengths [50], coalescent priors for rooted trees [12,22,49] and rate heterogeneity among sites [12,42]. Moreover, we hypothesize that using a proper neural network architecture for prior and variational parameterizations could help capture such complex correlations. Considering that our variational models estimate the phylogenetic parameters given a fixed tree, it would be interesting to investigate how we could implement adaptable prior models with a fully Bayesian phylogenetic inference approach such as VBPI [49], which applies a uniform prior on the tree topology. VBPI represents the variational posterior of the tree topology with a flexible distribution (subsplit Bayesian networks [47]) and a structured amortization of the branch lengths over tree topologies, which could make implementing the adaptable prior models a stimulating challenge. Last but not least, we envision assessing the performance and convergence of the VP models with empirical datasets as done in previous variational phylogenetic applications [7,12,22,48].

References

1. Alfaro, M.E., Holder, M.T.: The posterior and the prior in Bayesian phylogenetics. Ann. Rev. Ecol. Evol. Syst. **37**(1), 19–42 (2006). https://doi.org/10.1146/annurev.ecolsys.37.091305.110021

2. Ayres, D.L., et al.: BEAGLE 3: improved performance, scaling, and usability for a high-performance computing library for statistical phylogenetics. Syst. Biol. **68**(6), 1052–1061 (2019). https://doi.org/10.1093/sysbio/syz020

3. Bishop, C.M.: Pattern Recognition and Machine Learning. Springer, New York (2006). https://link.springer.com/book/9780387310732

4. Brown, J.M., Hedtke, S.M., Lemmon, A.R., Lemmon, E.M.: When trees grow too long: investigating the causes of highly inaccurate Bayesian branch-length estimates. Syst. Biol. **59**(2), 145–161 (2010). https://doi.org/10.1093/sysbio/syp081

5. Carpenter, B., et al.: Stan: a probabilistic programming language. J. Stat. Softw. **76**(1), 1–32 (2017). https://doi.org/10.18637/jss.v076.i01

6. Cohn, I., El-Hay, T., Friedman, N., Kupferman, R.: Mean field variational approximation for continuous-time Bayesian networks. J. Mach. Learn. Res. **11**(93), 2745–2783 (2010). http://jmlr.org/papers/v11/cohn10a.html

7. Dang, T., Kishino, H.: Stochastic variational inference for Bayesian phylogenetics: a case of CAT model. Mol. Biol. Evol. **36**(4), 825–833 (2019)

8. Fabreti, L.G., Höhna, S.: Bayesian inference of phylogeny is robust to substitution model over-parameterization. bioRxiv, pp. 2022–02 (2022). https://doi.org/10.1101/2022.02.17.480861

9. Felsenstein, J.: Evolutionary trees from DNA sequences: a maximum likelihood approach. J. Mol. Evol. **17**(6), 368–376 (1981). https://doi.org/10.1007/BF01734359

10. Fisher, A.A., Hassler, G.W., Ji, X., Baele, G., Suchard, M.A., Lemey, P.: Scalable Bayesian phylogenetics. Philos. Trans. R. Soc. B Biol. Sci. **377**(1861) (2022). https://doi.org/10/grqt53

11. Fortuin, V.: Priors in Bayesian deep learning: a review. Int. Stat. Rev. (2022). https://doi.org/10.1111/insr.12502. arXiv:2105.06868

12. Fourment, M., Darling, A.E.: Evaluating probabilistic programming and fast variational Bayesian inference in phylogenetics. PeerJ **7**(12), e8272 (2019). https://doi.org/10.7717/peerj.8272

13. Fourment, M., Magee, A.F., Whidden, C., Bilge, A., Matsen, F.A., Minin, V.N.: 19 dubious ways to compute the marginal likelihood of a phylogenetic tree topology. Syst. Biol. **69**(2), 209–220 (2020). https://doi.org/10.1093/sysbio/syz046. arXiv: 1811.11804

14. Hasegawa, M., Kishino, H., Yano, T.: Dating of the human-ape splitting by a molecular clock of mitochondrial DNA. J. Mol. Evol. **22**(2), 160–174 (1985). https://doi.org/10.1007/BF02101694

15. Hoffman, M.D., Blei, D.M., Wang, C., Paisley, J.: Stochastic variational inference. J. Mach. Learn. Res. **14**(40), 1303–1347 (2013). http://jmlr.org/papers/v14/hoffman13a.html

16. Hoffman, M.D., Johnson, M.J.: ELBO surgery: yet another way to carve up the variational evidence lower bound. In: Advances in Approximate Bayesian Inference. Neurips Workshop, Barcelona, Spain (2016). http://approximateinference.org/2016/accepted/HoffmanJohnson2016.pdf

17. Huelsenbeck, J.P., Larget, B., Miller, R.E., Ronquist, F.: Potential applications and pitfalls of Bayesian inference of phylogeny. Syst. Biol. **51**(5), 673–688 (2002). https://doi.org/10.1080/10635150290102366

18. Huelsenbeck, J.P., Ronquist, F.: Bayesian Analysis of Molecular Evolution Using MrBayes, pp. 183–226. Springer New York (2005). https://doi.org/10.1007/0-387-27733-1_7

19. Jojic, V., Jojic, N., Meek, C., Geiger, D., Siepel, A., Haussler, D., Heckerman, D.: Efficient approximations for learning phylogenetic HMM models from data. Bioinformatics **20**(Suppl. 1), 161–168 (2004). https://doi.org/10.1093/bioinformatics/bth917

20. Jordan, M.I., Ghahramani, Z., Jaakkola, T.S., Saul, L.K.: An introduction to variational methods for graphical models. Mach. Learn. **37**(2), 183–233 (1999). https://doi.org/10.1023/A:1007665907178

21. Jukes, T.H., Cantor, C.R.: Evolution of protein molecules. In: Munro, H.H. (ed.) Mammalian Protein Metabolism, vol. III, pp. 21–132. Academic Press, New York (1969). https://doi.org/10.1016/B978-1-4832-3211-9.50009-7

22. Ki, C., Terhorst, J.: Variational phylodynamic inference using pandemic-scale data. Mol. Biol. Evol. **39**(8) (2022). https://doi.org/10.1093/molbev/msac154

23. Kimura, M.: A simple method for estimating evolutionary rates of base substitutions through comparative studies of nucleotide sequences. J. Mol. Evol. **16**(2), 111–120 (1980). https://doi.org/10.1007/BF01731581

24. Kingma, D.P., Ba, J.: Adam: a method for stochastic optimization. In: Bengio, Y., LeCun, Y. (eds.) 3rd International Conference on Learning Representations, ICLR 2015, San Diego, CA, USA, 7–9 May 2015, Conference Track Proceedings (2015). https://arxiv.org/abs/1412.6980

25. Kingma, D.P., Welling, M.: Auto-encoding variational Bayes. In: Proceedings of the International Conference on Learning Representations (2014). https://arxiv.org/abs/1312.6114

26. Kingma, D.P., Welling, M.: An introduction to variational autoencoders. Found. Trends Mach. Learn. **12**(4), 307–392 (2019). https://doi.org/10.1561/2200000056

27. Kolaczkowski, B., Thornton, J.W.: Effects of branch length uncertainty on Bayesian posterior probabilities for phylogenetic hypotheses. Mol. Biol. Evol. **24**(9), 2108–2118 (2007). https://doi.org/10.1093/molbev/msm141

28. Krishnan, R., Liang, D., Hoffman, M.: On the challenges of learning with inference networks on sparse, high-dimensional data. In: Storkey, A., Perez-Cruz, F. (eds.) Proceedings of the Twenty-First International Conference on Artificial Intelligence and Statistics. Proceedings of Machine Learning Research, vol. 84, pp. 143–151. PMLR (2018). https://proceedings.mlr.press/v84/krishnan18a.html

29. Kruskal, W.H., Wallis, W.A.: Use of ranks in one-criterion variance analysis. J. Am. Stat. Assoc. **47**(260), 583–621 (1952). https://doi.org/10.2307/2280779

30. Nascimento, F.F., Reis, M.D., Yang, Z.: A biologist's guide to Bayesian phylogenetic analysis. Nat. Ecol. Evol. **1**(10), 1446–1454 (2017). https://doi.org/10.1038/s41559-017-0280-x

31. Nelson, B.J., Andersen, J.J., Brown, J.M.: Deflating trees: improving Bayesian branch-length estimates using informed priors. Syst. Biol. **64**(3), 441–447 (2015). https://doi.org/10.1093/sysbio/syv003

32. Paszke, A., et al.: PyTorch: an imperative style, high-performance deep learning library. In: Wallach, H., Larochelle, H., Beygelzimer, A., d'Alché-Buc, F., Fox, E., Garnett, R. (eds.) Advances in Neural Information Processing Systems 32, pp. 8024–8035. Curran Associates, Inc. (2019). https://arxiv.org/abs/1912.01703

33. Posada, D., Crandall, K.A.: Felsenstein phylogenetic likelihood. J. Mol. Evol. **89**(3), 134–145 (2021). https://doi.org/10.1007/s00239-020-09982-w

34. Ranganath, R., Gerrish, S., Blei, D.: Black box variational inference. In: Kaski, S., Corander, J. (eds.) Proceedings of the Seventeenth International Conference on Artificial Intelligence and Statistics. Proceedings of Machine Learning Research, vol. 33, pp. 814–822. PMLR, Reykjavik, Iceland (2014). https://proceedings.mlr.press/v33/ranganath14.html

35. Rannala, B., Zhu, T., Yang, Z.: Tail paradox, partial identifiability, and influential priors in Bayesian branch length inference. Mol. Biol. Evol. **29**(1), 325–335 (2012). https://doi.org/10.1093/molbev/msr210

36. Rezende, D.J., Mohamed, S., Wierstra, D.: Stochastic backpropagation and approximate inference in deep generative models. In: Xing, E.P., Jebara, T. (eds.) Proceedings of the 31st International Conference on Machine Learning. Proceedings of Machine Learning Research, vol. 32, pp. 1278–1286. PMLR, Bejing, China (2014). https://proceedings.mlr.press/v32/rezende14.html

37. Spielman, S.J., Wilke, C.O.: Pyvolve: a flexible python module for simulating sequences along phylogenies. PLoS ONE **10**(9), 1–7 (2015). https://doi.org/10.1371/journal.pone.0139047

38. Tavaré, S.: Some probabilistic and statistical problems in the analysis of dna sequences. In: Lectures on Mathematics in the Life Sciences, vol. 17, no. 2, pp. 57–86 (1986)

39. Tomczak, J., Welling, M.: VAE with a VampPrior. In: Storkey, A., Perez-Cruz, F. (eds.) Proceedings of the Twenty-First International Conference on Artificial Intelligence and Statistics. Proceedings of Machine Learning Research, vol. 84, pp. 1214–1223. PMLR (2018). https://proceedings.mlr.press/v84/tomczak18a.html

40. Wexler, Y., Geiger, D.: Variational upper bounds for probabilistic phylogenetic models. In: Speed, T., Huang, H. (eds.) RECOMB 2007. LNCS, vol. 4453, pp. 226–237. Springer, Heidelberg (2007). https://doi.org/10.1007/978-3-540-71681-5_16

41. Yang, Z.: Estimating the pattern of nucleotide substitution. J. Mol. Evol. **39**(1), 105–111 (1994). https://doi.org/10.1007/BF00178256

42. Yang, Z.: Among-site rate variation and its impact on phylogenetic analyses. Trends Ecol. Evol. **11**(9), 367–372 (1996). https://doi.org/10.1016/0169-5347(96)10041-0

43. Yang, Z.: Maximum likelihood estimation on large phylogenies and analysis of adaptive evolution in human influenza virus A. J. Mol. Evol. **51**(5), 423–432 (2000). https://doi.org/10.1007/s002390010105

44. Yang, Z., Rannala, B.: Branch-length prior influences Bayesian posterior probability of phylogeny. Syst. Biol. **54**(3), 455–470 (2005). https://doi.org/10.1080/10635150590945313

45. Zhang, C.: Improved variational Bayesian phylogenetic inference with normalizing flows. In: Larochelle, H., Ranzato, M., Hadsell, R., Balcan, M.F., Lin, H. (eds.) Advances in Neural Information Processing Systems, vol. 33, pp. 18760–18771. Curran Associates, Inc. (2020). https://proceedings.neurips.cc/paper/2020/hash/d96409bf894217686ba124d7356686c9-Abstract.html

46. Zhang, C., Bütepage, J., Kjellström, H., Mandt, S.: Advances in variational inference. IEEE Trans. Pattern Anal. Mach. Intell. **41**(8), 2008–2026 (2019). https://doi.org/10/ggmzgz

47. Zhang, C., Matsen, F.A.: Generalizing tree probability estimation via Bayesian networks. In: Advances in Neural Information Processing Systems 2018-Decem(NeurIPS), pp. 1444–1453 (2018). https://proceedings.neurips.cc/paper/2018/file/b137fdd1f79d56c7edf3365fea7520f2-Paper.pdf. arXiv: 1805.07834

48. Zhang, C., Matsen IV, F.A.: Variational Bayesian phylogenetic inference. In: International Conference on Learning Representations (2019). https://openreview.net/pdf?id=SJVmjjR9FX
49. Zhang, C., Matsen IV, F.A.: A variational approach to Bayesian phylogenetic inference. arXiv preprint arXiv:2204.07747 (2022). https://arxiv.org/abs/2204.07747
50. Zhang, C., Rannala, B., Yang, Z.: Robustness of compound Dirichlet priors for Bayesian inference of branch lengths. Syst. Biol. **61**(5), 779–784 (2012). https://doi.org/10.1093/sysbio/sys030

The Asymmetric Cluster Affinity Cost

Sanket Wagle[1]($\boxtimes$), Alexey Markin[2], Paweł Górecki[3], Tavis Anderson[2],
and Oliver Eulenstein[1]($\boxtimes$)

[1] Department of Computer Science, Iowa State University, Ames, IA 50011, USA
{swagle,oeulenst}@iastate.edu
[2] Virus and Prion Research Unit, National Animal Disease Center, USDA-ARS,
Ames, IA 50010, USA
[3] Faculty of Mathematics, Informatics and Mechanics, University of Warsaw,
Warsaw, Poland

Abstract. Tree comparison costs are sophisticated tools used to compare the results of different phylogenetic hypotheses and reconstruction methods and to evaluate the robustness of a tree to data perturbations. The Robinson-Foulds distance is a widely used measure for comparing the topologies of two trees, but it is highly sensitive to tree error. Consequently, tree differences may be over-estimated, leading to incorrect inference. An approach to overcome this shortcoming is the Cluster Affinity distance, which is a refinement of the Robinson-Foulds distance. These distances are symmetric and thus designed to compare the same type of trees. However, it is common to compare different types of trees, such as gene trees compared with species trees, or the integration of different datasets into a supertree: these comparisons are inherently asymmetric. Here, we introduce the asymmetric Cluster Affinity cost, a relaxation of the original Affinity cost to compare heterogeneous trees. We demonstrate that the characteristics of this cost are similar to the symmetric Cluster Affinity distance. Further, for the asymmetric affinity cost we describe efficient algorithms, derive the exact diameters, and use these to standardize the cost to be applicable in practice.

Keywords: Cluster Affinity · Phylogenetic tree · Supertrees

1 Introduction

Phylogenetics is the study of the evolutionary relationships among or within evolutionary entities, such as species or genes, and it relies on the computational inference of evolutionary trees or "phylogenies". These trees represent hypotheses of the relationships among different groups, and their inference can be based on data from DNA sequences, protein amino acid sequences, or morphological characters [24].

Different input data sets and tree inference methods can lead to incongruent hypotheses of evolutionary relationships. The current approach to resolving these conflicts is through tree comparison costs that evaluate the fit of different trees to a given dataset [1,8,16,22]. If the true tree is known, the different inferred trees can be compared, and those that describe the true tree topology,

© The Author(s), under exclusive license to Springer Nature Switzerland AG 2023
K. Jahn and T. Vinař (Eds.): RECOMB-CG 2023, LNBI 13883, pp. 131–145, 2023.
https://doi.org/10.1007/978-3-031-36911-7_9

or close to the true tree topology, can be identified [17]. Additionally, cost measures can provide a strategy to rank the trees generated from different input data sets and tree inference methods to help identify incongruence (e.g., [18]) that may have utility in phylogenomic studies [15,23]. Further, by evaluating the comparison costs of different phylogenetic analyses and datasets, areas of uncertainty and the robustness of phylogenetic inference can be assessed [12]. Therefore, tree comparison costs can be applied in comparative phylogenetics and facilitate phylogenomic studies through providing a metric for assessing tree errors and tree incongruence [12].

Tree comparison costs are sophisticated measuring tools, exhibiting distinct properties and characteristics, such as their (i) distribution, (ii) sensitivity to small changes in the tree structure caused by errors or noise in the data, and (iii) computability. Making measurement tools available with proven characteristics sensitive to the specific analysis task is important for comparative phylogenetics, and they may also have utility in other tree-focused research areas. For example, they may be used in epidemiology for comparing transmission trees and virus genealogies [9], to determine horizontal gene transfer events [5], and in natural language processing to compare and determine an aggregate parse tree [10]. Consequently, the development and improvement of tree comparison tools is a mature and highly active research area in computational and comparative phylogenetics.

To that effect, multiple phylogenetic costs have been considered and analyzed to compare phylogenetic trees. The Robinson-Foulds (RF) metric is widely used for comparing the topology of two trees. It measures the difference between the clusters of two trees and provides a distance score based on each tree's unique number of clusters. This metric is efficiently computable, which is desirable for large-scale trees and for comparing many trees, as the computational time does not increase excessively with the size of the trees or the number of trees being compared. However, the RF distance is frequently criticized for its high sensitivity to tree difference. This sensitivity means that even small differences in the trees can result in a large RF distance, which can make it difficult to distinguish between meaningful and random differences. This property of the RF metric can lead to over-interpretation of the results and incorrect conclusions about the similarity of the trees being compared [11,20]. The *cluster affinity distance* [13] alleviates this shortcoming by considering the symmetric distance between the clusters in each tree, allowing it to have a wider distribution and be resistant to small changes in the trees, which makes it preferable to the RF distance.

An additional motivation for phylogenetic distance metrics is comparing the trees from tree-construction methods based on genomic information to known species trees or supertrees. A supertree may involve different types of evolutionary information (i.e., morphological and molecular) from different numbers of taxa, derived from different analyses, that are subsequently combined into a single tree [2]. Consequently, the relationship between the trees is inherently asymmetric, and since most phylogenetic distance measures are symmetric, it is unlikely that they can accurately represent the asymmetric relationship between the trees. Asymmetry in tree comparison frequently appears in phylogenomic studies. When comparing a gene tree to a species tree, a gene tree is often not fully resolved due to low branch support or low phylogenetic signal in the alignment. In this case, the clusters present in the species tree do not have to be present in the gene tree, and therefore asymmetric tree comparison costs are often used for species tree evaluation [14,21]. Asymmetry in phylogenomics also appears due to gene duplications (i.e., paralogous genes) or horizontal gene transfer, and assessing trees objectively can improve the robustness of phylogenomic studies [12].

To address asymmetry in phylogenomic analyses, we propose an asymmetric Cluster Affinity (CA) cost based on finding the minimum cost for a cluster in the source tree for all the clusters in the target tree and formulating an efficient algorithm for the same. We derive the diameter for the Cluster Affinity cost and, since the Cluster Affinity cost is asymmetric, measure its asymmetricity by defining a separation diameter as the maximum difference between the two directions in which the Cluster Affinity cost can be computed.

Related Work. The Robinson-Foulds (RF) distance [16] is a standard metric for comparing phylogenetic trees due to its simplicity and interpretability. There were many attempts to generalize RF and overcome its shortcomings. The most notable of them are the (i) Matching Split (MS) and Matching Cluster (MC) distances [3,4,11] that find a minimum weight perfect matching between the sets of splits or clusters of the two trees, and (ii) the information-theoretic generalized RF [19]. The CA distance later relaxed the matching requirement in the MC distance, thus, allowing multiple clusters from one tree to 'map' towards a single cluster in another tree. This relaxation allowed for a substantially faster distance computation while preserving the key properties of the MC distance: the robustness to tree error and wide distribution range (i.e., an ability to discriminate between trees). The CA distance between two rooted trees T_1 and T_2 maps each cluster in T_1 to the most similar cluster in T_2 and vice versa. The mapping cost from T_1 to T_2 is then the total number of changes required to change the T_1 clusters to the respective mapped clusters in T_2. The CA distance is then the arithmetic mean between the mapping costs from T_1 to T_2 and in reverse.

Another key property of tree comparison measures is the *diameter* – the maximum distance between two trees over a fixed taxon set of size n. Understanding the diameter is crucial to make tree distances comparable for different pairs of trees by normalizing the distance with the diameter. Additionally, normalization ensures that distances are comparable across different tree comparison measurements. For RF, the diameter is not difficult to establish, but the exact diameters of MS, MC, and CA distances remain an open problem (although [3,4,13] provide bounds).

Our Contribution. In this work, we further relax the Cluster Affinity cost, present efficient algorithms for computing the cost and changes to the cost in response to the NNI and SPR tree edit operations, and derive an exact diameter for the cost. We relax the cost in two ways. First, we allow the cost to be asymmetric by only considering the CA mapping cost from T_1 to T_2; the asymmetry appears naturally when comparing different types of trees, e.g., a gene tree to a species tree in a phylogenomic analysis or an estimated tree to the ground truth in simulation/convergence studies. Second, we allow clusters in T_1 to map to *trivial* clusters in T_2 (i.e., leaves and the root cluster), which was prohibited in the original definition [13].

We present algorithms to (i) compute the cluster affinity cost between two trees in $O(n^2)$ time and (ii) to efficiently conduct a Nearest Neighbor Interchange (NNI) [6] search using the CA cost with $O(n \log n)$ time per NNI after a single $O(n^2)$ preprocessing step. Then, Subtree Prune and Regraft (SPR) search can be conducted efficiently using an NNI-graph presented in [7]. As NNI and SPR search strategies are a standard for species tree inference heuristics [2], these algorithms make the CA cost directly applicable for phylogenomic inference.

Further, we prove that the CA diameter for trees with n leaves is $\lceil \frac{n^2-2n}{4} \rceil$. Finally, we introduce the concept of a *separation diameter* that measures how asymmetric a cost is; that is, the maximum value of $|CA(T_1, T_2) - CA(T_2, T_1)|$ for all trees T_1 and T_2. We prove that the CA cost is significantly impacted by the topology of T_1, and the separation diameter for CA is in the order of $\Theta(n^2)$. This result implies that, in practice, CA costs need to be normalized by a diameter specific to the topology of T_1. While the exact topology-specific diameter remains an open research problem, we provide a practical upper bound on that diameter, supported by our theoretical results.

Finally, we demonstrate that our relaxed CA cost preserves the desirable distribution properties and robustness to error from the original CA formulation.

2   Asymmetric Cluster Affinity Cost

A phylogenetic tree T over a taxon set M is a rooted binary tree where each leaf is bijectively labeled with the elements from M. The vertex set and edge set of T are denoted by $V(T)$ and $E(T)$, respectively. By $L(T)$ we denote the set of all leaves of T. Note, in Sect. 2.3 we extend the labeling to also include non-bijective settings.

An edge (u, v) from $E(T)$ is directed from u to v, where v is a *child* of u, and u is the unique *parent* of v. For a vertex $w \in V(T)$, $\text{Ch}(w)$ is the set of all children of w. If two distinct vertices u and v have the same parent, they are *siblings* of each other. We also define $T(v)$ as the subtree of T rooted at v.

We define the *height* of a tree T as the edge-length of a longest path, i.e., the path containing the maximum number of edges from the tree's root to a leaf node in the tree, and is denoted by $h(T)$.

For two sets A and B, we define $A \Delta B$ as the symmetric difference between the two sets A and B. That is, $A \Delta B = (A \setminus B) \cup (B \setminus A)$.

A set of leaves $L(T(v))$ is called a cluster of the node v and is denoted by c_v. Note that we identify the leaves in a phylogenetic tree with the respective labels(taxa). A tree G can be represented by a set of clusters $C(G) = \{c_i | i \in V(G)\}$.

For convenience, we assume throughout the text that M is a taxon set, G, S, and T are trees over M, and $c \subseteq M$ is a cluster over M.

Definition 1 (Cluster Affinity (CA) cost [13]). Cluster Affinity cost *from c to S is* $d(c, S) := \min_{x \in C(S)} |c \triangle x|$, *and* Cluster Affinity cost *from G to S is* $d(G, S) := \sum_{x \in C(G)} d(x, S)$, *and the* Symmetric Cluster Affinity cost *between G and S is* $d_{sym}(G, S) := \frac{d(G,S)+d(S,G)}{2}$.

Definition 2 (Diameter of cost function). *The diameter of a cost function between trees is the maximum value that the cost function can have over all trees over the same leaf set.*

2.1   Tree Edit Operations

We define two classic tree edit operations for rooted trees, namely the Subtree Prune and Regraft (SPR) operation and the Nearest Neighbor Interchange (NNI) operation [6].

Let T be a phylogenetic tree and let $e = (u, v)$ be an edge in $E(T)$ and $w \in V(T)$. Let T' be the rooted binary tree obtained by deleting e and then adjoining a edge f between v and the component C_u that contains u in one of the two following ways:

1. If w is the root of T, we create a new vertex u' and a new edge from u' to w. Then, we adjoin f between u' and v and suppress the degree two vertex u. Then, u' becomes the new root for the tree T'.
2. Otherwise, we create a new vertex u' which subdivides the edge whose bottom not is w in C_u and adjoining f between u' and v. Then, we either suppress the degree-two vertex u or if u is the root of T, delete u and the edge incident with u, making the other end-vertex of the edge the root.

The *NNI operations* are SPR operations where the subtree is pruned close to its regrafting position as follows. For a non-root node v from T, let $NNI(T, v)$ be the SPR operation with the unique edge $e = (u, v)$ from T and w being the sibling of u.

2.2   Algorithm for the Cluster Affinity Cost

We describe an $O(n^2)$ algorithm for computing the CA cost between two trees on n leaves.

Theorem 1. *For fixed trees G, T and a cluster c from G, $d(c, T)$ can be computed in $O(n)$ time. Consequently, $d(G, T)$ can be obtained in $O(n^2)$ time.*

Proof. For a node $v \in V(T)$ such that $Ch(v) = \{x, y\}$, $|c_v| = |c_x| + |c_y|$ and $|c \cap c_v| = |c \cap c_x| + |c \cap c_y|$. Hence, since T is a tree, we can compute $|c_v|$ and $|c \cap c_v|$ via a post-order traversal on T in $O(n)$ time for all $v \in V(T)$. Thus we can compute $d(c, T)$ in $O(n)$ time using the following formula:

$$d(c, T) = |c| + \delta_c(Rt(T)),$$

where $\delta_c(v)$ is a recursive function given by

$$\delta_c(v) := \begin{cases} -1 & \text{if } v \in L(T) \cap c, \\ 1 & \text{if } v \in L(T) \setminus c, \\ \min(|c_v| - 2|c \cap c_v|, \min_{u \in Ch(v)} \delta_c(u)) & \text{otherwise.} \end{cases}$$

This completes the proof. $\qquad\qquad\qquad\qquad\qquad\qquad\qquad\qquad\qquad\qquad\square$

2.3   Cluster Affinity for Multi-copy Gene Trees

Note that the computational procedure from Sect. 2.2 can be directly applied to the case of comparison of a multi-copy gene tree (i.e., a gene tree with paralogous genes) with a species tree. For a multi-copy gene tree with leaves labeled by species and a fixed node in that tree, we define the cluster as the set of unique species labels below the node.

2.4   NNI Search Using Cluster Affinity Heuristic

We present an algorithm for the efficient NNI tree space traversal.

Theorem 2. *Given a cluster c and a tree T^0, let $T^1, T^2, T^3, \dots$ be a sequence of trees, such that T^i is obtained from T^{i-1} by a single NNI operation. Then, after $O(n)$ preprocessing on T^0, one can compute $d(c, T^i)$ for each $i > 0$ in $O(\log n)$ time.*

Proof. In the previous section, we showed that computing $d(c, T^0)$ can be performed in $O(n)$ time. During this computation, we obtain $|c\Delta c_v|$ values for each $v \in V(T^0)$. For convenience, let $d_v := |c\Delta c_v|$. We then place all d_v values in a binary min-heap. Recall that obtaining the minimum value from a binary min-heap can be performed in $O(1)$ time, and changing an element's value can be performed in $O(\log n)$ time. Building a min-heap requires $O(n)$ time.

Now, we show how to compute $d(c, T^1)$ and update the min-heap in $O(\log n)$ time. Then, computation of $d(c, T^2)$, $d(c, T^3)$, etc. follows the same algorithm. Let $T^1 = NNI(T^0, u)$, where $u \neq Rt(T)$. Let v be the parent of u, w be the sibling of u, and x be the sibling of v. Note that the only cluster that changes after the NNI is c_v. The new cluster in T^1 is $c' = c_w \cup c_x$. Note that we can compute $|c\Delta c'|$ in constant time $|c\Delta c'| = d_w + d_x - |c|$.

Updating d_v in the binary heap by removing the old value and then replacing it with $|c\Delta c'|$, can be done in $O(\log n)$ time. Then, querying the minimum value from the min-heap will give us $d(c, T^1)$ in constant time. $\qquad\qquad\square$

2.5   Diameter of the Cluster Affinity Cost

In this section, we derive the diameter for the asymmetric cluster affinity cost. However, first we require a few additional definitions to obtain the diameter of the Cluster Affinity cost.

A *rooted caterpillar tree* is a rooted tree T, where each internal node has at least one leaf child. We define a caterpillar C_n using the standard nested parenthesis notation as $(n, (n-1, \ldots (2, 1) \ldots))$, where the leaves are numbers. A *perfectly balanced tree* is a rooted tree T where each leaf is at the same distance in the number of edges from the root. *The cherry of a caterpillar tree* is the smallest subtree of the caterpillar tree, which contains two leaf nodes. For a tree T with a leaf set M and a node $v \in V(T)$ let $\tau_T(v) := \min(|M| - |c_v|, |c_v| - 1)$ and $\tau(T) := \sum_{v \in V(T)} \tau(v)$.

Definition 3 (Caterpillar-extend). *For a non-caterpillar tree T let v be an internal node in T such that v has two children u and t, $T(u)$ and $T(t)$ are two caterpillars such that $|L(T(u))| \geq |L(T(t))| \geq 2$. By $T \to T'$ we denote the caterpillar-extend operation $T' = \mathrm{NNI}(T, l)$, where l is a leaf-child of t. See Fig. 1 for an example of the caterpillar-extend operation.*

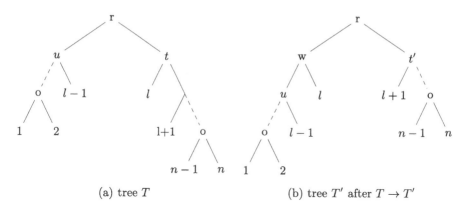

(a) tree T (b) tree T' after $T \to T'$

Fig. 1. Caterpillar-extend operation on tree T.

Lemma 1. *For every node v in a tree T, $d(c_v, S) \leq \tau_T(v)$ and $d(T, S) \leq \tau(T)$ for all trees S with the same leaf set as T.*

Proof. Consider a leaf l on tree S such that $l \in c_v$. Hence $|c_v \Delta \{l\}| = |c_v| - 1$. Similarly, for the cluster at the root of the tree S, we have $|c_v \Delta M| = |M| - |c_v|$. Since $d(c_v, S) = \min_{x \in C(S)} |c_v \Delta x|$ and both clusters $\{l\}$ and M exist in $C(S)$, $d(c_v, S) \leq \min(|c_v| - 1, |M| - |c_v|)$. Recall that $\tau_T(v) = \min(|c_v| - 1, |M| - |c_v|)$ and $\tau(T) = \sum_{v \in V(T)}(\tau(v))$. Hence $d(c_v, S) \leq \tau_T(v)$ and $\forall T', d(T, T') \leq \tau(T)$ $\square$

Lemma 2. *If $T \to T'$, then $\tau(T) \leq \tau(T')$.*

Proof. Let v, u and t be the nodes from the caterpillar-extend operation definition. Note, that $|c_u| \geq |c_t|$ and $\tau(T) = \sum_{i \in V(T)} \tau_T(i)$.

Let c_w represent the new cluster that is formed in T'. Note that there are only two different clusters between T and T', namely, c_t and c_w. Hence, if $\tau_{T'}(w) \geq \tau_T(t)$ then $\tau(T) \geq \tau(T')$. Since $|c_u| \geq |c_t|$, $|c_t| \leq \frac{|M|}{2}$ and $\tau_T(t) = |c_t| - 1$. Then, there are two cases for c_w. If $|c_w| \leq \frac{|M|}{2}$, we have $\tau_{T'}(w) = |c_w| - 1 = |c_u| > \tau_T(t)$. Otherwise, $|c_w| > \frac{|M|}{2}$ and it follows from $|M| \geq |c_t| + |c_u|$, that $\tau_T(w) = |M| - |c_w| = |M| - |c_u| - 1 \geq \tau_T(t)$. We conclude $\tau_{T'}(w) \geq \tau_T(t)$, which implies $\tau(T) \geq \tau(T')$.   □

Corollary 1. *Any maximal sequence of caterpillar-extend operations that starts in a tree T terminates in a caterpillar tree T^*. Moreover, $\tau(T^*)$ is maximal in the set of all trees of fixed size, and it does not depend on T^*, as long as T^* is a caterpillar.*

Lemma 3. *For any n, $\tau(C_n) = d(C_n, \bar{C}_n)$ where $\bar{C}_n = (1, (2, \ldots (n, n-1) \ldots))$.*

Proof. Consider a cluster c from C_n. For any cluster r in $\bar{C}_n$, associated with the vertex $v \in V(\bar{C}_n)$, such that $|c \cap r| \geq 1$ and $|r| > 1$, $|r| \geq n - |c|$ since there are $n - |c|$ taxa that are above c in C_n. More precisely, if $|c \cap r| \geq 1$ and $|r| > 1$, then $|r| = n - |c| + |c \cap r|$. Thus for a cluster r such that $|c \cap r| \geq 1$ and $|r| > 1$,

$$|c \Delta r| = |c| + n - |c| + |c \cap r| - 2|c \cap r| = n - |c \cap r|.$$

Note that $|c \cap r| = |c|$ if r is the cluster of the root. Also, note that there is always at least one leaf node l in $\bar{C}_n$ such that $l \in c$ and thus $|c \Delta \{l\}| = |c| - 1$. For all the remaining cases, when $r \cap c = \emptyset$, we have $|c \Delta r| = |c| + |r|$. Hence we have,

$$d(c, \bar{C}_n) = \min(|c| - 1, n - |c|) = \tau(c).$$

Hence,

$$d(C_n, \bar{C}_n) = \sum_{v \in V(C_n)} d(c_v, \bar{C}_n) = \sum_{v \in V(C_n)} \tau(c_v) = \tau(C_n).$$

This completes the proof.   □

Theorem 3 (CA cost diameter). *The maximum Cluster Affinity cost between two trees of size n is $\left\lceil \frac{n^2 - 2n}{4} \right\rceil$.*

Proof. We show that the diameter is $\tau(T^*)$ where T^* is a caterpillar. Let T and S be two trees over the same set of leaves M and let $|M| = n$. Then, by Lemma 1 $d(T, S) \leq \tau(T)$. Next, we transform T into a caterpillar T^* by a sequence of caterpillar-extend operations. By Lemma 2, $\tau(T) \leq \tau(T^*)$. By Corollary 1, $\tau(T^*)$ is maximal and does not depend on T^* as long as T^* is a caterpillar. We showed that for any pair of trees, T, S, $d(T, S)$ is bounded by the value $\tau(T^*)$. Since, $\tau(T^*)$ is reached by two caterpillars, by Lemma 3, we conclude that $\tau(T^*)$ is maximal. It remains to derive the exact value:

$$\tau(T^*) = \sum_{c \in C(T)} \min(n - |c|, |c| - 1) = \sum_{j=\lceil \frac{n}{2}+1 \rceil}^{n-1} (n-j) + \sum_{i=2}^{\lceil \frac{n}{2} \rceil} (i-1)$$

$$= \sum_{j=\lceil \frac{n}{2}+1 \rceil}^{n} (n-j) + \sum_{i=1}^{\lceil \frac{n}{2} \rceil} (i-1) = n * \frac{n}{2} - \sum_{j=1}^{n}(j) + \sum_{j=1}^{\lceil n/2 \rceil}(j) - \left\lceil \frac{n}{2} \right\rceil + \sum_{i=1}^{\lceil n/2 \rceil}(i)$$

$$= \frac{n^2}{2} - \frac{n(n+1)}{2} - \left\lceil \frac{n}{2} \right\rceil + 2 * \sum_{i=1}^{\lceil n/2 \rceil}(i) = \frac{n^2}{2} - \frac{n(n+1)}{2} - \left\lceil \frac{n}{2} \right\rceil + \frac{n}{2}(\frac{n}{2} + 1)$$

$$= \frac{n^2}{4} - \left\lceil \frac{n}{2} \right\rceil = \left\lceil \frac{n^2 - 2n}{4} \right\rceil.$$

Hence, the diameter for the Cluster Affinity cost is $\left\lceil \frac{n^2-2n}{4} \right\rceil$.   □

2.6   Separation Diameter

We define the separation cost for Cluster Affinity cost as the absolute difference between the Cluster Affinity costs between the two trees S and G. That is $\sigma(G, S) = |d(G, S) - d(S, G)|$. Similarly, let the *separation diameter* of the Cluster Affinity cost be the diameter of the separation cost for the Cluster Affinity cost. Below we show how to derive a bound for the separation diameter for the Cluster Affinity cost.

Lemma 4. *The separation diameter of the Cluster Affinity cost is bounded above by* $\left\lceil \frac{n^2-2n}{4} \right\rceil$.

Proof. It follows immediately from Theorem 3.   □

Lemma 5. *For any $n = 2^m$, there exists a perfectly balanced tree P_n such that* $d(C_n, P_n) = \tau(C_n) = \frac{n^2-2n}{4}$.

Proof. We construct a perfectly balanced tree P_n such that each non-trivial cluster in P_n is of the form $\{i, \ldots, j, n-j+1, \ldots, n-i+1\}$ for some $1 \leq i \leq j \leq \frac{n}{2}$. We define the labeling in P_n as follows: let the ith cherry in P_n in the prefix order have leaf labels i and $n - i + 1$ for $1 \leq i \leq n/2$.

Consider a cluster $c = \{1, 2, \ldots, k\}$ in C_n and a non-leaf cluster $p \in P_n$. Then $p = \{i, \ldots j, n-j+1, \ldots, n-i+1\}$ for some $1 \leq i \leq j \leq n$. We have two cases as follows:

1. **Case 1:** if $i \leq k \leq j$, then $|c \cap p| = k - i + 1$ and thus $|c \cap p| \leq \frac{|p|}{2}$. Hence, $|c \triangle p| = |c| + |p| - 2(|c \cap p|) \geq |c| \geq \tau_G(c)$.

2. **Case 2:** if $n-j+1 \le k \le n-i+1$ then $|c \cap p| = k-(n-j+1)+1+(j-i+1) = k - n + 2j - i + 1$. Thus,

$$
\begin{aligned}
|c \Delta p| &= k + 2(j - i + 1) - 2(k - n + 2j - i + 1) \\
&\ge (n - k) + (n - 2(\frac{n}{2})) \\
&= n - k = \tau_G(c).
\end{aligned}
$$

For all the other cases, while $|p|$ increases, $|c \cap p|$ remains constant, and hence they can be ignored. Hence, for every cluster $c \in C(C_n)$, $d(c, P_n) = \tau_{C_n}(c)$. Thus, $d(C_n, P_n) = \tau(C_n) = \frac{n^2 - 2n}{4}$. □

Lemma 6. *For a perfectly balanced tree T of size $n = 2^m$, $\tau(T) = n \log n - 3n + 2$.*

Proof. Let c be an internal cluster apart from the root cluster such that $c \in C(T)$ and c is at a height h from the root and hence $|c| = 2^h$. Since T is a perfectly balanced tree, $|c| \le \frac{n}{2}$ and thus $\tau(c) = |c| - 1$. Thus, for a level at height h, the total contribution of level h nodes to $\tau(T) = (2^h - 1) \cdot 2^{m-h}$. Hence,

$$
\begin{aligned}
\tau(T) = \sum_{c \in C(T)} \tau_T(c) &= \sum_{h=0}^{m-1} (2^h - 1) \cdot 2^{m-h} \\
&= n(m - 1 - 2 \cdot \frac{n-1}{n}) = n \log_2 n - 3n + 2.
\end{aligned}
$$

Thus $\tau(T) = n \log n - 3n + 2$. □

Lemma 7. *For every n, there exists a caterpillar tree G and a tree S such that the separation diameter $\sigma(G, S) \ge \lceil \frac{n^2 - 4n}{16} \rceil - n \log_2 n - 3n + 2$.*

Proof. It follows from Lemma 6 and Lemma 1, that $d(S, G)) \le \tau(S) = n \log_2 n - 3n + 2$ and hence, $\sigma(G, S) \ge \lceil \frac{n^2 - 2n}{4} \rceil - (n \log_2 n - 3n + 2)$ when $n = 2^m$ for some m.

For the remaining n's, we prove it by construction. Let $S_n = (n, (n - 1, (n - 2, \ldots (n - n' + 1, P_{n'}) \ldots)))$ where $n' = 2^{\lfloor \log_2 n \rfloor}$ and $P_{n'}$ is a perfectly balanced tree of the size n' from the proof of Lemma 6. Since all clusters of size larger than n' are present in C_n and $n > n' \ge \frac{n}{2}$, we have $\sigma(C_n, S_n) \ge \sigma(C_{n'}, P_{n'}) \ge \lceil \frac{(n')^2 - 2n'}{4} \rceil - n' \log_2 n' - 3n' + 2 \ge \lceil \frac{n^2 - 4n}{16} \rceil - n \log_2 n - 3n + 2$. □

Theorem 4. *The separation diameter is $\Theta(n^2)$.*

Proof. From Lemma 7, $\sigma(G, S) \ge \lceil \frac{n^2 - 4n}{16} \rceil - n \log_2 n - 3n + 2$ and hence $\sigma(G, S) = \Omega(n^2)$. Since the Cluster Affinity distance is bound by $O(n^2)$, the separations affinity distance is also $O(n^2)$. Hence, the separation diameter is $\Theta(n^2)$. □

3   Empirical Study

We study the characteristics of the relaxed and asymmetric Cluster Affinity cost and compare it to the asymmetric (one-sided) RF cost, also known as the false negative rate, using simulated datasets. We define the one-sided RF between trees T_1 and T_2 as the number of clusters in T_1 that are not present in T_2.

3.1   CA Displays a Broad Distribution Range

Table 1 shows the descriptive statistics of the (asymmetric) CA, RF costs on pairs of random trees generated under the birth-death process with a birth rate of 1.0 and a death rate of 0.5.

We observe that CA has a broader range (standard deviation and the min-max range) than RF and the mean/median of CA is significantly less skewed towards the maximum. These results are similar to the comparison between the original (symmetric) Cluster Affinity distance and RF [13]; thus, demonstrating that our relaxation maintained the key properties of the original CA distance.

Additionally, we quantified the CA asymmetry between the tree pairs. For a pair of trees (T_1, T_2) we computed the separation cost $(|d(T_1, T_2) - d(T_2, T_1)|)$ and normalized by the maximum observed CA cost in our dataset. The average separation cost was 0.057 with a maximum value of 0.270 and a standard deviation of 0.041. That is, the asymmetry between birth-death trees was 6% on average and 27% in the worst case.

We observed that for a pair of trees with $n \in \{100, 1000\}$, it required an average of 1.48ms and 16.96ms to compute the asymmetric RF distance, while it required an average of 123.34ms and 13,594.96ms to compute the cluster affinity cost. This is due to the higher computational complexity and resolution of the cluster affinity cost.

Table 1. Asymmetric CA and RF distribution statistics for trees with $n \in \{100, 1000\}$ leaves. All values are normalized by the maximum observed RF or CA respectively.

Leaves		RF	Affinity cost
100	Mean	0.997	0.816
	SD	0.004	0.056
	Median	1.000	0.810
	Min	0.969	0.687
1000	Mean	0.999	0.883
	SD	4.857×10^{-4}	0.036
	Median	1.000	0.878
	Min	0.996	0.807

3.2 CA Is Robust to Minor Tree Edit Operations

We demonstrate the CA is more robust to tree edit operations (and hence the error) than RF. Our experimental setup follows [13] for comparison between the original CA and RF distances.

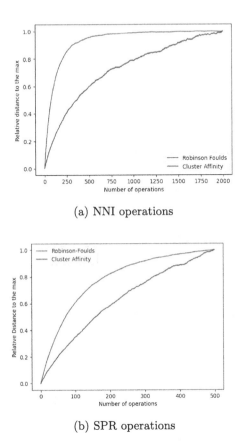

(a) NNI operations

(b) SPR operations

Fig. 2. Distances by edit operations: average RF and Affinity distances of 100 trees on 100 leaves as a function of the number of consecutive tree edit operations and the ratio to the diameter.

Dataset. We generated a profile P consisting of 100 random trees on 100 leaves, where each tree in P was generated using the birth-death model with birth rate 1.0 and death rate 0.5. We generated two profiles $Q(i) := \{q(i)_1, \ldots, q(i)_{100}\}$ for every $i \in \{1, \ldots, 2000\}$ for the NNI operation and $i \in \{1, \ldots, 500\}$ for the SPR operation. The profiles were generated as follows:

1. Given a tree edit operation, the initial profile $Q(1)$ is set to profile P. The range is set to 2000 for NNI and 500 for SPR.

2. For each $i \in \{2, \ldots, r\}$ the profile $Q(i + 1)$ is generated from profile $Q(i)$. The tree $q(i + 1)_j$ is created by applying the input tree edit operation to tree $q(i + 1)_j$ for each $j \in \{1, \ldots, 100\}$, where each edge on which the operation is performed is chosen uniformly and independently at random.

Experimental Setting. Distances were computed between tree pairs $q(1)_j$ and $q(i)_j$ averaged over all $j \in \{1, \ldots, 100\}$, under the RF cost and Cluster Affinity cost for every $i \in \{1, \ldots, r\}$ using the profiles that were generated for each of the edit operations. Similarly, the maximum of the distances between the tree pairs $q(1)_j$ and $q(i)_j, \forall j \in \{1, \ldots, 100\}$ was computed to finally compute the ratio of the averages to the observed maximum distances.

Results. The respective RF and Cluster Affinity costs between $q(1)$ and $q(i)$ over consecutive NNI and SPR operations are shown in Fig. 2. We observe that for both NNI and SPR operations, the RF cost tends to increase almost exponentially. For NNI operations, the RF cost saturates after approximately 400 operations, while it does not saturate but steadily increases with SPR operations. On the other hand, for the Cluster Affinity cost, we observe that it increases slowly for each consecutive operation.

4 Conclusion

A major goal in phylogenetics is to reconstruct a true hypothesis on the evolutionary history of an organism. Approaches that merge disparate datasets and gene trees into larger species trees and supertrees can provide greater resolution for evolutionary inference. However, to achieve this, methods are required to estimate uncertainty and identify when and where conflicts in different gene trees and datasets occur. In this study, we propose an asymmetric Cluster Affinity (CA) cost based on finding the minimum cost for a cluster in the source tree for all the clusters in the target tree. We present efficient algorithms for the CA cost and derive its diameter. Given that the CA cost is asymmetric, we measure its asymmetricity by defining a separation diameter as the maximum difference between the two directions in that Cluster Affinity cost can be computed.

Acknowledgements. We thank the reviewers for their constructive and valuable comments. This work was supported in part by the U.S. Department of Agriculture (USDA) Agricultural Research Service (ARS project number 5030-32000-231-000-D, and 5030-32000-231-095-S). The funders had no role in study design, data collection and interpretation, or the decision to submit the work for publication. Mention of trade names or commercial products in this article is solely for the purpose of providing specific information and does not imply recommendation or endorsement by the USDA. USDA is an equal opportunity provider and employer. PG was supported by the grant of National Science Centre 2017/27/B/ST6/02720.

References

1. Allen, B.L., Steel, M.: Subtree transfer operations and their induced metrics on evolutionary trees. Ann. Comb. **5**, 1–15 (2001)
2. Bininda-Emonds, O.R.: Phylogenetic Supertrees: Combining Information to Reveal the Tree of Life, vol. 4. Springer, Dordrecht (2004). https://doi.org/10.1007/978-1-4020-2330-9
3. Bogdanowicz, D., Giaro, K.: Matching split distance for unrooted binary phylogenetic trees. IEEE/ACM Trans. Comput. Biol. Bioinf. **9**(1), 150–160 (2011)
4. Bogdanowicz, D., Giaro, K.: On a matching distance between rooted phylogenetic trees. Int. J. Appl. Math. Comput. Sci. **23**(3), 669–684 (2013)
5. Bogdanowicz, D., Giaro, K.: Comparing phylogenetic trees by matching nodes using the transfer distance between partitions. J. Comput. Biol. **24**(5), 422–435 (2017)
6. Bordewich, M., Semple, C.: On the computational complexity of the rooted subtree prune and regraft distance. Ann. Comb. **8**, 409–423 (2005). https://doi.org/10.1007/s00026-004-0229-z
7. Chaudhary, R., Burleigh, J.G., Eulenstein, O.: Efficient error correction algorithms for gene tree reconciliation based on duplication, duplication and loss, and deep coalescence. BMC Bioinform. **13**, 1–10 (2012)
8. Estabrook, G.F., McMorris, F., Meacham, C.A.: Comparison of undirected phylogenetic trees based on subtrees of four evolutionary units. Syst. Zool. **34**(2), 193–200 (1985)
9. Giardina, F., Romero-Severson, E.O., Albert, J., Britton, T., Leitner, T.: Inference of transmission network structure from HIV phylogenetic trees. PLoS Comput. Biol. **13**(1), e1005316 (2017)
10. Kulkarni, A., Sabetpour, N., Markin, A., Eulenstein, O., Li, Q.: CPTAM: constituency parse tree aggregation method. In: SDM (2022)
11. Lin, Y., Rajan, V., Moret, B.M.: A metric for phylogenetic trees based on matching. IEEE/ACM Trans. Comput. Biol. Bioinf. **9**(4), 1014–1022 (2011)
12. Lozano-Fernandez, J.: A practical guide to design and assess a phylogenomic study. Genome Biol. Evol. **14**(9), evac129 (2022)
13. Moon, J., Eulenstein, O.: The cluster affinity distance for phylogenies. In: Cai, Z., Skums, P., Li, M. (eds.) ISBRA 2019. LNCS, vol. 11490, pp. 52–64. Springer, Cham (2019). https://doi.org/10.1007/978-3-030-20242-2_5
14. Page, R.D.M.: Modified mincut supertrees. In: Guigó, R., Gusfield, D. (eds.) WABI 2002. LNCS, vol. 2452, pp. 537–551. Springer, Heidelberg (2002). https://doi.org/10.1007/3-540-45784-4_41
15. Prum, R.O., et al.: A comprehensive phylogeny of birds (Aves) using targeted next-generation DNA sequencing. Nature **526**(7574), 569–573 (2015)
16. Robinson, D.F., Foulds, L.R.: Comparison of phylogenetic trees. Math. Biosci. **53**(1–2), 131–147 (1981)
17. Russo, C., Takezaki, N., Nei, M.: Efficiencies of different genes and different tree-building methods in recovering a known vertebrate phylogeny. Mol. Biol. Evol. **13**(3), 525–536 (1996)
18. Shen, X.X., Steenwyk, J.L., Rokas, A.: Dissecting incongruence between concatenation-and quartet-based approaches in phylogenomic data. Syst. Biol. **70**(5), 997–1014 (2021)
19. Smith, M.R.: Information theoretic generalized Robinson-Foulds metrics for comparing phylogenetic trees. Bioinformatics **36**(20), 5007–5013 (2020)

20. Steel, M.A., Penny, D.: Distributions of tree comparison metrics-some new results. Syst. Biol. **42**(2), 126–141 (1993)
21. Swenson, M.S., Suri, R., Linder, C.R., Warnow, T.: An experimental study of quartets MaxCut and other supertree methods. In: Moulton, V., Singh, M. (eds.) WABI 2010. LNCS, vol. 6293, pp. 288–299. Springer, Heidelberg (2010). https://doi.org/10.1007/978-3-642-15294-8_24
22. Waterman, M.S., Smith, T.F.: On the similarity of dendrograms. J. Theor. Biol. **73**(4), 789–800 (1978)
23. Wickett, N.J., et al.: Phylotranscriptomic analysis of the origin and early diversification of land plants. Proc. Natl. Acad. Sci. **111**(45), E4859–E4868 (2014)
24. Yang, Z., Rannala, B.: Molecular phylogenetics: principles and practice. Nat. Rev. Genet. **13**(5), 303–314 (2012)

The K-Robinson Foulds Measures
for Labeled Trees

Elahe Khayatian[1], Gabriel Valiente[2], and Louxin Zhang[1(✉)]

[1] Department of Mathematics, National University of Singapore,
Singapore 119076, Singapore
elahe.khayatian@u.nus.edu, matzlx@nus.edu.sg
[2] Department of Computer Science, Technical University of Catalonia,
Barcelona, Spain
gabriel.valiente@upc.edu

Abstract. Investigating the mutational history of tumor cells is important for understanding the underlying mechanisms of cancer and its evolution. Now that the evolution of tumor cells is modeled using labeled trees, researchers are motivated to propose different measures for the comparison of mutation trees and other labeled trees. While the Robinson-Foulds distance is widely used for the comparison of phylogenetic trees, it has weaknesses when it is applied to labeled trees. Here, k-Robinson-Foulds dissimilarity measures are introduced for labeled tree comparison.

Keywords: phylogenetic trees · mutation trees · Robinson-Foulds distance · k-Robinson-Foulds dissimilarity

1 Introduction

Trees in biology are a fundamental concept as they depict the evolutionary history of entities. These entities may consist of organisms, species, proteins, genes or genomes. Trees are also useful for healthcare analysis and medical diagnosis. Introducing different kinds of tree models has given rise to the question about how these models can be efficiently compared for evaluation. This question has led to defining a dissimilarity measure in the space of targeted trees. For example, mutation/clonal trees are introduced to model tumor evolution, in which nodes represent cellular populations and are labeled by the gene mutations carried by those populations [11,19]. The progression of tumors varies among different patients; additionally, information about such variations is significant for cancer

G. Valiente—Supported by the Ministerio de Ciencia e Innovación (MCI), the Agencia Estatal de Investigación (AEI) and the European Regional Development Funds (ERDF) through project METACIRCLE PID2021-126114NB-C44, also supported by the European Regional Development Fund (FEDER).
L. Zhang—Supported partially by Singapore MOE Tier 1 grant R-146-000-318-114.

K. Jahn and T. Vinař (Eds.): RECOMB-CG 2023, LNBI 13883, pp. 146–161, 2023.
https://doi.org/10.1007/978-3-031-36911-7_10

treatment. Therefore, dissimilarity measures for mutation trees have become a focus of recent research.

In prior studies on species trees, several measures have been introduced to compare two phylogenetic trees. Some examples of such distances are Robinson-Foulds distance (RF) [17], Nearest-Neighbor Interchange (NNI) [13,16], Quartet distance [6], and Path distance [20,21]. Although these distances have been widely used for phylogenetic trees, they are defined based on the assumption that the involved trees have the same label sets. Moreover, only leaves of phylogenetic trees are labeled. Thus, these distances are not useful for comparing trees with different label sets or trees in which all the nodes are labeled.

To get around some limitations of the above-mentioned distances in the comparison of mutation trees, researchers have introduced new measures for mutation trees. Some of these measures are Common Ancestor Set distance (CASet) [5], Distinctly Inherited Set Comparison distance (DISC) [5], and Multi-Labeled Tree Dissimilarity measure (MLTD) [11]. Even though these distances allow for comparing clonal trees efficiently, they are defined based on the assumption that no mutation can occur more than once during a tumor history, and once a mutation appears in one stage, it will not get lost in the next stages [5,11]. Therefore, these distances are not useful for comparing the trees used to model the tumor evolution where mutations can occur multiple times and disappear after being gained.

In addition to the three measures mentioned above, a group of other dissimilarity measures have been introduced for the comparison of mutation trees, including Parent-Child Distance (PCD) [8] and Ancestor-Descendant Distance (AD) [8]. These measures are metric only in the space of 1-labeled mutation trees, in which each node is labeled by exactly one mutation. These distances are again defined based on the assumption that no mutation can occur more than once during the tumor evolution.

Apart from the measures mentioned above, another recently proposed distance is the generalized RF distance (GRF) [14,15]. This measure allows for the comparison of phylogenetic trees, phylogenetic networks, mutation and clonal trees. An important point about this distance is that its value depends on the intersection between clusters or clones of trees. However, this intersection does not contribute to the RF distance. In fact, if two clusters or clones of two trees are different, their contribution to the RF distance is 1; otherwise, it is 0. Hence, the generalized RF distance has a better resolution than the RF distance. However, it is defined based on the assumption that two distinct nodes in each tree are labeled by two disjoint sets [14].

In addition to the generalized RF distance, there are some other generalizations of the RF distance, such as Bourque distance [9]. This distance is effective for comparing mutation trees whose nodes are labeled by non-empty sets. However, like the above distances, it does not allow for multiple occurrences of mutations during the tumor history [9]. Other generalization of the RF distance have also been proposed for gene trees [1,2].

The above-mentioned measures are not able to quantify similarity or difference of some tree models. Two instances of such models are the Dollo [7] and the Camin-Sokal model [3]. The reason behind the inadequacy of the measures for these models is that it is possible for mutations to get lost after they are gained in the Dollo model, and a mutation can occur more than once during the tumor history in the Camin-Sokal model [14]. Hence, some measures are needed to mitigate the problem. To the best of our knowledge, Triplet-based Distance (TD) [4] is the only measure introduced so far to resolve the issue. The distance is useful for comparing mutation trees whose nodes are labeled by non-empty sets; additionally, it allows for multiple occurrences of mutations during the tumor history and losing a mutation after it is gained [4]. Thus, the measure is applicable to the broader family of trees in which two nodes of a tree may have non-disjoint sets of labels. Nevertheless, it is not able to compare those trees in which there is a node whose label has more than one copy of a mutation.

Although no tree model has been introduced so far that allows for more than one copy of a mutation in the label of a single node, current models can be extended to deal with copy number of mutations. For example, the constrained k-Dollo model [18] takes the variant read count and the total read count of each mutation in each cell, derived from single-cell DNA sequencing data, as input; then, based on three thresholds for the variant read count, the total read count, and the variant allele frequency, it decides whether a mutation is present or absent in a cell or it is missing [18]. Alternatively, the model can consider the exact frequency numbers to show the multiplicity of each mutation in each cell. This motivates us to define and evaluate k-RF dissimilarity measures that can be used to compare pairs of labeled trees whose nodes are labeled by non-empty multisets.

The rest of the paper consists of six sections. Section 2 introduces the basic concepts and notations. In Sect. 3, we define some dissimilarity measures for 1-labeled unrooted and rooted trees. In Sect. 4, we present several mathematical properties of the k-RF measures. We also examine the frequency distribution of the pairwise k-RF scores in the space of 1-labeled unrooted and rooted trees. In Sect. 5, we extend definitions of the measures to multiset-labeled unrooted and rooted trees. In Sect. 6, we demonstrate that k-RF measures perform well when used for clustering trees. Finally, in Sect. 7, a brief summary of the paper's aim, the important results, and open lines for future work are mentioned.

2   Concepts and Notations

A graph consists of a set of nodes and a set of edges that are each an unordered pair of distinct nodes, whereas a directed graph consists of a set of nodes and a set of directed edges that are each an ordered pair of distinct nodes.

Let G be a (directed) graph. We use $V(G)$ and $E(G)$ to denote its node and edge set, respectively. Let $u, v \in V(G)$ such that $(u, v) \in E(G)$. We say that u and v are adjacent, the edge (u, v) is incident to u, and u is an end of (u, v). The degree of v is defined as the number of edges incident to v. The nodes of degree one are called the *leaves*. We use $Leaf(G)$ to denote the leaf set of G. Non-leaf nodes are called *internal nodes*. Furthermore, if G is directed, the indegree (resp., outdegree) of v is defined as the number of edges (x, y) such that $y = v$ (resp., $x = v$). An edge is said to be *internal* if its two ends are both internal nodes.

A *path* of length k from u to v consists of a sequence of nodes $u_0, u_1, \ldots, u_k$ such that $u_0 = u$, $u_k = v$ and $(u_{i-1}, u_i) \in E(G)$ for each i. The *distance* from u to v is the smallest length of a path from u to v, denoted as $d_G(u, v)$. Note that if G is undirected, $d_G(u, v) = d_G(v, u)$ for all $u, v \in V(G)$. The *diameter* of G, denoted as $\mathrm{diam}(G)$, is defined as $\max_{u,v \in V(G)} d_G(u, v)$. The diameter of a directed graph is set to ∞ if there is no directed path from one node to another.

2.1   Trees

A tree T is a graph in which there is exactly one path from every node to any other node. It is *binary* if every internal node is of degree 3. It is a *line tree* if every internal node is of degree 2. Each line tree has exactly two leaves.

2.2   Rooted Trees

A rooted tree is a directed tree with a specific root node where the edges are oriented away from the root. In a rooted tree, there is exactly one edge entering u for every non-root node u, and thus there is a unique path from its root to every other node.

Let T be a rooted tree and $u, v \in V(T)$. If $(u, v) \in E(T)$, v is called a child of u and u is called the parent of v. In general, for $u \neq v$, if u belongs to the unique path from $root(T)$ to v, v is said to be a descendant of u, and u is said to be an ancestor of v. We use $C_T(u)$, $A_T(u)$ and $D_T(u)$ to denote the set of all children, ancestors and descendants of u, respectively. Note that $u \notin A_T(u)$ and $u \notin D_T(u)$.

We say that $v \in V(T)$ is in level n if the length of the path from the root of T to v is n. The depth of T, denoted as $depth(T)$, is the maximum level a node can have.

A binary rooted tree is a rooted tree in which the root is of indegree 0 and outdegree 2, and every other internal node is of indegree 1 and outdegree 2.

A rooted line tree is a rooted tree in which each internal node has only one child.

A rooted caterpillar tree is a tree in which every internal node has at most one child that is internal.

2.3   Labeled Trees

Let L be a set and $\mathbb{P}(L)$ be the set of all subsets of L. A tree or rooted tree T is labeled with the subsets of L if T is equipped with a function $\ell : V(T) \to \mathbb{P}(L)$ such that $\cup_{v \in V(T)} \ell(v) = L$, and $\ell(v) \neq \emptyset$ for every $v \in V(T)$. In particular, if $\ell(v)$ contains exactly one element for each $v \in V(T)$ and ℓ is one-to-one, T is said to be a 1-labeled tree on L. In addition, if T is 1-labeled on L, then for $C \subseteq V(T)$, $L(C)$ is defined as $L(C) = \{x \in L \mid \exists w \in C : \ell(w) = \{x\}\}$.

Two (rooted) labeled trees S and T are isomorphic, denoted by $S \cong T$, if there is a bijection $\phi : V(S) \to V(T)$ such that $\ell(\phi(v)) = \ell(v)$ for every $v \in V(S)$ and $(u, v) \in E(S)$ if and only if $(\phi(u), \phi(v)) \in E(T)$. Moreover, ϕ maps the root of S to the root of T if S and T are rooted.

2.4 Phylogenetic and Mutation Trees

Let X be a finite taxon set. A phylogenetic tree (resp., rooted phylogenetic tree) on X is a binary tree (resp., binary rooted tree) in which the leaves are uniquely labeled by the taxa of X and the internal nodes are unlabeled.

A mutation tree on a set M of mutated genes is a rooted tree whose nodes are labeled by the subsets of M.

2.5 Dissimilarity Measures for Trees

Let $\mathcal{T}$ be a set of trees. A dissimilarity measure on $\mathcal{T}$ is a symmetric real function $d : \mathcal{T} \times \mathcal{T} \to \mathbb{R}^{\geq 0}$. Intuitively, a dissimilarity measure satisfies the property that the more similar two trees are, the lower their measure value is. A pseudometric on $\mathcal{T}$ is a dissimilarity measure that satisfies the triangle inequality condition. Finally, a metric (distance) on $\mathcal{T}$ is a pseudometric d such that $d(S, T) = 0$ if and only if $S \cong T$.

3   The k-RF Dissimilarity Measures for 1-Labeled Trees

In this section, we propose k-RF dissimilarity measures for 1-labeled unrooted and rooted trees.

3.1 The k-RF Measures for 1-Labeled Trees

Let X be a set of labels and let T be a 1-labeled tree over X. Each $e = (u, v) \in E(T)$ induces the following two-part partition on X:

$$P_T(e) = \{L(B_e(u)), L(B_e(v))\}, \tag{1}$$

where

$$B_e(u) = \{w \mid d_T(w, u) < d_T(w, v)\}, B_e(v) = \{w \mid d_T(w, v) < d_T(w, u)\}. \tag{2}$$

Next, we define:

$$\mathcal{P}(T) = \{P_T(e) \mid e \in E(T)\}. \tag{3}$$

Let S and T be two 1-labeled trees. Recall that the RF distance $d_{RF}(S, T)$ between S and T is defined as:

$$d_{RF}(S, T) = |\mathcal{P}(S) \triangle \mathcal{P}(T)|. \tag{4}$$

Example 1. Consider the 3 1-labeled trees in Fig. 1. We have $d_{\mathrm{RF}}(S,T) = 4$ because $P_T(e_1), \ldots, P_T(e_6)$ are:

$$\{\{a,b,c,d,e,f\}, \{g\}\}, \ \{\{a,b,c,d,f\}, \{e,g\}\}, \ \{\{a,b,c,d\}, \{e,f,g\}\},$$
$$\{\{a,e,f,g\}, \{b,c,d\}\}, \ \{\{a,c,d,e,f,g\}, \{b\}\}, \ \{\{a,b,d,e,f,g\}, \{c\}\},$$

respectively, and $P_S(\bar{e}_1), \ldots, P_S(\bar{e}_6)$ are:

$$\{\{a,b,c,d,f,g\}, \{e\}\}, \ \{\{a,b,c,d,e,g\}, \{f\}\}, \ \{\{a,b,c,d\}, \{e,f,g\}\},$$
$$\{\{a,e,f,g\}, \{b,c,d\}\}, \ \{\{a,b,d,e,f,g\}, \{c\}\}, \ \{\{a,c,d,e,f,g\}, \{b\}\},$$

respectively. In addition, $d_{\mathrm{RF}}(\acute{S},T) = 12$ even if the two trees have the same topology and only one node is labeled differently.

The above example indicates that the RF measure cannot capture local similarity (and difference) for some 1-labeled trees. One popular dissimilarity measure for sets is the Jaccard distance. It is obtained by dividing the size of the symmetric difference of two sets by the size of their union. Two 1-labeled trees are identical if and only if they have same edges. Therefore, we propose to use $|E(S)\triangle E(T)|$ and its generalization to measure the dissimilarity for 1-labeled trees S and T.

Let k be a non-negative integer and let T be a 1-labeled tree. Each edge $e = (u,v)$ induces the following pair of subsets of labels:

$$P_T(e, k) = \{L(B_e(u,k)), L(B_e(v,k))\}, \tag{5}$$

where $B_e(x, k) = \{w \in B_e(x) \mid d_T(w,x) \leqslant k\}$ for $x = u, v$. Clearly, $B_e(u, \infty) = B_e(u)$ and $B_e(u, 0) = \{u\}$. Next, we define:

$$\mathcal{P}_k(T) = \{P_T(e, k) \mid e \in E(T)\}. \tag{6}$$

Definition 1. *Let $k \geqslant 0$ and let S and T be two 1-labeled trees. The k-RF dissimilarity measure between S and T is defined as:*

$$d_{k\text{-RF}}(S,T) = |\mathcal{P}_k(S)\triangle\mathcal{P}_k(T)|. \tag{7}$$

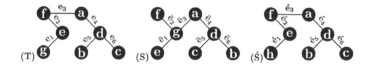

Fig. 1. Three 1-labeled trees.

Example 1 (cont.) We have $d_{1\text{-RF}}(\acute{S}, T) = 4$, as $P_T(e_i, 1)$ $(1 \leqslant i \leqslant 6)$ are:

$$\{\{g\}, \{e, f\}\}, \quad \{\{e, g\}, \{a, f\}\}, \quad \{\{e, f\}, \{a, d\}\},$$
$$\{\{a, f\}, \{b, c, d\}\}, \quad \{\{b\}, \{a, c, d\}\}, \quad \{\{c\}, \{a, b, d\}\},$$

respectively, and $P_{\acute{S}}(\acute{e}_i, 1)$ $(1 \leqslant i \leqslant 6)$ are:

$$\{\{h\}, \{e, f\}\}, \quad \{\{e, h\}, \{a, f\}\}, \quad \{\{e, f\}, \{a, d\}\},$$
$$\{\{a, f\}, \{b, c, d\}\}, \quad \{\{b\}, \{a, c, d\}\}, \quad \{\{c\}, \{a, b, d\}\},$$

respectively. We also have $d_{1\text{-RF}}(S, T) = 8$. The 1-RF dissimilarity measure captures the difference of the trees better than the RF distance for the trees.

3.2   The k-RF Measures for 1-Labeled Rooted Trees

Let $k \geqslant 0$ be an integer and let T be a 1-labeled rooted tree. For a node $w \in V(T)$, we define $B_k(w)$ and $D_k(w)$ as:

$$B_k(w) = \{x \in V(T) \mid \exists y \in A_T(w) \cup \{w\} : d(y, w) + d(y, x) \leqslant k\}, \quad (8)$$
$$D_k(w) = \{w\} \cup \{x \in D_T(w) \mid d(w, x) \leqslant k\}. \quad (9)$$

For each $e = (u, v) \in E(T)$, we define $P_T(e, k)$ as:

$$P_T(e, k) = (L(D_k(v)), L(B_k(u) \setminus D_k(v))), \quad (10)$$

which is an ordered pair of label subsets. Note that the first subset of $P_T(e, k)$ contains the labels of the descendants that are at distance at most k from v, whereas the second subset contains the labels of the other nodes around the edge e within a distance of k. Next, we define:

$$P_k(T) = \{P_T(e, k) \mid e \in E(T)\}. \quad (11)$$

Definition 2. *Let $k \geqslant 0$ and let S and T be two 1-labeled rooted trees. Then, the k-RF dissimilarity measure between S and T is defined as:*

$$d_{k\text{-RF}}(S, T) = |P_k(S) \triangle P_k(T)|. \quad (12)$$

Example 2. In Fig. 2 there are two 1-labeled rooted trees S and T. In T, $P_T(e_i, 1)$ $(1 \leqslant i \leqslant 7)$ are the following ordered pairs of label subsets:

$$(\{f, h\}, \{b, d\}), \quad (\{c, f, g\}, \{b, h\}), \quad (\{c\}, \{f, g, h\}), \quad (\{g\}, \{c, f, h\}),$$
$$(\{a, d, e\}, \{b, h\}), \quad (\{a\}, \{b, d, e\}), \quad (\{e\}, \{a, b, d\}).$$

In S, $P_S(\bar{e}_i, 1)$ $(1 \leqslant i \leqslant 7)$ are the following ordered pairs of label subsets:

$$(\{b, d\}, \{c, f\}), \quad (\{a, d, e\}, \{b, c\}), \quad (\{a\}, \{b, d, e\}), \quad (\{e\}, \{a, b, d\}),$$
$$(\{f, g, h\}, \{b, c\}), \quad (\{g\}, \{c, f, h\}), \quad (\{h\}, \{c, f, g\}).$$

Therefore, $d_{1\text{-RF}}(S, T) = 8$.

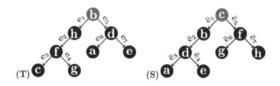

Fig. 2. Two 1-labeled rooted trees in Example 2, for which the 1-RF score is 8.

4   Characterization of k-RF for 1-Labeled Trees

In order to evaluate the k-RF dissimilarity measures, we first reveal some mathematical properties of the k-RF measures; we then present experimental results on the frequency distribution of the measures.

4.1   Mathematical Properties

All the propositions given in this subsection are proved in the Supplementary Document.

Proposition 1. *Let S and T be two 1-labeled trees.*

(a) *Let $|L(S) \cap L(T)| \leqslant 2$ and $|E(T)| \geqslant 2$. For any $k \geqslant 1$, $d_{k\text{-RF}}(S,T) = |E(S)| + |E(T)|$.*

(b) *Assume that $L(S) \neq L(T)$. Then, for $k < \min\{diam(T), diam(S)\}$, we have $k + 1 \leqslant d_{k\text{-RF}}(S,T) \leqslant |E(S)| + |E(T)|$.*
   In addition, if $k \geqslant \min\{diam(T), diam(S)\}$ and $|L(S)| = |L(T)|$, we have $d_{k\text{-RF}}(S,T) = |E(S)| + |E(T)|$.

(c) *Renaming each node with its label, we have $d_{0\text{-RF}}(S,T) = |E(S) \triangle E(T)|$.*

(d) *If $k \geqslant \max\{diam(S), diam(T)\} - 1$, then $d_{k\text{-RF}}(S,T) = d_{RF}(S,T)$.*

Proposition 2. *Let $k \geqslant 0$ be an integer. The k-RF dissimilarity measure satisfies the non-negativity, symmetry and triangle inequality conditions. It also satisfies that $d_{k\text{-RF}}(S,T) = 0$ if and only if $S \cong T$. In other words, k-RF is a metric in the space of 1-labeled trees.*

Proposition 3. *Let S and T be two 1-labeled trees with n nodes and $k \geqslant 0$. Then, $d_{k\text{-RF}}(S,T)$ can be computed in $O(kn)$ time.*

The above facts also hold for 1-labeled rooted trees (Theorems 1 and 2, in the Supplementary Document).

4.2   The Distribution of k-RF Scores

We examined the distribution of the k-RF dissimilarity scores for 1-labeled unrooted and rooted trees with the same label set and with different label sets.

The distribution of the frequency of the pairwise k-RF scores in the space of 6-node 1-labeled unrooted and rooted trees are presented in Fig. 3 for each

Table 1. The number of pairs of 1-labeled 6-node unrooted (top) and rooted (bottom) trees that have c common labels and have 1-RF score d for $c = 3, 4, 5$ and $d = 2, 4, 6, 8, 10$.

c \ 1-RF	2	4	6	8	10
3	0	0	0	3,072	1,676,544
4	0	0	432	16,800	1,662,384
5	0	340	3,720	53,100	1,622,456

c \ 1-RF	2	4	6	8	10
3	0	0	0	79,872	60,386,304
4	0	0	7,776	419,136	60,039,264
5	0	4,080	65,760	1,310,880	59,085,456

k from 0 to 4. Recall that 4-RF is actually the RF distance. The frequency distribution for the RF distance in the space of phylogenetic trees is known to be Poisson [20]. It seems also true that the pairwise 0-RF and 4-RF scores have a Poisson distribution in the space of 6-node 1-labeled unrooted and rooted trees. However, the distribution of the pairwise k-RF scores is unlikely Poisson when $k = 1, 2, 3$. This fact was also observed on 1-labeled trees of other small sizes (data shown in the Supplementary Document).

We examined 1,679,616 (respectively, 60,466,176) pairs of 6-node 1-labeled unrooted (respectively, rooted) trees (respectively, rooted tress) such that the trees in each pair have c common labels, with $c = 3, 4, 5$. Table 1 shows that the most of the pairs have a largest dissimilarity score of 10.

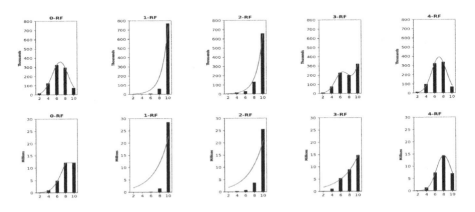

Fig. 3. The distribution of the frequency of the k-RF scores for the 1-labeled 6-node unrooted (top row) and rooted (bottom row) trees. In each barplot, the x-axis represents k-RF scores and the y-axis represents the number of tree pairs for which the k-RF is equal to a given score.

5   A Generalization to Multiset-Labeled Trees

In this section, we extend the measures introduced in Sect. 3 to multiset-labeled unrooted and rooted trees.

5.1   Multisets and Their Operations

A multiset is a collection of elements in which an element x can occur one or more times [10]. The set of all distinct elements appearing in a multiset A is denoted by $\text{Supp}(A)$. Each multiset A can be represented by $A = \{x^{m_A(x)} \mid x \in \text{Supp}(A)\}$, where $m_A(x)$ is the number of occurrences of x. In this paper, we simply represent A by the monomial $x_1^{m_A(x_1)} \ldots x_n^{m_A(x_n)}$ for multiset instances, where x_i^1 is simplified to x_i for each i.

Let A and B be two multisets. We say A is a multi-subset of B, denoted by $A \subseteq_m B$, if for every $x \in \text{Supp}(A)$, $m_A(x) \leqslant m_B(x)$. In addition, we say that $A = B$ if $A \subseteq_m B$ and $B \subseteq_m A$. Furthermore, the union, sum, intersection, difference, and symmetric difference of A and B are respectively defined as follows:

- $A \cup_m B = \{x^{\max\{m_A(x), m_B(x)\}} \mid x \in \text{Supp}(A) \cup \text{Supp}(B)\}$;
- $A \uplus_m B = \{x^{m_A(x) + m_B(x)} \mid x \in \text{Supp}(A) \cup \text{Supp}(B)\}$;
- $A \cap_m B = \{x^{\min\{m_A(x), m_B(x)\}} \mid x \in \text{Supp}(A) \cap \text{Supp}(B)\}$;
- $A \setminus_m B = \{x^{m_A(x) - m_B(x)} \mid x \in \text{Supp}(A) : m_A(x) > m_B(x)\}$;
- $A \triangle_m B = (A \cup_m B) \setminus_m (A \cap_m B)$;

where $m_X(x)$ is defined as 0 if $x \notin \text{Supp}(X)$ for $X = A, B$.

Let L be a set and $\mathbb{P}_m(L)$ be the set of all multi-subsets on L. A tree T is labeled with the multi-subsets of L if T is equipped with a function $\ell : V(T) \to \mathbb{P}_m(L)$ such that $\cup_{v \in V(T)} \text{Supp}(\ell(v)) = L$ and $\ell(v) \neq \emptyset$, for every $v \in V(T)$. For $C \subseteq V(T)$ and $x \in L$, we define $L_m(C)$ and $m_T(x)$ as follows:

$$L_m(C) = \uplus_{v \in C} \ell(v); \tag{13}$$

$$m_T(x) = \sum_{v \in V(T)} m_{\ell(v)}(x). \tag{14}$$

5.2   The RF and k-RF Measures for Multiset-Labeled Trees

Let T be a multiset-labeled tree. Then, each edge $e = (u, v)$ of T induces a pair of multisets

$$P_T(e) = \{L_m(B_e(u)), L_m(B_e(v))\}, \tag{15}$$

where $L_m()$ is defined in Eq. (13), and $B_e(u)$ is defined in Eq. (2). Note that Eq. (15) is obtained from Eq. (1) by replacing $L()$ with $L_m()$.

Remark 1. In a multiset-labeled tree T, two edges may induce the same multi-set pair as shown in Fig. 4. Hence, $\mathcal{P}(T)$ in Eq. (3) is a multiset in general.

We use Eq. (15), Eq. (3) and Eq. (4) to define the RF-distance for multiset-labeled trees by replacing $\triangle$ with $\triangle_m$ in Eq. (4).

Fig. 4. A multiset-labeled tree in which different edges define the same label multi-subset pairs. Here, $P_T(e_2) = P_T(e_3) = \{abc, a^2b^2c\}$.

Let $k \geqslant 0$. We use Eq. (5), Eq. (6), and Eq. (7) to define the k-RF dissimilarity measures for multiset-labeled trees by replacing $L()$ with $L_m()$ in Eq. (5) and replacing $\triangle$ with $\triangle_m$ in Eqn. (7).

Example 3. Consider the multiset-labeled trees S, $\acute{S}$, and T in Fig. 5. $\mathcal{P}_k(T), \mathcal{P}_k(S)$ and $\mathcal{P}_k(\acute{S})$ for $k = 0, 1, \infty$ are summarized in Table 2. We obtain:

$$d_{0\text{-RF}}(T, \acute{S}) = 2; \quad d_{1\text{-RF}}(T, \acute{S}) = 6; \quad d_{\text{RF}}(T, \acute{S}) = 12;$$
$$d_{0\text{-RF}}(S, \acute{S}) = 10; \quad d_{1\text{-RF}}(S, \acute{S}) = 12; \quad d_{\text{RF}}(S, \acute{S}) = 12.$$

It is not hard to see that both $d_{0\text{-RF}}(T, \acute{S})$ and $d_{1\text{-RF}}(T, \acute{S})$ reflect the local similarity of the two multiset-labeled trees better than $d_{\text{RF}}(T, \acute{S})$.

Fig. 5. Three multiset-labeled trees.

Table 2. The edge-induced unordered pairs of multisets in the three trees in Fig. 5 for $k = 0, 1, \infty$.

Tree	$\mathcal{P}_0(\ )$	$\mathcal{P}_1(\ )$	$\mathcal{P}_\infty(\ )$
T	$\{c^2, e^2\}, \{c, e^2\},$ $\{ac, c\}, \{ac, d\},$ $\{ab^2, d\}, \{cd, d\}$	$\{c^2, ce^2\}, \{ab^2cd^2, ac^2\},$ $\{ac^2, c^2e^2\}, \{ab^2, ac^2d^2\},$ $\{acd, ce^2\}, \{a^2b^2cd, cd\}$	$\{a^2b^2c^3d^2e^2, c^2\}, \{a^2b^2c^3d^2, c^2e^2\},$ $\{a^2b^2c^2d^2, c^3e^2\}, \{ab^2cd^2, ac^4e^2\},$ $\{ab^2, ac^5d^2e^2\}, \{a^2b^2c^4de^2, cd\}$
S	$\{c^2, e\}, \{ce, e\},$ $\{ac, e\}, \{ac, d\},$ $\{abc, d\}, \{bd, d\}$	$\{ac^2e^2, c^2\}, \{a^2bc^2d, bd\},$ $\{ab^2cd^2, ace\}, \{ac^3e, ce\},$ $\{acd, c^3e^2\}, \{abc, abcd^2\}$	$\{a^2b^2c^3d^2e^2, c^2\}, \{a^2b^2c^2d^2, c^3e^2\},$ $\{ab^2cd^2, ac^4e^2\}, \{a^2b^2c^4e, ce\},$ $\{abc, abc^4d^2e^2\}, \{a^2bc^5de^2, bd\}$
$\acute{S}$	$\{c^2, e^2\}, \{c, e^2\},$ $\{ac, c\}, \{ac, d\},$ $\{b^3, d\}, \{cd, d\}$	$\{c^2, ce^2\}, \{ac^2, c^2e^2\},$ $\{acd, ce^2\}, \{ac^2d^2, b^3\},$ $\{ac^2, b^3cd^2\}, \{ab^3cd, cd\}$	$\{ab^3c^3d^2e^2, c^2\}, \{ab^3c^3d^2, c^2e^2\}$ $\{ab^3c^2d^2, c^3e^2\}, \{ac^4e^2, b^3cd^2\},$ $\{ac^5e^2d^2, b^3\}, \{ab^3c^4e^2d, cd\}$

5.3   The k-RF Measures for Multiset-Labeled Rooted Trees

Let $k \geqslant 0$ be an integer. We use Eq. (10), Eq. (11), and Eq. (12) to define k-RF dissimilarity measures for multiset-labeled rooted trees by replacing $L()$ with $L_m()$ in Eq. (10) and replacing $\triangle$ with $\triangle_m$ in Eq. (12).

Proposition 4. *Let $k \geqslant 0$ be an integer. The k-RF dissimilarity measure satisfies the non-negativity, symmetry, and triangle inequality conditions. Hence, k-RF is a pseudometric for each k in the space of multiset-labeled trees.*

Remark 2. The k-RF dissimilarity measure does not necessarily satisfy the condition that $d_{k\text{-RF}}(S, T) = 0$ implies $S \cong T$. For example, for the multiset-labeled trees S and T in Fig. 6, $\mathcal{P}_2(S) = \mathcal{P}_2(T)$, which consist of $\{a^2d^2, b\}$, $\{abd, ad\}$, and $\{a, abd^2\}$. Hence, k-RF is not a metric in general.

Proposition 5. *Let $k \geqslant 0$ and S and T be two (rooted) trees whose nodes are labeled by $L(S)$ and $L(T)$, respectively. Then, the $d_{k\text{-RF}}(S, T)$ can be computed in $O(n(k + |L(S)| + |L(T)|))$ if the total multiplicity of each label is upper bounded by a constant.*

Remark 3. The parts (b) and (d) of Proposition 1, with the same proof, hold for multiset-labeled (rooted) trees.

5.4   Correlation Between the k-RF and the Other Measures

Using the Pearson correlation, we compared the k-RF measures with the three existing measures CASet∩, DISC∩ [5], and the generalized RF distance (GRF) [14] in the space of set-labeled trees for different k from 0 to 28.

Firstly, we conducted the correlation analysis in the space of mutation trees with the same label set. Using a method reported in [9], we generated a simulated dataset containing 5,000 rooted trees in which the root was labeled with 0 and the other nodes were labeled by the disjoint subsets of $\{1, 2, \ldots, 29\}$, where the trees have different number of nodes. We computed all $\binom{5,000}{2}$ pairwise scores for CASet∩, DISC∩, GRF and k-RF. We then computed the Pearson correlation between k-RF and the other measures, which is summarized in the left dotplot in Fig. 7.

Our results show that CASet∩, DISC∩ and GRF were all positively correlated with each k-RF measure. We observed the following facts:

Fig. 6. Two multiset-labeled trees, for which the 2-RF dissimilarity score is 0.

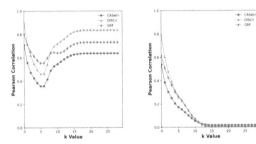

Fig. 7. Pearson correlation between the k-RF and the existing measures: CASet∩, DISC∩, and GRF. The analyses were conducted on rooted trees with the same label set (left) and with different but overlapping label sets (right).

- The GRF and k-RF measures had the largest Pearson correlation for each $k <$ 8, whereas the DISC∩ and k-RF measures had the largest Pearson correlation for each $k \geqslant 8$.
- The 5-RF and 6-RF were less correlated to CASet∩, DISC∩ and GRF than other k-RF measures.
- The Pearson correlation between k-RF and each of CASet∩ and DISC∩ increased when k went from 6 to 15.
- The Pearson correlation did not change when $k \geqslant 19$. In fact, k-RF became the RF distance when $k \geqslant 19$, as the maximum diameter of the trees in the dataset was 20.

Secondly, we conducted the Pearson correlation analysis on the trees with different but overlapping label sets. The dataset was generated by the same method and was a union of 5 groups of rooted trees, each of which contained 200 trees over the same label set. We computed the dissimilarity scores for each tree in the first family and each tree in other groups and then computed the Pearson correlation between different measures. Again, all the dissimilarity measures were positively correlated, but less correlated than in the first case; see Fig 7 (right).

The right dotplot of Fig. 7 shows that the k-RF and DISC∩ had the largest Pearson correlation for k from 1 to 9, and the k-RF and the CASet∩ has the largest Pearson correlation for $k \geqslant 10$. Moreover, all the Pearson correlations decreased when k changed from 1 to 15. This trend was not observed in the first case.

6   Validating the k-RF Measures by Clustering Trees

A clustering experiment was designed in order to demonstrate which of the k-RF measures, CASet∩, DISC∩, and GRF distances, is good at grouping trees according to their label sets, as follows.

We generated 5 tree families each containing 50 trees by repeatedly calling the program reported in [9]. The nodes in each tree family were labeled by the subsets of a 30-label set. The label sets used for different tree families were different, but overlapping. As the nodes were labeled by disjoint subsets, each different label between the label sets of two trees induces at least d different pairs, where d is the degree of the node with the label. Thus, a large number of different elements between the label sets could make the trees more distinguishable by the k-RF measures. Therefore, the label sets used for the different tree families differed in only one label.

We then computed the pairwise dissimilarity scores for all 250 trees in the groups using each measure; we clustered the 250 trees into k clusters using the k-means algorithm, where $k = 2, \ldots, 57$; and we assessed the quality of each clustering using the Silhouette score [12].

As Fig. 8 illustrates, neither of the CASet∩, DISC∩, and GRF distances were able to recognize the exact number of families. However, CASet∩ had the highest Silhouette score when the number of clusters was 5, compared to DISC∩, GRF, and the k-RF measures for $k = 0, 1, \ldots, 12$. In addition, the figure shows that the k-RF measures for $k = 12, \ldots, 19$ could recognize the correct number of families. Moreover, the Silhouette score of the k-RF measures at 5 increased when k increased from 8 to 19. This interesting observation could be the result of the point that with the rise of k, the number of pairs of trees with the highest possible k-RF score may increase, making families more recognizable. Note that the existence of such pairs is guaranteed when the minimum diameter of trees is at most k (see Remark 3). Thus, as the minimum diameter of the trees was 8, the increase occurred from $k = 8$ onward.

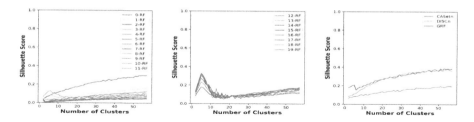

Fig. 8. The Silhouette scores of clustering 250 rooted trees by using k-RF measures for $k = 0, 1, \ldots, 11$ (left), k-RF measures for $k = 12, \ldots, 19$ (middle), and the three measures CASet∩, DISC∩, and GRF (right).

7   Conclusions

Introducing an efficient measure for the comparison of mutation trees is a significant task. Therefore, this paper aims to define a new set of dissimilarity measures, namely the k-RF measures. These measures allow for the comparison of any two trees whose nodes are labeled by (not necessarily the same) multisets; thus, they are applicable to mutation trees as well.

We first presented some results about the k-RF measures for 1-labeled unrooted and rooted trees; then, we extended the results to trees in which the nodes are labeled with multisets. For 1-labeled trees, the k-RF measure is proved to be a metric. Then, we examined the frequency distribution of the measures and observed that in the space of 1-labeled unrooted and rooted trees with $n \geqslant 4$ nodes, 0-RF and $(n-2)$-RF scores seem to have a Poisson distribution while the distribution of other k-RF measures does not appear to be Poisson.

After evaluating the measures for 1-labeled trees, we examined the trees whose nodes are labeled by multisets. For these trees, each k-RF measure is a pseudometric. Then, using mutation trees with the same and different label sets, we evaluated correlations of CASet∩, DISC∩, and GRF with each k-RF measure for $0 \leqslant k \leqslant 28$. We observed that each k-RF measure is positively correlated with each of the other three measures. Moreover, The correlation values were generally higher when the measures were used to compare mutation trees with the same label set. Finally, we demonstrated that some k-RF measures are better for clustering mutation trees according to their label sets than the CASet∩, DISC∩, and GRF measures; additionally, we found that for mutation trees with minimum diameter of n, the k-RF measures with $k > n$ are better able to cluster the trees, compared to the one with $k \leqslant n$.

Future work includes how to apply the k-RF measures to designing tree inference algorithms like GraPhyC [8] and also how to infer the exact frequency distribution of the k-RF measures for each $k \geqslant 1$. It is also interesting to investigate the generalization of RF-distance for clonal trees [14].

Supplementary Information

The computer program for the k-RF can be downloaded from https://github.com/Elahe-khayatian/k-RF-measures.git. Supplementary material for this paper can be found at https://zenodo.org/badge/latestdoi/516782294

References

1. Briand, S., Dessimoz, C., El-Mabrouk, N., Lafond, M., Lobinska, G.: A generalized Robinson-Foulds distance for labeled trees. BMC Genom. **21**, 1–13 (2020)
2. Briand, S., Dessimoz, C., El-Mabrouk, N., Nevers, Y.: A linear time solution to the labeled Robinson-Foulds distance problem. Syst. Biol. **71**, 1391–1403 (2022)
3. Camin, J.H., Sokal, R.R.: A method for deducing branching sequences in phylogeny. Evol. **19**, 311–326 (1965)
4. Ciccolella, S., Bernardini, G., Denti, L., Bonizzoni, P., Previtali, M., Vedova, G.D.: Triplet-based similarity score for fully multilabeled trees with poly-occurring labels. Bioinformatics **37**, 178–184 (2021)
5. DiNardo, Z., Tomlinson, K., Ritz, A., Oesper, L.: Distance measures for tumor evolutionary trees. Bioinformatics **36**, 2090–2097 (2020)
6. Estabrook, G.F., McMorris, F.R., Meacham, C.A.: Comparison of undirected phylogenetic trees based on subtrees of four evolutionary units. Syst. Zool. **34**, 193–200 (1985)

7. Farris, J.S.: Phylogenetic analysis under Dollo's law. Syst. Biol. **26**, 77–88 (1977)
8. Govek, K., Sikes, C., Oesper, L.: A consensus approach to infer tumor evolutionary histories. In: Proceedings of ACM International Conference on Bioinformatics, Computational Biology and Health Informatics, pp. 63–72. ACM Press, New York (2018)
9. Jahn, K., Beerenwinkel, N., Zhang, L.: The Bourque distances for mutation trees of cancers. Algorithms Mol. Biol. **16**, 1–15 (2021)
10. Jürgensen, H.: Multisets, heaps, bags, families: what is a multiset? Math. Struct. Comput. Sci. **30**, 139–158 (2020)
11. Karpov, N., Malikic, S., Rahman, M.K., Sahinalp, S.C.: A multi-labeled tree dissimilarity measure for comparing clonal trees of tumor progression. Algorithms Mol. Biol. **14**, 1–18 (2019)
12. Kaufman, L., Rousseeuw, P.J.: Finding Groups in Data: An Introduction to Cluster Analysis. John Wiley & Sons, Hoboken (2009)
13. Li, M., Tromp, J., Zhang, L.: On the nearest neighbour interchange distance between evolutionary trees. J. Theor. Biol. **182**, 463–467 (1996)
14. Llabrés, M., Rosselló, F., Valiente, G.: A generalized Robinson-Foulds distance for clonal trees, mutation trees, and phylogenetic trees and networks. In: Proceedings of 11th ACM International Conference on Bioinformatics, Computational Biology and Health Informatics. ACM Press, NY (2020)
15. Llabrés, M., Rosselló, F., Valiente, G.: The generalized Robinson-Foulds distance for phylogenetic trees. J. Comput. Biol. **28**, 1–15 (2021)
16. Robinson, D.F.: Comparison of labeled trees with valency three. J. Comb. Theory **11**, 105–119 (1971)
17. Robinson, D.F., Foulds, L.R.: Comparison of phylogenetic trees. Math. Biosci. **53**, 131–147 (1981)
18. Sashittal, P., Zhang, H., Iacobuzio-Donahue, C.A., Raphael, B.J.: Condor: Tumor phylogeny inference with a copy-number constrained mutation loss model. bioRxiv (2023)
19. Schwartz, R., Schäffer, A.A.: The evolution of tumour phylogenetics: principles and practice. Nat. Rev. Genet. **18**, 213–229 (2017)
20. Steel, M.A., Penny, D.: Distributions of tree comparison metrics: some new results. Syst. Biol. **42**, 126–141 (1993)
21. Williams, W.T., Clifford, H.T.: On the comparison of two classifications of the same set of elements. Taxon **20**, 519–522 (1971)

Bounding the Number of Reticulations in a Tree-Child Network that Displays a Set of Trees

Yufeng Wu[1]([⊠]) and Louxin Zhang[2]

[1] Department of Computer Science and Engineering, University of Connecticut, Storrs, CT 06269, USA
yufeng.wu@uconn.edu
[2] Department of Mathematics and Center for Data Science and Machine Learning, National University of Singapore, Singapore 119076, Singapore
matzlx@nus.edu.sg

Abstract. Phylogenetic network is an evolutionary model that uses a rooted directed acyclic graph (instead of a tree) to model an evolutionary history of species in which reticulate events (e.g., hybrid speciation or horizontal gene transfer) occurred. Tree-child network is a kind of phylogenetic network with structural constraints. Existing approaches for tree-child network reconstruction can be slow for large data. In this paper, we present several computational approaches for bounding from below the number of reticulations in a tree-child network that displays a given set of rooted binary phylogenetic trees. Through simulation, we demonstrate that the new lower bounds on the reticulation number for tree-child networks can practically be computed for large tree data. The bounds can provide estimates of reticulation for relatively large data.

Keywords: Phylogenetic network · Tree-child network · Algorithm · Phylogenetics

1 Introduction

Phylogenetic network is an emerging evolutionary model for several complex evolutionary processes, including recombination, hybrid speciation, horizontal gene transfer and other reticulate events [11,13]. On the high level, phylogenetic network is a leaf-labeled rooted acyclic digraph. Different from phylogenetic tree model, a phylogenetic network can have nodes (called reticulate nodes) with in-degrees of two or larger. The presence of reticulate nodes greatly complicates the application of phylogenetic networks. The number of possible phylogenetic networks even with a small number of reticulate nodes is very large [6]. A common computational task related to an evolutionary model is the inference of the model (tree or network) from data. A set of phylogenetic trees is a common data for phylogenetic inference. An established research problem on phylogenetic networks is inferring a phylogenetic network as the *consensus* of multiple phylogenetic trees where the network satisfies certain optimality conditions [5,10].

© The Author(s), under exclusive license to Springer Nature Switzerland AG 2023
K. Jahn and T. Vinař (Eds.): RECOMB-CG 2023, LNBI 13883, pp. 162–178, 2023.
https://doi.org/10.1007/978-3-031-36911-7_11

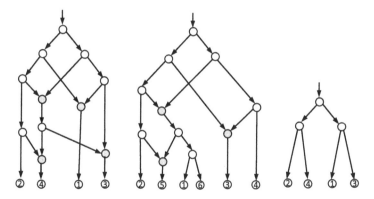

Fig. 1. An arbitrary phylogenetic network with four reticulate nodes (filled circles) on taxa 1, 2, 3, 4 (left), a tree-child network (middle) and a phylogenetic tree (right). Note: the tree on the right is displayed in the network to the left.

Each phylogenetic tree is somehow "contained" (or "displayed") in the network. The problem of inferring a phylogenetic network from a set of phylogenetic trees is called the network reconstruction problem. We refer to the recent surveys [16,24] for the mathematical relation between trees and networks.

The network reconstruction problem has been actively studied recently in computational biology. There are two types of approaches for this problem: unconstrained network reconstruction and constrained network reconstruction. Unconstrained network reconstruction [4,14,21,22] aims to reconstructing a network without additional topological constraints. While such approaches infer more general networks, they are often slow and difficult to scale to large data. Constrained network reconstruction imposes some type of topological constraints on the inferred network. Such constraints simplify the network structure and often lead to more efficient algorithms. There are various kinds of constraints studied in the literature. One popular constraint is requiring simplified cycle structure in networks (e.g., so-called galled tree [12,18,23].

Another topological constraint, the so-called tree-child property [1], has been studied actively recently. A phylogenetic network is tree-child if every non-leaf node has at least one child that is of in-degree one. This property implies that every non-leaf node is connected to some leaf through a path that is not affected by the removal of any reticulate edge (edge going into a reticulate node; see Fig. 1). A main benefit of tree-child network is that it can have more complex structure than say galled trees, and is therefore potentially more applicable. While tree-child networks have complex structure, they can efficiently be enumerated and counted by a simple recurrence formula [15,25] and so may likely allow faster computation for other tasks. There is a parametric algorithm for determining whether a set of multiple trees can be displayed in a tree-child network simultaneously [17].

Given a phylogenetic network $\mathcal{N}$, we say a phylogenetic tree T (with the same set of taxa as $\mathcal{N}$) is *displayed* in $\mathcal{N}$ if T can be obtained by (i) first deleting all but one incoming edges at each reticulate node of $\mathcal{N}$ (this leads to a tree), and then (ii) removing the degree-two nodes so that the resulting tree becomes a phylogenetic tree. As an example, in Fig. 1, the tree on the right is displayed in the network on the left. Given a set of phylogenetic trees $\mathcal{T}$, we want to reconstruct a tree-child network such that it displays *each* tree $T \in \mathcal{T}$ and its so-called reticulation number (denoted as $\text{TCR}_{\min}$) is the smallest among all such tree-child networks. Here, reticulation number is equal to the number of reticulate edges minus the number of reticulate nodes. $\text{TCR}_{\min}$ is called the tree-child reticulation number of $\mathcal{T}$. There exists no known polynomial-time algorithm for computing the exact $\text{TCR}_{\min}$ for multiple trees.

Since computing the exact tree-child reticulation number $\text{TCR}_{\min}$ of multiple trees is challenging, heuristics for estimating the range of $\text{TCR}_{\min}$ have been developed. Existing heuristics aim at finding a tree-child network with the number of reticulation that is as close to $\text{TCR}_{\min}$ as possible. At present, the best heuristics is ALTS [26]. ALTS can construct near-parsimonious tree-child networks for data that is infeasible for other existing methods. However, a main downside of ALTS is that it is a heuristic and so how close a network reconstructed by ALTS to the optimal one is unknown. Moreover, ALTS still cannot work on large data (say 50 trees with 100 taxa, and with relatively large number of reticulations).

We can view the network reconstruction heuristics as providing an *upper bound* to the reticulation number. In order to gain more information on the reticulation number, a natural approach is computing a *lower bound* on the reticulation number. Such lower bounds, if practically computable, can provide information on the range of the reticulation number. In some cases, if a lower bound matches the heuristically computed upper bound for some data, we can actually know the exact reticulation number [21]. Computing a tight lower bound on reticulation number, however, is not easy: to derive a lower bound one has to consider *all* possible networks that display a set of trees $\mathcal{T}$; in contrast, computing an upper bound on reticulation number of $\mathcal{T}$ only requires one feasible network. For unconstrained networks with multiple trees, the only known non-trivial lower bound is the bound computed by PIRN [21]. While this bound performs well for relatively small data, it is computationally intensive to compute for large data. For tree-child networks, we are not aware of any published non-trivial lower bounds.

In this paper, we present several lower bounds on $\text{TCR}_{\min}$. By simulation, we show that these lower bounds can be useful estimates of $\text{TCR}_{\min}$.

Background on tree-child network

Throughout this paper, when we say network, we refer to tree-child network in which reticulate nodes can have two or more incoming reticulate edges (unless otherwise stated). Edges in the network that are not reticulate edge are called tree edges. Note that the network may not be binary. Trees are assumed to be rooted binary trees on the same taxa.

The Tree-Child Property. A phylogenetic network is *tree-child* if every non-leaf node has at lease one child that is a tree node. In Fig. 1, the middle phylogenetic network is tree-child, whereas the left network is not, in which both the parent u of the leaf 4 and the parent v of the leaf 3 are reticulate and the node

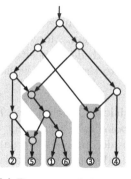 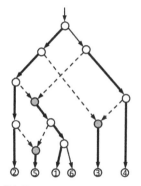

   (a) Decompose into trees   (b) Decompose into paths

Fig. 2. Illustration of the decomposition of a phylogenetic network with k reticulate nodes and n leaves. In this example, $k = 3$ and $n = 6$. Part 2(a): decompose into $k + 1$ disjoint tree components. The tree component rooted at the network root is highlighted in green; other tree components each rooted at a recirculate node are in blue. Part 2(b): decompose into n paths (each path appears in a tree component; ordered by the leaf labels). Reticulate edges: dashed lines. Edges in paths: thick lines. Tree edges not on paths: thin lines.

right above u has u and v as its child. One important property about tree-child network is that there is a directed path (with only tree edges) from any node to some leaf (see e.g. [26]).

Network Decomposition. Consider a phylogenetic network $\mathcal{N}$ with k reticulate nodes. Let the root of $\mathcal{N}$ be r_0 and the k reticulate nodes be $r_1, r_2, \cdots, r_k$. For each i from 0 to k, r_i and its descendants that are connected to r_i by a path consisting of only tree edges induces a subtree of $\mathcal{N}$. Such $k + 1$ subtrees are called the *tree components* of N [9]. Note that the tree components are disjoint and the node set of N is the union of the node sets of these tree components (see Fig. 2). Network decomposition is a powerful technique for studying the tree-child networks [3,6] and other network classes [7] (see [24] for a survey).

Path Decomposition. The network decomposition for a tree-child network leads to a set of trees, where the trees are connected by reticulate edges. We can further decompose each tree component into *paths* as follows. Suppose a tree component contains p leaves. Further suppose the leaves are ordered in some way. We create a path for each leaf sequentially. Let a be the current leaf. We create a path of edges from a node as close to the root of the tree as possible, and down to a. We then remove *all* edges starting at a path and ending at a different path. This procedure (called path decomposition) is illustrated in Fig. 2(b), where the path creation follows the numerical order of the leaves. Note that path decomposition is a valid decomposition of a network $\mathcal{N}$: each node in $\mathcal{N}$ belongs to a unique path after decomposition. This is because only edges (not nodes) are removed during the above procedure. Also note that path decomposition depends on the ordering on nodes: suppose we trace two paths backward to the root; when two paths meet, the path ending at an earlier node continues and the path to a later node ends. This implies that the ordering of leaves affect the outcome of the path decomposition. Moreover, a path starts at either a reticulate node or a tree node in path decomposition. At least one incoming edge is needed to connect the path to the rest of the network (unless the path starts from the root of the network).

Displaying Trees and Path Decomposition. When a tree T is displayed in $\mathcal{N}$, there are edges in $\mathcal{N}$ that form a topologically equivalent tree (possibly with degree-two nodes) as T. Now, when $\mathcal{N}$ is decomposed into paths, to display T, we need to *connect* the paths by using (either tree or reticulate) edges *not* belonging to the paths. Intuitively, tree edges connect the paths in a fixed way while reticulate edges lead to different topology of paths. That is, to display different trees, we need to connect the paths using reticulate edges. This simple property is the foundation of the lower bounds we are to describe soon.

Recall that to display a tree, we need to make choices for each reticulate edge whether to keep or discard. This choice is called the *display choice* for this reticulate edge.

2   Lower Bounds on the Tree-Child Reticulation Number by Integer Programming

In this section, we present several practically computable lower bounds on the tree-child reticulation numbers. These bounds are derived based on the decomposition of tree-child networks.

2.1   TCLB$_1$: A Simple Lower Bound

Recall that any tree-child network with n taxa $\{1, 2, \ldots, n\}$ can be decomposed by path decomposition into n simple paths (possibly in different ways), where each path starts with some network node and ends at a taxon. Now we consider a specific network $\mathcal{N}$ and a specific decomposition of $\mathcal{N}$ into n paths P_i $(1 \leq i \leq n)$. Each P_i starts from nodes in the network and ends at taxon i. We say P_i and P_j are connected if some node within P_i is connected by an edge to the start node of P_j or vice versa. We define a binary variable $C_{i,j}$ to indicate whether or not P_i and P_j are connected for each pair of i and j such that $1 \leq i < j \leq n$. Note that $C_{i,j}$ is for a specific network $\mathcal{N}$ and a specific path decomposition of $\mathcal{N}$. For an example, in the path decomposition in Fig. 2(b), We have:

$$C_{1,2} = C_{1,4} = C_{1,5} = C_{1,6} = C_{2,3} = C_{2,4} = C_{2,5} = C_{3,4} = 1$$

and $C_{i,j} = 0$ for other index pairs. Note that each of these $C_{i,j} = 1$ corresponds to a specific (tree or reticulate) edge *not* inside paths. We have the following simple observation.

Lemma 1. *Let $\mathcal{T}$ be a set of trees. Let $C(\mathcal{N}) = \sum_{1 \leq i < j \leq n} C_{i,j}$ where $C_{i,j}$ is for an arbitrary path decomposition of a network $\mathcal{N}$ that displays $\mathcal{T}$. Then,*

$$\mathrm{TCLB} \triangleq min_{\mathcal{N}} C(\mathcal{N}) - n + 1 \leq \mathrm{TCR}_{\min}.$$

Note that TCLB may well underestimate $\text{TCR}_{\min}$ because $C_{i,j}$ are binary and there can be more than one edges connecting two paths in a path decomposition of the optimal network. We only give an informal proof here. See [26] for proofs on related properties.

Proof. Note that TCLB is taking the minimum over all possible networks $\mathcal{N}$ for an input tree set $\mathcal{T}$. $C(\mathcal{N})$ counts the number of pairs of simple paths P_i and P_j that are connected. If $C_{i,j} = 1$, there must be at least one edge between P_i and P_j. Also note that all the n simple paths must be connected which means exactly $n - 1$ of $C_{i,j}$ are equal to 1 in order to connect these n simple paths. Any additional $C_{i,j} = 1$ implies one reticulation. We consider all tree-child networks $\mathcal{N}$ that display the given trees. For each $\mathcal{N}$, there may exist different path decomposition. The above argument works for any valid path decomposition $\mathcal{N}$. That is, an arbitrary path decomposition of $\mathcal{N}$ gives a lower bound on the number of reticulations in $\mathcal{N}$. This establishes the lower bound. $\square$

While Lemma 1 leads to a lower bound, TCLB is hard to compute because it needs to consider all possible networks $\mathcal{N}$ that displays the given trees $\mathcal{T}$. We now show that we can practically compute a weaker bound TCLB_1, which bounds from below TCLB and thus $\text{TCR}_{\min}$.

We consider a binary tree $T \in \mathcal{T}$. (Our bounds can be generalized to non-binary trees.) The following lemma illustrates one structural property of tree-child network when displaying a *subtree* T_1 of T. Assume T_1 is rooted at node v. Let $S(v)$ be the set of taxa under the node v. Since T_1 is also displayed in $\mathcal{N}$, there exists some non-path edges (i.e., edges not on the paths in the path decomposition) which connects the paths, one path for each leaf in $S(v)$, that displays T_1. Let $v(\mathcal{N})$ be the node in $\mathcal{N}$ that is the root of the displayed subtree in $\mathcal{N}$. We say T_1 is displayed at node $v(\mathcal{N})$.

Lemma 2. *Let $\mathcal{N}$ be a tree-child network displaying $\mathcal{T}$ and let T_1 be a subtree rooted at v of T and be displayed at a node $v(\mathcal{N})$. Then, for any path decomposition, $v(\mathcal{N})$ is on some path. That is, we can always trace from a taxon from $S(v)$ upwards in $\mathcal{N}$ and reach $v(\mathcal{N})$ by following only path edges for the path decomposition.*

Proof. By the tree-child property, there is a leaf a that can be reached from $v(\mathcal{N})$ following only tree edges. Thus, $v(\mathcal{N})$ and a must be inside the same tree component (recall path component is obtained by further decomposition of some network decomposition into trees). Therefore, no matter how path composition is performed, there is always a leaf a' where $v(\mathcal{N})$ is on the path ending at a'. $\square$

Lemma 3. *Let v be an internal node of $T \in \mathcal{T}$ with v_1 and v_2 as its children. In a tree-child network $\mathcal{N}$ displaying $\mathcal{T}$, for any path decomposition of $\mathcal{N}$,*

$$\sum_{i \in S(v_1)} \sum_{j \in S(v_2)} C_{i,j} \geq 1$$

Proof. First $T \in \mathcal{T}$ is displayed in $\mathcal{N}$. Then there exist edges of $\mathcal{N}$ that connect the paths in a path decomposition to form T (otherwise T cannot be displayed in $\mathcal{N}$). So suppose we trace these edges to locate the two subtrees rooted at v_1 and v_2. By Lemma 2, there are nodes $v_1(\mathcal{N})$ and $v_2(\mathcal{N})$ in $\mathcal{N}$ where the two subtrees are displayed at, and are on some paths (denoted as P and P_j respectively). Here, j is a taxon and $j \in S(v_2)$. When there are multiple such nodes for displaying a same subtree, we choose the one that is closest to the root of $\mathcal{N}$.

Now there is a node $v(\mathcal{N})$ in the network where the subtree of T rooted at v is displayed. Again by Lemma 2, $v(\mathcal{N})$ is on a path P_i for some leaf i. This implies either $v_1(\mathcal{N})$ or $v_2(\mathcal{N})$ is on P_i too. Without loss of generality, suppose $v_1(\mathcal{N})$ is on P_i. Then there must exist an edge between the path to i and P_j and $i \in S(v_1)$. This is because (i) there exists a path in $\mathcal{N}$ from $v(\mathcal{N})$ to $v_2(\mathcal{N})$ that is taken to display T in $\mathcal{N}$; (ii) this path can have only a single edge; if not, then there exists at least a node v_3 *not* on P or P_j (recall $v_2(\mathcal{N})$ is the one closest to the root among all choices for $v_2(\mathcal{N})$); (iii) let v_3 be on a decomposed path P_k (which connects to a leaf k; but this violates the assumption that $v_1(\mathcal{N})$ and $v_2(\mathcal{N})$ display two subtrees of v. This implies $C_{i,j} = 1$. We don't know which i and j for the network $\mathcal{N}$. Nonetheless, there exists some $i \in S(v_1)$ and $j \in S(v_2)$ where $C_{i,j} = 1$. $\qquad \square$

Lemma 3 leads to the following lower bound TCLB_1.

Proposition 1. *Let $C_{i,j}$ be binary variables for $1 \leq i < j \leq n$. Let $\text{TCLB}_1 = min(\sum_{1 \leq i < j \leq n} C_{i,j}) - n + 1$ where $C_{i,j}$ satisfies the following constraint: for any internal node v of a tree $T \in \mathcal{T}$ with two children v_1 and v_2, the condition stated in Lemma 3 is satisfied. Then TCLB_1 is a lower bound on the tree-child reticulation number.*

As an example, consider the tree on the right in Fig. 1. We have the following constraints: $C_{2,4} \geq 1, C_{1,3} \geq 1, C_{1,2} + C_{1,4} + C_{2,3} + C_{3,4} \geq 1$. When there are multiple trees, we create such constraints for each tree. TCLB_1 takes the minimum over all choices of $C_{i,j}$ that satisfy all the constraints.

While we don't know how to efficiently compute TCLB_1, it is straightforward to apply integer linear programming formulation (ILP) to compute TCLB_1. Our experience in using ILP modelling shows that TCLB_1 can usually be computed efficiently (in practice) even for large data: for 100 binary trees with 100 taxa, it usually takes less than one second even using a very basic ILP solver.

2.2   TCLB$_2$: A Stronger Lower Bound

We now present techniques to strengthen it to obtain a stronger lower bound called TCLB_2. We start with a stronger version of Lemma 3. We need a special kind of path decomposition, called *ordered path decomposition*, of a network $\mathcal{N}$. An ordered path decomposition is a path decomposition where its paths can be arranged in a total order, and *all* reticulate and non-path tree edges are oriented in one direction relative to this total order. Such ordered path decomposition always exists. To see this, recall that $\mathcal{N}$ is a digraph. Thus, all components of $\mathcal{N}$ obtained by network decomposition can be arranged in a total order. Then we can obtain a tree decomposition by decomposing each component into paths. This leads to a tree decomposition where paths are linearly ordered from left to right and all reticulate edges and all non-path tree edges are oriented from left to right.

We now consider an ordered path decomposition. We let $f(v)$ be the taxon in $S(v)$ that is ordered the first among all the taxa in $S(v)$. That is, $f(v)$ is the taxon under node v that is ordered the first among all the taxa (leaves) under v.

Lemma 4. *Let v_1 and v_2 be the two children of node v of some tree. Then,*

$$C_{f(v_1),f(v_2)} = 1$$

Proof. Recall the proof of Lemma 3. When we trace the subtree rooted at v_1, the root of this subtree must be located within the simple path for $f(v_1)$. This is because the network is acyclic and the simple paths are ordered as in the specific path decomposition. Recall that all reticulate and non-path edges are oriented from left to right. So when we trace edges in a bottom up order (starting from leaves), we must reach the node (i.e. $f(v_1)$) that is ordered the first (i.e., the leftmost). The situation for v_2 is similar. Thus, by the same reason as in Lemma 3, $P_{f(v_1)}$ and $P_{f(v_2)}$ must be connected.   □

Lemma 4 leads to a stronger lower bound TCLB$_2$. This is because if $C_{i,j}$ values satisfy the conditions in Lemma 4, they also satisfy the conditions in Lemma 3.

Let $\mathcal{O}^*$ be the total order of the n taxa in an ordered path decomposition. We let $B(\mathcal{O}^*) = \sum_{i,j} C_{i,j}$, where $C_{i,j} = 1$ if Lemma 4 specifies which two taxa i and j must have $C_{i,j} = 1$, when we consider *all* internal nodes of each tree in $\mathcal{T}$. If i and j are not forced by Lemma 4, $C_{i,j} = 0$. That is, $B(\mathcal{O}^*)$ is fully decided if $\mathcal{O}^*$ is given. By Lemma 4, $B(\mathcal{O}^*) - n + 1$ is a lower bound on TCR$_{\min}$. One technical difficulty is that we don't know $\mathcal{O}^*$ for $\mathcal{N}$. Nonetheless, we can derive a lower bound on TCR$_{\min}$ by taking the *minimum* over all possible $\mathcal{O}$. Thus, we have the following observation.

Proposition 2. TCLB$_2 \overset{\Delta}{=} \min_{\mathcal{O}}(B(\mathcal{O}) - n + 1)$ *is a lower bound on the tree-child reticulation number.*

Naively, to compute TCLB$_2$, we have to consider all possible total orders of the taxa. Enumerating all possible total orders of n taxa is infeasible even for relatively small n value. To develop a practically computable bound, we again apply ILP. Due to the lack of space, we provide the details on the ILP formulation in the full version of this paper.

3 Cherry Bound: Analytical Lower Bound in Terms of the Number of Distinct Cherries in the Given Trees

There is no known polynomial time algorithm for computing the lower bounds in Sect. 2. A natural research question is developing good lower bounds that are polynomial time computable. In the following, we describe an analytical lower bound (called cherry bound) on tree-child reticulation number. Compared with the ILP-computed bounds in Sect. 2, cherry bound is much easier to compute. However, experience shows that cherry bound tends to be weaker than the ILP-computed bounds. Cherry bound is expressed in terms of the number of distinct cherries in given trees $\mathcal{T}$. Here, a cherry is a two-leaf subtree in some $T \in \mathcal{T}$. We let C be the number of *distinct* cherries in the given trees $\mathcal{T}$.

Consider a tree-child network $\mathcal{N}$ with r reticulate nodes that displays $\mathcal{T}$, where $|\mathcal{T}| \geq 2$. Note that a reticulate node in $\mathcal{N}$ has two or more incoming edges. We let n_R be the total number of reticulate edges of $\mathcal{N}$. That is, n_R is equal to the sum of in-degrees of each reticulate node. The reticulation number R of $\mathcal{N}$ is equal to $n_R - r$.

Now suppose we collapse common cherries in $\mathcal{T}$. Here, a common cherry is present in *each* of the trees in $\mathcal{T}$. We collapse such common cherry into a single (new) taxon and repeat until there is no common cherry left. Note that this step is identical to common subtree collapsing, which is a preprocessing step commonly practiced in phylogenetic network construction. Collapsing identical subtrees in given set of trees is a common practice for computing $R_{\min}$ (see, e.g., [13,26]). So in the following, we assume there is no common cherry in $\mathcal{T}$.

Since cherry is a subtree of two leaves in $\mathcal{T}$, each cherry needs to be displayed in $\mathcal{N}$ by obtaining a tree T (through making display choices for reticulate edges) where T displays this cherry. One can view the process of obtaining T is traversing certain nodes of $\mathcal{N}$. We have the following observation.

Lemma 5. *To obtain a cherry in $\mathcal{T}$, we need to traverse either the tail or the head of some reticulate edge in $\mathcal{N}$. That is, displaying a cherry must depend on the choices we make about which reticulate edges to keep for displaying a tree.*

Proof. Suppose displaying a cherry in $\mathcal{N}$ can be achieved by following a path that doesn't contain either the head or the tail of some reticulate edge. Then for any display choice (keep or discard) we make for reticulate edges, such path leading to the cherry that is always present. So, this cherry is a common cherry in $\mathcal{T}$, which contradicts our assumption of no common cherry. □

By Lemma 5, each cherry in the given phylogenetic trees is related to the display choices in $\mathcal{N}$. It is obvious that a cherry displayed in a tree must also be displayed in the network $\mathcal{N}$. Therefore we consider the cherries displayed in the network $\mathcal{N}$. Suppose we add reticulate edges one by one to the network. Adding a reticulate edge can lead to new cherries to be displayed in the network. The more distinct cherries there are, the more reticulation is needed. We now make this more precise by establishing an upper bound on the number of distinct cherries that can be displayed by adding a single reticulate edge, which is an edge entering a reticulate node. Note that displaying a cherry can involve more than one reticulate edge. Suppose there are R reticulations and so there are at least $2R$ reticulate edges.

Lemma 6. *Selecting a reticulate edge e_r to display a tree in a network $\mathcal{N}$ can add at most 2 distinct cherries.*

Proof. Recall a cherry is a size-two subtree and is so displayed in the network $\mathcal{N}$. To display a cherry in $\mathcal{N}$, there are a set of tree or reticulate edges of $\mathcal{N}$ that connect the two taxa of the cherry when displaying choices are made. We refer these edges as the cherry display of this cherry. We classify the cherries into two cases based on the types of edges in a cherry display.

Type 1. The cherry display contains at least one reticulate edge. That is, keeping a reticulate edge can only generate a type-1 cherry.

Type-2. The cherry display contains only tree edges. That is, a type-2 cherry is only related to discarding (but not keeping) some reticulate edges.

We now argue that keeping a reticulate edge e_r can only generate at most one type-1 cherry and at most one type-2 cherry. To see this, we first consider the case of keeping e_r. We call a taxon a a *tree*-taxon under an ancestor node v if a can be reached from v by following a simple path with only tree edges, i.e. a is a descendant of v in the tree component containing v. Due to the property of tree-child network, at most one type-1 cherry can be obtained by keeping e_r: there must be only one tree-taxon a below the destination of e_r, and one tree-taxon b below the other child of the source of e_r, and keeping e_r can only create a single distinct cherry (a, b). Note that otherwise, no cherry can be formed by keeping e_r. If e_r is kept, we have to remove its twin reticulate edges e'_r, this may display another cherry in the tree component containing the source node of e'_r, which is of type-2.

Therefore, we conclude that at most two distinct cherries can be associated with a reticulate edge. □

By Lemma 5, each distinct cherry in $\mathcal{T}$ is associated the display choices of some reticulate edge. By Lemma 6, one reticulate edge can lead to at most two distinct cherries. So $2n_R \geq C$. Note that reticulation number $R = n_R - r$ and $n_R \geq 2r$ (there are at least two reticulate edges per reticulate node). So, $R = n_R - r \geq \frac{n_R}{2}$. So,

$$R \geq \frac{n_R}{2} \geq \frac{C}{4}$$

Proposition 3. *[Cherry bound two on reduced trees] Let C be the number of distinct cherries in a set of trees $\mathcal{T}$ which have no common cherries. We let $\mathrm{TCLB}_0 = \frac{C}{4}$ (called the cherry bound). Then $\mathrm{TCR}_{\min} \geq \mathrm{TCLB}_0$ (i.e., TCLB_0 is a lower bound).*

Note that cherry bound is also valid when we restrict to binary tree-child networks. We have derived another lower bound based on the number of distinct cherries. Due to the space limit, we omit the details here.

4   Results

We have implemented the lower bounds in the program PIRN, which is down-loadable from https://github.com/yufengwudcs/PIRN. To compute the $TCLB_1$ and $TCLB_2$ bounds, PIRN uses *GLPK*, an open-source ILP solver by default. While GLPK can practically compute $TCLB_1$ for most data we tested, it becomes slow for computing $TCLB_2$ for relatively large data. Our experience shows that $TCLB_2$ can be practically computed using Gurobi, a more power-ful ILP solver, even the data becomes relatively large. However, Gurobi is not open-source. In order to support Gurobi, PIRN outputs the ILP formulation in a file which can be loaded into Gurobi so that $TCLB_2$ can be computed in an interactive way. The results we presented below were computed using Gurobi in this interactive approach.

4.1   Simulation Data

To test the performance of lower bounds, we use the simulation data analyzed in [26]. The simulation data were generated using the approach first developed in [21]. Briefly, we first produced reticulate networks using a simulation scheme similar to the well-known coalescent simulation backwards in time. At each step, there are two possible events: (a) lineage merging (which corresponds to specia-tion), and (b) lineage splitting (which corresponds to reticulation). The relative frequency of these two events (denoted as r) influences the level of reticulation in the simulated network: a larger r will lead to more reticulation events in simulation. The following lists the simulation parameters (Table 1).

Table 1. *A list of parameters and their default values used in the simulation.*

Description	Symbol	Simulated values (default: boldface)
Number of taxa	n	**10**, 20, 50
Reticulation level	r	1.0, **3.0**, 5.0
Number of gene trees	K	10, 50

We used the average over ten replicate data for each simulation settings. The following three lower bounds (all developed in this paper) were evaluated:

1. $TCLB_0$: the cherry bound
2. $TCLB_1$: the practically computable bound by ILP.
3. $TCLB_2$: slower to compute by ILP but usually more accurate bound.

In order to measure the accuracy of lower bounds, ideally we want to compare with the exact tree-child reticulation number. However, these methods tend to be slow for the data we tested. Therefore, we use the following two heuristic upper bounds instead as a rough estimate on tree-child reticulation number.

1. ALTS. This method calculates a heuristic upper bound on tree-child reticulation number.
2. PIRNs. Note: PIRNs outputs a unconstrained network. Since the output network may not be optimal, its reticulation number can occasionally be smaller than the computed lower bounds for tree-child reticulation number. But this is rare.

We use the following statistics for benchmarking various methods.

1. Average value of the (lower/upper) bounds.
2. For each lower bound, the average percentage of differences between a lower bound LB and the ALTS bound UB_a: $\frac{UB_a - LB}{UB_a}$.
3. Running time (in seconds).

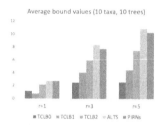 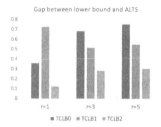

(a) Average lower/upper bound values

(b) Average gap between lower and the ALTS bound (normalized by ALTS)

Fig. 3. *Closeness of three lower bounds (TCLB$_0$, TCLB$_1$ and TCLB$_2$) on 10 trees over 10 taxa under three reticulation levels $r = 1, 3, 5$. Two upper bounds, ALTS and PIRNs, are used for comparison. Part 3(a) shows the average lower/upper bound values. Part 3(b) shows the gap between each of the tree lower bounds and the ALTS bound (divided by the ALTS bounds).*

Figure 3 shows the performance of the tree lower bounds, TCLB$_0$, TCLB$_1$ and TCLB$_2$ on relatively small data (ten gene trees over ten taxa). Our results show that TCLB$_2$ clearly outperforms the other two lower bounds in terms of accuracy. At lower reticulation level ($r = 1$), the gap between TCLB$_2$ and ALTS is only a little over 10%. At higher reticulation levels, the gap between TCLB$_2$ and ALTS is larger but is still much smaller than the other two lower bounds. Recall that ALTS is restricted to tree-child network while PIRNs works with unconstrained networks.

We also examined the closeness of the lower bounds on larger data. We simulated 50 gene trees with varying number of taxa: 10, 20 and 50. Our results (Fig. 4) show that TCLB$_2$ still performs the best among the three lower bounds in term of the accuracy.

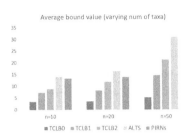
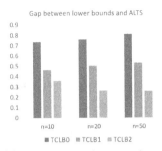

(a) Average lower/upper bound values for larger data

(b) Average gap between lower and the ALTS bound (normalized by ALTS)

Fig. 4. *Performance of three lower bounds (TCLB$_0$, TCLB$_1$ and TCLB$_2$) on larger data. Reticulation level: $r = 3$. 50 gene trees. Vary the number of taxa (n): 10, 20 and 50. Two upper bounds, ALTS and PIRNs, are used for comparison. PIRNs is too slow for $n = 50$, and no result is given for this setting. Part 4(a) shows the average lower/upper bound values. Part 4(b) shows the gap between each of the tree lower bounds and the ALTS bound (divided by the ALTS bounds).*

Time to Compute the Bounds. Figure 5 shows the running time to compute the bounds. We vary the reticulation levels (which may lead to networks with different number of reticulations), and also the number of taxa. Our results show that computing TCLB$_2$ takes longer time than the other two bounds. All lower bounds are faster to compute than the two upper bounds. ALTS is more efficient than PIRNs, while the ALTS bounds tend to be larger than the PIRNs bounds. PIRNs cannot be applied on large data (say $n = 50$). ALTS also appears to be close to its practical range when $n = 50$: there is one instance where ALTS failed to complete the computation by exhausting the memory in a Linux machine with 64 G memory).

More on Large Data. TCLB$_0$ can be easily computed for large data because it is based on simple properties of input trees and can be easily computed in polynomial time. While we don't have a polynomial time algorithm for computing TCLB$_1$, our experience shows that TCLB$_1$ can usually be easily computed even when only an open source ILP solver such as GLPK is used. This can be seen from Fig. 5.

TCLB$_2$ can be practically computed using a state-of-the-art ILP solver such as Gurobi for moderately large data (e.g., 50 gene trees with 50 taxa). As an example, on a dataset with 50 trees (each with 50 taxa), a lower bound of 16 is computed within a few seconds using Gurobi. The TCLB$_1$ bound of 8 blue was computed in a fraction of seconds even with an open source ILP solver. PIRNs took 10 h to compute a unconstrained network with 20 reticulations. ALTS took over 10 min to find a tree-child network with 23 reticulations. While the lower bound doesn't match the best upper bound, the lower bound can provide a range of the solution for large data. We note that Gurobi usually computes TCLB$_2$

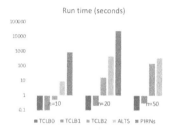

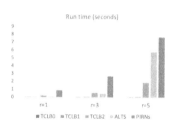

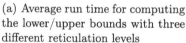

(a) Average run time for computing the lower/upper bounds with three different reticulation levels

(b) Average run time for computing the lower/upper bounds with varying numbers of taxa

Fig. 5. *Running time (in seconds) to compute three lower bounds (*TCLB$_0$, TCLB$_1$ *and TCLB$_2$). Two upper bounds, ALTS and PIRNs, are used for comparison. Part 5(a) shows the average run time (in seconds) for on 10 trees over 10 taxa under three reticulation levels $r = 1, 3, 5$. Part 5(b) shows the average run time (in seconds) for varying numbers of taxa (n): 10, 20 and 50 (reticulation level fixed at $r = 3$ and 50 gene trees).*

much faster than GLPK. Unless the data is small (say with 10 taxa or less), we recommend to use Gurobi.

To test its scalability, we simulated 50 gene trees with 100 taxa. TCLB$_1$ can still be practically computed in less than one second even using GLPK. TCLB$_2$ can be computed using Gurobi, but in a long time. As an example, it took over 10 h for obtaining TCLB$_2 = 48$ on a dataset with 50 simulated tree over 100 taxa. In contrast, TCLB$_1 = 37$ and TCLB$_0 = 15$. Our experience shows that for very large data, the difference between TCLB$_1$ and TCLB$_2$ is not very large. Therefore, TCLB$_1$ can provide a quick estimate on the reticulation number since it can be practically computed for large data, In fact, TCLB$_1$ is perhaps the only practical method that can provide a reasonable strong estimate on reticulation for large data. We are not aware of any other existing approaches for estimating either a lower or upper bound that can be computed for the large simulated data we use here. Here, the large dataset mentioned here has 50 gene trees, 100 taxa and is simulated using reticulation parameter $r = 3.0$ (which can lead to a tree-child reticulation number of over 40).

4.2 Real Biological Data

To evaluate how well our bounds work for real biological data, we test our methods on a grass dataset. The dataset was originally from the Grass Phylogeny Working Group [8] and has been analyzed by a number of papers on phylogenetic networks. There are some variations in the exact form of data, depending on the preprocessing steps performed. The grass data we analyze here have five trees over 14 taxa. Earlier analyses focus on calculating the so-called subtree prune and regraft distances between pairs of these trees [2, 19, 20]. The first attempt for reconstructing phylogenetic network for all five trees is [21]. In [21], the (unconstrained) reticulation number of these fives tree are known to be between 11 (lower bound) and 13 (upper bound). The upper bound was improved to 12 by PIRNs [14]. Regarding to tree-child reticulation number, ALTS found a tree-child network with 13 reticulations. No non-trivial lower bounds for tree-child reticulation number for these five grass trees are known before.

We compute the three lower bounds on the five grass trees. The cherry bound $TCLB_0$ is 2, while the fast ILP bound $TCLB_1$ is 3. These two bounds can be calculated very fast but obviously the bounds are not very precise. It takes 75 seconds to compute $TCLB_2$ using Gurobi, which gives a lower bound of 11. This matches the lower bound in [21]. Note that the lower bound in [21] is based on pairwise distances between the five trees, and takes much longer time to compute: when the number of tree increases, that bound becomes more difficult to compute. Although $TCLB_2$ just provides the same bound as [21], it is close to the currently best upper bound (13). Our results show that $TCLB_2$ can indeed produce good estimates on tree-child reticulation number.

5   Conclusion

Our results show that the lower bounds (especially $TCLB_0$ and $TCLB_1$) are faster to compute than existing upper bounds (namely ALTS) on large data. Our results show that there are trade-offs in accuracy and efficiency when computing lower bounds. The $TCLB_2$ bound is the most accurate, but is also the slowest to compute. The simple cherry bound is very easy to calculate but usually is not very accurate. For large trees, the fast ILP-based $TCLB_1$ bound may be a good choice to obtain quick estimate on tree-child reticulation number. We note that upper bound heuristics such as ALTS can construct a plausible phylogenetic network for the given gene trees, while lower bounds only provide a range of the reticulation number. Still, our lower bounds can provide quick estimate about the reticulation level of a set of phylogenetic trees for large data which is beyond the current feasibility range of existing upper bound methods.

The tree-child network model often allows faster computation. The lower bounds on tree-child reticulation number are much faster to compute than lower bounds [21] on the general reticulation number. There are a number of open questions about lower bounds for tree-child reticulation number. For example, is there a polynomial time algorithm for computing the $TCLB_1$ bound? Can one develop a new lower bound that has better (or similar) accuracy as $TCLB_2$ and is faster to compute?

Acknowledgments. Research is partly supported by U.S. NSF grants CCF-1718093 and IIS-1909425 (to YW) and Singapore MOE Tier 1 grant R-146-000-318-114 (to LZ). The work was started while YW was visiting the Institute for Mathematical Sciences of National University of Singapore in April 2022, which was partly supported by grant R-146-000-318-114.

References

1. Cardona, G., Rossello, F., Valiente, G.: Comparison of tree-child phylogenetic networks. IEEE/ACM Trans Comput. Biol. Bioinf. **6**(4), 552–569 (2009)
2. Bordewich, M., Linz, S., John, K.S., Semple, C.: A reduction algorithm for computing the hybridization number of two trees. Evol. Bioinf. **3**, 86–98 (2007)
3. Cardona, G., Zhang, L.: Counting and enumerating tree-child networks and their subclasses. J. Comput. Syst. Sci. **114**, 84–104 (2020)
4. Chen, Z., Wang, L.: Algorithms for reticulate networks of multiple phylogenetic trees. IEEE/ACM Trans. Comput. Biol. Bioinf. **9**(2), 372–384 (2012)
5. Elworth, R.A.L., Ogilvie, H.A., Zhu, J., Nakhleh, L.: Advances in computational methods for phylogenetic networks in the presence of hybridization. In: Warnow, T. (ed.) Bioinformatics and Phylogenetics. CB, vol. 29, pp. 317–360. Springer, Cham (2019). https://doi.org/10.1007/978-3-030-10837-3_13
6. Fuchs, M., Yu, G.-R., Zhang, L.: On the asymptotic growth of the number of tree-child networks. Eur. J. Comb. **93**, 103278 (2021)
7. Gambette, P., Gunawan, A.D.M., Labarre, A., Vialette, S., Zhang, L.: Locating a tree in a phylogenetic network in quadratic time. In: Przytycka, T.M. (ed.) RECOMB 2015. LNCS, vol. 9029, pp. 96–107. Springer, Cham (2015). https://doi.org/10.1007/978-3-319-16706-0_12
8. Grass Phylogeny Working Group: Phylogeny and subfamilial classification of the grasses (poaceae). Ann. Mo. Bot. Gard. **88**, 373–457 (2001)
9. Gunawan, A.D., DasGupta, B., Zhang, L.: A decomposition theorem and two algorithms for reticulation-visible networks. Inf. Comput. **252**, 161–175 (2017)
10. Gunawan, A.D., Rathin, J., Zhang, L.: Counting and enumerating galled networks. Disc. Appl. Math. **283**, 644–654 (2020)
11. Gusfield, D.: ReCombinatorics: The Algorithmics of Ancestral Recombination Graphs and Explicit Phylogenetic Networks. MIT press, Cambridge (2014)
12. Gusfield, D., Eddhu, S., Langley, C.: The fine structure of galls in phylogenetic networks. Informs J. Comput. **16**(4), 459–469 (2004)
13. Huson, D.H., Rupp, R., Scornavacca, C.: Phylogenetic Networks: Concepts, Algorithms and Applications. Cambridge University Press, Cambridge (2010)
14. Mirzaei, S., Wu, Y.: Fast construction of near parsimonious hybridization networks for multiple phylogenetic trees. IEEE/ACM Trans. Comput. Biol. Bioinf. **13**, 565–570 (2016)
15. Pons, M., Batle, J.: Combinatorial characterization of a certain class of words and a conjectured connection with general subclasses of phylogenetic tree-child networks. Sci. Rep. **11**(1), 1–14 (2021)
16. Steel, M.: Phylogeny: discrete and random processes in evolution. SIAM (2016)
17. van Iersel, L., Janssen, R., Jones, M., Murakami, Y., Zeh, N.: A practical fixed-parameter algorithm for constructing tree-child networks from multiple binary trees. Algorithmica **84**(4), 917–960 (2022)
18. Wang, L., Zhang, K., Zhang, L.: Perfect phylogenetic networks with recombination. J. Comput. Biol. **8**(1), 69–78 (2001)
19. Wu, Y.: A practical method for exact computation of subtree prune and regraft distance. Bioinformatics **25**(190–196), 2009 (2009)
20. Wu, Y., Wang, J.: Fast computation of the exact hybridization number of two phylogenetic trees. In: Borodovsky, M., Gogarten, J.P., Przytycka, T.M., Rajasekaran, S. (eds.) ISBRA 2010. LNCS, vol. 6053, pp. 203–214. Springer, Heidelberg (2010). https://doi.org/10.1007/978-3-642-13078-6_23

21. Wu, Y.: Close lower and upper bounds for the minimum reticulate network of multiple phylogenetic trees. Bioinformatics (supplement issue for ISMB 2010 proceedings) **26**, 140–148 (2010)
22. Wu, Y.: An algorithm for constructing parsimonious hybridization networks with multiple phylogenetic trees. J. Comput. Biol. **20**, 792–804 (2013)
23. Zhang, L.: On tree-based phylogenetic networks. J. Comput. Biol. **23**(7), 553–565 (2016)
24. Zhang, L.: Clusters, trees, and phylogenetic network classes. In: Warnow, T. (ed.) Bioinformatics and Phylogenetics. CB, vol. 29, pp. 277–315. Springer, Cham (2019). https://doi.org/10.1007/978-3-030-10837-3_12
25. Zhang, L.: Generating normal networks via leaf insertion and nearest neighbor interchange. BMC Bioinf. **20**(20), 1–9 (2019)
26. Zhang, L., Abhari, N., Colijn, C., Wu, Y.: A fast and scalable method for inferring phylogenetic networks from trees via aligning the ancestor sequences of taxa. In: To be Presented at RECOMB 2023 Conference (2022)

Finding Agreement Cherry-Reduced Subnetworks in Level-1 Networks

Kaari Landry[1]($\boxtimes$), Olivier Tremblay-Savard[1], and Manuel Lafond[2]

[1] University of Manitoba, Winnipeg, MB, Canada
landryk1@cs.umanitoba.ca
[2] Université de Sherbrooke, Sherbrooke, QC, Canada

Abstract. Phylogenetic networks are increasingly being considered as better suited to represent the complexity of the evolutionary relationships between species. One class of phylogenetic networks that has received a lot of attention recently is the class of orchard networks, which is composed of networks that can be reduced to a single leaf using cherry reductions. Cherry reductions, also called cherry-picking operations, remove either a leaf of a simple cherry (sibling leaves sharing a parent) or a reticulate edge of a reticulate cherry (two leaves whose parents are connected by a reticulate edge). In this paper, we present a fixed-parameter tractable algorithm to solve the problem of finding a maximum agreement cherry-reduced subnetwork (MACRS) between two rooted binary level-1 networks. This is the first exact algorithm proposed to solve the MACRS problem. As proven in earlier work, there is a direct relationship between finding an MACRS and calculating a distance based on cherry operations. As a result, the proposed algorithm also provides a distance that can be used for the comparison of level-1 networks.

Supplementary material for this paper can be found on arXiv.org.

Keywords: Cherry operations · Graphs and networks · Trees · Network problems · Algorithm design and analysis · Biology and genetics · Phylogenetic Networks

1 Introduction

Phylogenetic trees have been used extensively throughout the years to represent simple evolutionary relationships between species. Because of this, many tools and techniques are readily available to efficiently build, compare and evaluate trees. Phylogenetic networks on the other hand are much better suited to represent more complex relationships, such as the ones resulting from hybridization, recombination and lateral gene transfer events [11]. In the last 15 years or so, bioinformatics research has focused increasingly on solving problems related to phylogenetic networks, such as network construction [1, 22–26, 29], minimum hybridization number [2, 3, 8–10, 15, 27], tree/network containment [16, 17, 28], and distance calculation between networks [5, 19, 21].

One crucial concept that has been shown to be a very useful tool in solving several of the important phylogenetic network problems mentioned above is the

K. Jahn and T. Vinař (Eds.): RECOMB-CG 2023, LNBI 13883, pp. 179–195, 2023.
https://doi.org/10.1007/978-3-031-36911-7_12

one of cherry-picking sequences [10, 20]. A cherry-picking sequence is made up of operations that can reduce a network by either removing one leaf of a simple (tree-like) cherry (*i.e.* two leaf siblings descending from the same parent vertex), or removing one reticulate edge of a reticulated cherry (two leaves whose parent vertices are connected by a reticulate edge). The concept of cherry-picking has been so valuable that it led to the definition of *orchard networks*, also known as cherry-picking networks, which are simply phylogenetic networks that can be reduced to a single leaf by cherry-picking operations [6, 17]. Recent work has been focusing on further characterizing and classifying different subtypes of orchard networks [13, 14, 18].

Lately, we have used a generalized definition of *cherry operations* to describe both cherry reductions (*i.e..* cherry picking) and cherry expansions (the reverse of a reduction, which adds a simple or reticulate cherry) [19]. We have then defined four novel distances between orchard networks that are based on cherry operations, with three of them being different formulations of an equivalent distance (construction, deconstruction and tail distances) and the fourth one (mixed distance) being a lower bound for the other three. In the process of describing these distances, the concept of a *maximum agreement cherry-reduced subnetwork* (MACRS – note that we replace cherry-picking used in [19] by cherry-reduced here for clarity) was defined to represent a network contained in both networks being compared that maximizes the number of vertices. We showed that finding an MACRS of two orchard networks was NP-hard, and this was analogous to the problem of calculating the three equivalent distances. Distance calculation between phylogenetic networks is useful in the context of phylogenetic network construction, as it provides a way to measure the discrepancy between a constructed network and a simulated or manually curated one, or any other network produced by an alternative method.

We present an exact fixed-parameter tractable (FPT) algorithm to compute an MACRS of two rooted binary level-1 networks that is exponential in the sum of reticulations present in both networks. More precisely, our algorithm runs in $O(3^r n^3)$, where r is the sum of reticulations and n represents the maximum number of vertices of the input networks. Our approach essentially consists of enumerating a certain set of subnetworks of the input networks in which all possible combinations of reticulation edges have been removed. Then, it makes use of a dynamic programming algorithm that finds whether there is an MACRS (and what it is, if it exists) or not between two level-1 networks in which reticulations that are remaining cannot be removed (we call this problem MACRS-SIMPLE). We prove that the initial MACRS problem can be solved by solving the MACRS-SIMPLE problem on all combinations of enumerated subnetworks.

It is worth noting that another important difference between the previous defining work on MACRS and this article is the definition of networks. Specifically, we allow leaves of the network to have multiple labels. In fact, we force all leaf labels to be conserved as the network is trimmed by cherry reductions by subsuming labels of a removed leaf onto its cherry sibling that remains. In this way, we keep a "memory" of reductions and this compressed representation of networks allows to restore all possible alternative network (bijective) leaf labelings from it.

Finally, we conclude the paper by discussing how the enumeration step could be optimized by considering the relationships between the reticulations of both input networks. We also briefly present a preliminary idea of how the proposed algorithm could be extended to higher level binary networks. Even though the proposed approach applies to orchard networks and not to general networks, the orchard network class actually contains network types that are of interest to the research community, such as the tree-child networks [4] and tree-sibling time-consistent networks [6]. The tree-child networks in particular, in addition to having been studied extensively in the literature, are biologically relevant, since all ancestral species (internal vertices) have a path that can go to a leaf using only tree vertices. This reflects the idea that ancestral species have descendants that will perdure through mutation and speciation events, and that hybridization events are not as common as speciation events [18].

2   Preliminaries

We first introduce the notions regarding networks, then proceed to defining cherry operations and our problem of interest.

2.1   Networks

A *phylogenetic network* $\mathcal{N}$, or a *network* for short, is an acyclic directed graph without vertices of in-degree and out-degree 1, and whose vertices and edges are denoted $V(\mathcal{N})$ and $E(\mathcal{N})$, respectively. We assume that all networks are binary. For $v \in V(\mathcal{N})$, we use v^- and v^+ to denote the in-degree and out-degree of v, respectively. The set $V(\mathcal{N})$ contains

- the *root* $\rho(\mathcal{N})$, which is the unique node satisfying $\rho(\mathcal{N})^- = 0$ and $\rho(\mathcal{N})^+ = 2$. In the case that $|V(\mathcal{N})| = 2$, $\rho(\mathcal{N})^+ = 1$;
- the *leaves* $L(\mathcal{N})$, which satisfy $l^- = 1$ and $l^+ = 0$ for all $l \in L(\mathcal{N})$;
- the *internal* vertices $V(\mathcal{N}) \setminus (L(\mathcal{N}) \cup \{\rho(\mathcal{N})\})$, which contains:
  - the *tree vertices* $T(\mathcal{N})$, which satisfy $v^- = 1$ and $v^+ = 2$ for all $v \in T(\mathcal{N})$;
  - the *reticulation vertices* $R(\mathcal{N})$, or simply *reticulations*, which satisfy $v^- = 2$ and $v^+ = 1$ for all $v \in R(\mathcal{N})$.

We use X to denote the set of all taxa. For our purposes, the leaves of a network $\mathcal{N}$ are labeled by one *or more* taxa. For $l \in L(\mathcal{N})$, we will use $X(l)$ to denote the set of taxa that label l. We require that $X(l) \neq \emptyset$, and that for any distinct leaves $l_1, l_2 \in L(\mathcal{N})$, $X(l_1) \cap X(l_2) = \emptyset$.

The edges directed into a reticulation vertex are called *reticulation edges*, denoted $E_R(\mathcal{N})$. For $v \in V(\mathcal{N})$, the out-neighbors of v are called its *children*. If v has a single in-neighbor, we denote it by $p(v)$ and call it the *parent* of v (if $v \in \{\rho(\mathcal{N})\} \cup R(\mathcal{N})$, then $p(v)$ is undefined). Vertices u and v are *siblings* if $p(u), p(v)$ are defined and $p(u) = p(v)$. When there is a directed path from vertex v to vertex u, we call v an *ancestor* of u and we call u a *descendant* of v. The descendants of v are denoted $reach(v, \mathcal{N})$ while its ancestors are denoted

$reach^-(v,\mathcal{N})$ (note that v itself is in both sets). The union of the labels in $reach(v,\mathcal{N}) \cap L(\mathcal{N})$ is denoted $X(v)$. We denote by $R(v)$ the set of reticulations in $reach(v,\mathcal{N})$.

Two networks $\mathcal{N}_1, \mathcal{N}_2$ are *weakly isomorphic* if there exists a bijection σ : $V(\mathcal{N}_1) \to V(\mathcal{N}_2)$ such that $(u,v) \in E(\mathcal{N}_1)$ if and only if $(\sigma(u), \sigma(v)) \in E(\mathcal{N}_2)$, and such that for each $l \in L(\mathcal{N}_1)$, $X(l) \cap X(\sigma(l)) \neq \emptyset$. For this we use the notation $\mathcal{N} \simeq \mathcal{N}'$. If, for each $l \in L(\mathcal{N}_1)$, $X(l) = X(\sigma(l))$, then we say $\mathcal{N}_1$ and $\mathcal{N}_2$ are *strongly isomorphic* which we denote by $\mathcal{N}_1 = \mathcal{N}_2$.

A network $\mathcal{N}$ may have only one edge whose endpoints are $\rho(\mathcal{N})$ and a leaf. Then $\mathcal{N}$ is a *single-leaf network* or *singleton*. We say $\rho(\mathcal{N})$ *roots* $\mathcal{N}$. If, for a vertex v, and for all vertices $v' \in reach(v,\mathcal{N})$, if every path from $\rho(\mathcal{N})$ to v' goes through v, then we say v roots the subnetwork below it.

While a network $\mathcal{N}$ is directed, there is an undirected version of $\mathcal{N}$ on the same vertex set and with an undirected edge $\{u,v\}$ present for every $(u,v) \in E(\mathcal{N})$ which we call the *underlying graph*. It is on this underlying graph that we identify the set of *biconnected components* of $\mathcal{N}$. Such a component is a maximal subgraph B that cannot be disconnected by the removal of an edge therein. Note that every individual leaf and some tree vertices alone constitute a biconnected component, we refer to such single vertex components as *trivial*, and all others as *non-trivial*. For a set of biconnected components $B_1...B_b$ on a network $\mathcal{N}$, a *bridge* is an edge (u,v) such that $u \in B_i$, $v \in B_j$ for any arbitrary $1 \leq i \neq j \leq b$.

The *level* of a network is the maximum number of reticulations across all biconnected components of a network. A level-k network has no biconnected component with more than k reticulations. A level-1 network has every biconnected component with either 0 or 1 reticulations. This does not limit the number of reticulations over the whole network, just in each biconnected component.

2.2   Cherries and Cherry Reductions

A *cherry* is a pair of leaves that are siblings or that have a reticulation edge joining their parents. More specifically, a pair $(x,y) \in L(\mathcal{N}) \times L(\mathcal{N})$ is called a *cherry* if either $p(x) = p(y)$, in which case (x,y) is called a *simple* cherry, or $p(x) \in R(\mathcal{N})$ and $(p(y), p(x)) \in E(\mathcal{N})$, in which case (x,y) is called a *reticulated* cherry.

Let $\mathcal{N}$ be a network and let (x,y) be a pair of vertices. Then *applying the cherry reduction* (x,y) on $\mathcal{N}$ creates a new network as follows:

– If (x,y) is a simple cherry of $\mathcal{N}$, then the (x,y) *reduction* consists of removing the leaf x and the edge $(p(x), x)$, suppressing the resulting node of in and out-degree 1 if any, and re-assigning $X(y) = X(y) \cup X(x)$. Note that the operation we introduce here differs from the cherry reduction operation described in previous work, where both the leaf x and the set $X(x)$ are deleted. The purpose of our new definition is to preserve a reference to which label *could* have been assigned to y. This is to say that the labels on a given leaf are interchangeable [19, Lemma 3].

- If (x, y) is a reticulated cherry of $\mathcal{N}$, then we remove the reticulation edge $(p(y), p(x))$ and the resulting vertices of in and out-degree 1 are suppressed. In this case, we say that the reticulation edge $(p(y), p(x))$ is *removed* by the cherry reduction (x, y).
- If (x, y) is not a cherry of $\mathcal{N}$, then $\mathcal{N}$ is unchanged.

The resulting graph is a network, and always has a cherry unless it is a singleton. In fact, this is the defining feature of the class of *orchard networks*: an orchard can be continually reduced by cherry reductions until it becomes a singleton [6,17].

See Fig. 1 for an illustration of the two cherry reduction operations, and the concepts of isomorphism.

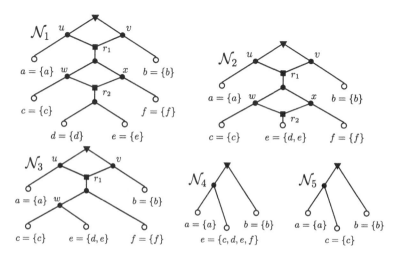

Fig. 1. In this figure, leaves are represented by open circles, tree vertices as filled circles, reticulations as filled squares, and the root of the network as a filled, inverted triangle. Network $\mathcal{N}_1$ is a level-1 network with $|R(\mathcal{N})| = 2$. $\mathcal{N}_1$ is a reticulation-trimmed subnetwork of $\mathcal{N}_1$ with respect to $F = \emptyset$. Network $\mathcal{N}_2 = \mathcal{N}_1\langle (d, e) \rangle$, where (d, e) is a simple cherry/reduction. Network $\mathcal{N}_3 = \mathcal{N}_2\langle (e, f) \rangle$ where (e, f) is a reticulated cherry/reduction. $\mathcal{N}_3$ is reticulation-trimmed subnetwork of $\mathcal{N}_1$ and of $\mathcal{N}_2$ with respect to $F = \{(x, r_2)\}$. Network $\mathcal{N}_4 = \mathcal{N}_3\langle (c, e) \cdot (f, e) \cdot (e, b) \rangle$ and is a reticulation-trimmed subnetwork of $\mathcal{N}_1$ and of $\mathcal{N}_2$ with respect to $F = \{(x, r_2), (v, r_1)\}$ or to $F = \{(w, r_2), (v, r_1)\}$. Network $\mathcal{N}_5 \simeq \mathcal{N}_4$, in fact, there are CSs that may lead to leaf e being any of leaves c, d, e, or f. Each of these networks would have the same label set on that leaf, and all are weakly isomorphic with $\mathcal{N}_5$. There is no reticulation-trimmed subnetwork of $\mathcal{N}_1$ or of $\mathcal{N}_2$ with respect to $F = \{(u, r_1)\}$ or to $F = \{(v, r_1)\}$.

Cherry reductions often occur in batches, and a sequence S of pairs of leaves is called a *cherry sequence* (*CS*). The number of elements in S is denoted $|S|$. The cherry at position i of a CS S is referred to by S_i. We use $\mathcal{N}\langle S \rangle$ to denote the network obtained from $\mathcal{N}$ by first applying cherry reduction S_1 on $\mathcal{N}$, then S_2 on the resulting network, and so on until $S_{|S|}$ is applied. Note that we allow

S to contain pairs that do not modify the network (e.g. non-cherries). The subsequence from (including) the first cherry to (excluding) the ith cherry in S is $S_{(0:i)}$. When a CS S reduces a network $\mathcal{N}$ to a singleton, then we say S is *complete* for $\mathcal{N}$. We assume networks are orchard networks hereafter.

Cherries on a network can be reduced in any order. We restate a theorem of [16] that we adapt to our formalism[1].

Theorem 1. *Let $\mathcal{N}$ be a network, let (x, y) be a cherry of $\mathcal{N}$, and let S be a CS that contains (x, y). Then there exists a CS S' such that $\mathcal{N}\langle S\rangle = \mathcal{N}\langle S'\rangle$, and whose first element is (x, y).*

2.3   Maximum Agreement Cherry-Reduced Subnetworks

For networks $\mathcal{N}$ and $\mathcal{N}'$, when there exists a CS S such that $\mathcal{N}\langle S\rangle \simeq \mathcal{N}'$, we say that $\mathcal{N}'$ is a *cherry-reduced subnetwork* (*CRS*) of $\mathcal{N}$, denoted by $\mathcal{N}' \subseteq_{cr} \mathcal{N}$. We can now define the main problem of focus.

The *Maximum Agreement Cherry-Reduced Subnetwork* (MACRS) problem.
Input: Two orchard networks $\mathcal{N}_1$ and $\mathcal{N}_2$
Find: A network $\mathcal{N}^*$ with the maximum number of vertices that satisfies $\mathcal{N}^* \subseteq_{cr} \mathcal{N}_1$ and $\mathcal{N}^* \subseteq_{cr} \mathcal{N}_2$

We call a solution $\mathcal{N}^*$ to the above problem an MACRS of $\mathcal{N}_1$ and $\mathcal{N}_2$.

3   An MACRS Algorithm on Level-1 Networks

We show that the MACRS problem can be solved in time $O(3^r n^3)$ for $n = \max(|V(\mathcal{N}_1)|, |V(\mathcal{N}_2)|)$, and $r = |R(\mathcal{N}_1)| + |R(\mathcal{N}_2)|$ on level-1 networks. We employ a two-step strategy. We first enumerate a number of inputs that have been specially reduced to a selected set of remaining reticulations. Second, these inputs are provided to a cubic time dynamic programming algorithm on an easier version of MACRS that uses only simple reductions. Because the number of special inputs is limited by 3^r, we get an FPT algorithm. MACRS is thus split into two subproblems. We first introduce them and show how they can be used to solve MACRS. The later sections then focus on each problem separately.

Let $\mathcal{N}$ be a network and let $F \subseteq E_R(\mathcal{N})$ be a subset of reticulation edges. We wish to generate all the maximal cherry-reduced subnetworks of $\mathcal{N}$ under the restriction that the reticulation edges removed by cherry operations coincide with F. Thus, we say that a network $\mathcal{N}'$ is a *reticulation-trimmed subnetwork* of $\mathcal{N}$ *with respect to F* if there exists a CS S such that $\mathcal{N}\langle S\rangle = \mathcal{N}'$, and such that $(u, v) \in F$ if and only if S contains a reticulated cherry reduction that removes (u, v), and S is of minimum length i.e. we require that there is no other CS S' with $|S'| < |S|$ that satisfies the same properties.

---

[1] Note that the authors prove the statement under the assumption that S is complete, and that leaves are single-labeled. However the proof is easy to adapt to our context.

Furthermore, we say that $\mathcal{N}'$ is a *reticulation-trimmed subnetwork* of $\mathcal{N}$ if there exists a set $F \subseteq E_R(\mathcal{N})$ such that $\mathcal{N}'$ is a reticulation-trimmed subnetwork of $\mathcal{N}$ with respect to F.

The RETICULATION-TRIMMED ENUMERATION problem:
Input: An orchard network $\mathcal{N}$.
Find: The set of all reticulation-trimmed subnetworks of $\mathcal{N}$.

Note that the size of the set of reticulation-trimmed subnetworks depends heavily on the network structure. For instance, it is possible to show that it is linear when all reticulations are arranged in a path, and exponential when all reticulations are independent (none is an ancestor of the other). It is possible to calculate the size of this set exactly by algorithmic means through an abstraction of the network structure. However, we reserve the analysis of the impact of this parameter on our algorithm for future work.

Once the set of edges to remove by reticulation reductions have been guessed, it remains to infer the set of non-reticulated cherry operations. A *simple CS* is a CS that contains only simple cherries. In this way, $R(\mathcal{N}) = R(\mathcal{N}\langle S\rangle)$ for any simple CS S. For networks $\mathcal{N}$ and $\mathcal{N}'$, when there exists a simple CS S such that $\mathcal{N}\langle S\rangle \simeq \mathcal{N}'$ we say that $\mathcal{N}'$ is a CRS-SIMPLE of $\mathcal{N}$. Note that owing to our definition of weak isomorphism, $\mathcal{N}\langle S\rangle \simeq \mathcal{N}'$ does not mean that S transforms $\mathcal{N}$ into $\mathcal{N}'$. A better intuition would rather be that after applying S on $\mathcal{N}$, we could choose one label in the label set of each leaf of $N\langle S\rangle$ and of $\mathcal{N}'$, such that the resulting networks would be isomorphic in the traditional sense.

The *Simple Maximum Agreement Cherry-Reduced Subnetwork* (MACRS-SIMPLE) problem.
Input: Two orchard networks $\mathcal{N}_1$ and $\mathcal{N}_2$.
Find: A network $\mathcal{N}^*$ with a maximum number of vertices such that $\mathcal{N}^*$ is a CRS-SIMPLE of $\mathcal{N}_1$ and a CRS-SIMPLE of $\mathcal{N}_2$.

A solution $\mathcal{N}^*$ to the above problem will be called a MACRS-SIMPLE of $\mathcal{N}_1$ and $\mathcal{N}_2$.

For the standard MACRS problem on networks $\mathcal{N}_1$ and $\mathcal{N}_2$, there is always a solution as long as $X(\mathcal{N}_1) \cap X(\mathcal{N}_2) \neq \emptyset$, however since reticulations cannot be removed by simple CS, the MACRS-SIMPLE problem may not have a solution (for instance when the two networks have a different number of reticulation vertices). We can now describe our main algorithm, where we assume that the MACRS-SIMPLE routine correctly returns an optimal solution.

Algorithm 1. MACRS Finder

    Input Two networks $\mathcal{N}_1$ and $\mathcal{N}_2$
    Output A MACRS of $\mathcal{N}_1$ and $\mathcal{N}_2$
1: $\tilde{\mathcal{N}} \leftarrow$ empty network
2: **for each** reticulation-trimmed subnetwork $\mathcal{N}_1'$ of $\mathcal{N}_1$ **do**
3:     **for each** reticulation-trimmed subnetwork $\mathcal{N}_2'$ of $\mathcal{N}_2$ **do**
4:         Let $\mathcal{N}'$ be a MACRS-SIMPLE of $\mathcal{N}_1'$ and $\mathcal{N}_2'$
5:         **if** $\mathcal{N}'$ exists and $|V(\mathcal{N}')| > |V(\tilde{\mathcal{N}})|$ **then** $\tilde{\mathcal{N}} \leftarrow \mathcal{N}'$
6:     **end for**
7: **end for**
8: return $\tilde{\mathcal{N}}$

An optimization technique is evident here: as we mentioned, there is only a solution to MACRS-SIMPLE $(\mathcal{N}_1, \mathcal{N}_2)$ when $|R(\mathcal{N}_1)| = |R(\mathcal{N}_2)|$ since only simple reductions will be performed. Thus, we need only test such pairs. We do not expect this optimization to provide any theoretical improvement in general and it is not included in the analysis of bounds presented here. It may, however, provide an improvement in practice, a test which we defer to future work.

In the remainder of this section, we focus on proving that this algorithm works correctly. We will deal with the complexity of the algorithm once we have dealt with the RETICULATION-TRIMMED ENUMERATION and MACRS-SIMPLE subproblems. We begin by showing that one can always obtain a subnetwork by first going through a reticulation-trimmed subnetwork, and then using only simple cherry reductions.

Lemma 1. *Let $\mathcal{N}$ be a network. Then for any $\mathcal{N}' \subseteq_{cr} \mathcal{N}$, there exists a reticulation-trimmed subnetwork $\mathcal{N}''$ of $\mathcal{N}$ and a simple CS S such that $\mathcal{N}''\langle S \rangle = \mathcal{N}'$.*

Theorem 2. *Algorithm 1 correctly finds a MACRS of $\mathcal{N}_1$ and $\mathcal{N}_2$.*

Proof. Let $\mathcal{N}^*$ be a MACRS of $\mathcal{N}_1$ and $\mathcal{N}_2$. Let $\tilde{\mathcal{N}}$ be the network returned by Algorithm 1. We first claim that if $\tilde{\mathcal{N}}$ is non-empty, it does satisfy $\tilde{\mathcal{N}} \subseteq_{cr} \mathcal{N}_1, \mathcal{N}_2$. To see this, note that every pair $\mathcal{N}_1', \mathcal{N}_2'$ of networks enumerated by Algorithm 1 satisfy $\mathcal{N}_1' \subseteq_{cr} \mathcal{N}_1$ and $\mathcal{N}_2' \subseteq_{cr} \mathcal{N}_2$, by the definition of reticulation-trimmed subnetworks. Moreover, if a MACRS-SIMPLE $\mathcal{N}'$ of $\mathcal{N}_1', \mathcal{N}_2'$ exists, then by transitivity, $\mathcal{N}' \subseteq_{cr} \mathcal{N}_1' \subseteq_{cr} \mathcal{N}_1$ and $\mathcal{N}' \subseteq_{cr} \mathcal{N}_2' \subseteq_{cr} \mathcal{N}_2$. Since $\tilde{\mathcal{N}}$ is one of those $\mathcal{N}'$, this proves our claim. $\square$

Let us now focus on the optimality of $\tilde{\mathcal{N}}$. First note that $|V(\mathcal{N}^*)| \geq |V(\tilde{\mathcal{N}})|$: if $\tilde{\mathcal{N}}$ is an empty network, this is obvious, and otherwise, by our above claim, $\tilde{\mathcal{N}}$ is a cherry-reduced subnetwork of $\mathcal{N}_1$ and $\mathcal{N}_2$ and thus cannot be larger than $\mathcal{N}^*$.

Let us now show that $|V(\mathcal{N}^*)| \leq |V(\tilde{\mathcal{N}})|$. By Lemma 1, there exists a reticulation-trimmed subnetwork $\mathcal{N}_1'$ of $\mathcal{N}_1$ (resp. $\mathcal{N}_2'$ of $\mathcal{N}_2$) such that $\mathcal{N}^*$ can be obtained from $\mathcal{N}_1'$ (resp. $\mathcal{N}_2'$) using only simple CSs. Thus, $\mathcal{N}^*$ is a CRS-SIMPLE of $\mathcal{N}_1'$ and $\mathcal{N}_2'$. Algorithm 1 will eventually enumerate $\mathcal{N}_1'$ and $\mathcal{N}_2'$ and find a MACRS-SIMPLE $\mathcal{N}'$ of them, which is of maximum size and thus has at least as many vertices as $\mathcal{N}^*$. Since the returned $\tilde{\mathcal{N}}$ is the $\mathcal{N}'$ of maximum size found by the algorithm, it follows that $|V(\mathcal{N}^*)| \leq |V(\tilde{\mathcal{N}})|$.

4   Subroutines

4.1   Enumerating the Set of Reticulation-Trimmed Subnetworks

We now show how to enumerate the set of all reticulation-trimmed subnetworks of a network $\mathcal{N}$ in time $O(3^{|R(\mathcal{N})|}|V(\mathcal{N})|)$. The reticulation-trimmed subnetworks are characterized by having no more reductions than what sufficiently removes the desired reticulation edges. Luckily, we will see that at most one such network can exist; we must only remove the complete subnetwork under both endpoints of the reduced reticulation edge. This is guaranteed possible by cherry reductions, assuming all reticulations below these endpoints have also been specified for removal. Algorithm 2 shows how to enumerate the relevant edges, and uses Algorithm 3 as a subroutine, which finds the reticulation-trimmed subnetwork with respect to a given edge set. We show that the reticulation-trimmed subnetwork of $\mathcal{N}$ with respect to $F \subseteq R(\mathcal{N})$ is uniquely defined in Lemma 4. We say that a set of edges F is *disjoint* if, for any two distinct edges $(u, v), (x, y) \in F$, $\{u, v\} \cap \{x, y\} = \emptyset$.

Algorithm 2. REDUCED-SET-FINDER

    Input A network $\mathcal{N}$
    Output The set of all reticulation-trimmed subnetworks of $\mathcal{N}$
1: **for each** $F \in \{\mathcal{P}(E_R(\mathcal{N})) : (a, b), (c, b) \notin F \text{ for any } a, b, c\}$ **do**
2:     $\mathbf{N} \leftarrow \mathbf{N} \cup \text{RT-SUBNET-MAKER}(\mathcal{N}, F)$
3: **end for**
4: **return** $\mathbf{N}$

Lemma 2. *Let $\mathcal{N}$ be a network, $F \subseteq E_R(\mathcal{N})$ be a set, and $\mathcal{N}'$ be a network that is a reticulation-trimmed subnetwork of $\mathcal{N}$ with respect to F. Then F is disjoint.*

Next, for $F \subseteq E_R(\mathcal{N})$, a topological sort of F is an ordering of its element such that for distinct edges $e_1, e_2 \in F$, if there is a path from a vertex of e_1 to a vertex of e_2 in $\mathcal{N}$, then e_1 comes later than e_2 in this ordering.

Lemma 3. *Let $\mathcal{N}$ be a network and $F \subseteq E_R(\mathcal{N})$ be a set such that there exists a reticulation-trimmed subnetwork of $\mathcal{N}$ with respect to F. Then there exists a topological sort of F.*

The next lemma is crucial, as it shows that reticulation-trimmed subnetworks with respect to a given F are either unique, or do not exist. This allows us to enumerate in reasonable time.

Lemma 4. *Let $\mathcal{N}$ be a network and let $F \subseteq E_R(\mathcal{N})$. Then there does not exist two non-strongly isomorphic reticulation-trimmed subnetworks of $\mathcal{N}$ with respect to F.*

For an example, given a network $\mathcal{N}$, of an $F \subseteq E_R(\mathcal{N})$ that does not admit a reticulation-trimmed subnetwork of $\mathcal{N}$, consider $\mathcal{N}$ with 2 reticulations, r_1, r_2 such that $r_1 \in reach^-(r_2)$. Choosing $F = \{(p_1, r_1)\}$, for p_1 chosen arbitrarily between r_1's parents, will not admit a reticulation-trimmed subnetwork since reticulation r_2 must have leaves below its endpoints to be in a cherry, but this choice of F has no corresponding reticulated reductions of r_1 making it impossible to construct a CS S that reduces only r_2.

We next describe Algorithm 3, which produces the reticulation-trimmed networks with respect to some given F, see Fig. 2 for an illustration.

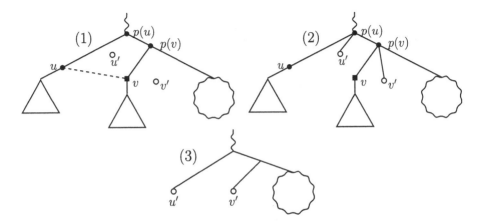

Fig. 2. In this figure, leaves are represented by open circles, tree vertices as filled circles, reticulations as filled squares. A subnetwork without reticulations is represented by a large open triangle, a subnetwork that may be reticulated is represented by a large open blob. This Figure shows an example of the operation of Algorithm 3, note how $R(u) \cup R(v) \setminus \{v\} = \emptyset$ in this example. Subnetwork under label (1) is an example network at line 7, the dotted line represents the removed reticulation edge (u, v) by line 5 and both leaves u' and v' have been constructed (leaf labels are not shown). The network under label (2) shows the state of network (1) at line 8 when edges $(p(u), u')$ and $(p(v), v')$ have been added. The network under label (3) shows the state of the network under (1) at line 9 when vertices in $reach(u, \mathcal{N}') \cup reach(v, \mathcal{N}')$ are removed.

Lemma 5. *Algorithm 3 on $(\mathcal{N}, F)$ returns the reticulation-trimmed subnetwork $\mathcal{N}'$ of $\mathcal{N}$ with respect to F if it exists, and NULL if not, and runs in time $O(|V(\mathcal{N})|)$.*

Theorem 3. *Algorithm 2 correctly enumerates all reticulation-trimmed subnetworks of a network $\mathcal{N}$, and runs in time $O(3^{|R(\mathcal{N})|}|V(\mathcal{N})|)$.*

Algorithm 3. RT-SUBNET-MAKER

    Input A network $\mathcal{N}$ and a disjoint set $F \subseteq E_R(\mathcal{N})$

    Output the reticulation-trimmed subnetwork of $\mathcal{N}$ with respect to F, or NULL if it does not exist

1: $\mathcal{N}' \leftarrow \mathcal{N}$

2: Find a topological sort F' of F

3: **for each** $(u, v) \in F'$ in order **do**

4:     **if** $R(u) \cup R(v) \setminus \{v\} = \emptyset$ **then**

5:         delete edge (u, v)

6:         construct leaf u' such that $X(u') = X(u)$

7:         construct leaf v' such that $X(v') = X(v)$

8:         add edges $(p(u), u')$ and $(p(v), v')$ to $\mathcal{N}'$

9:         remove all vertices in $reach(u, \mathcal{N}') \cup reach(v, \mathcal{N}')$

10:    **else**

11:        return NULL

12:    **end if**

13: **end for**

14: return $\mathcal{N}'$

Proof. It is already proved (Lemma 2) that non-disjoint F does not admit a reticulation-trimmed subnetwork, so it is correct to filter those. The remaining correctness follows from the exhaustive nature of the construction of all F and by the correctness of Algorithm 3.   □

    As for the time complexity, filtering non-disjoint F implies a threefold choice on each reticulation (we either include one, or none of its incoming edges, but not both by disjointness). Thus the size of the set is $O(3^{|R(\mathcal{N})|})$. Recalling that Algorithm 3 can be implemented in time $O(|V(\mathcal{N})|)$, the total runtime for Algorithm 2 is in $O(3^{|R(\mathcal{N})|}|V(\mathcal{N})|)$.

4.2 An Algorithm for MACRS-SIMPLE

A dynamic programming algorithm that solves the MACRS-SIMPLE problem in cubic time is given and proved in this section.

    Assume we have networks $\mathcal{N}_1$ and $\mathcal{N}_2$ as input to the MACRS-SIMPLE problem. We assume that we have computed the set of biconnected components of $\mathcal{N}_1$ and $\mathcal{N}_2$ in a preprocessing step, along with the bridge edges. This can be done in time $O(|V(\mathcal{N})|)$, see [7]. Since the networks considered are level-1, each biconnected component B contains exactly one vertex u that has no in-neighbor in B, and exactly one vertex r that has no out-neighbor in B. If B is trivial, then $u = r$, and otherwise r is a reticulation vertex and there are two edge-disjoint paths from u to r in B [12]. We refer to these two paths as *component paths*. The vertex u will be called the *root* of B and denoted $\rho(B)$, and r will be called the *bottom* of B. We let $\mathcal{B}_1$ be the set of biconnected components of $\mathcal{N}_1$ and $\mathcal{B}_2$ be the set of biconnected components of $\mathcal{N}_2$. Finally for $i \in \{1, 2\}$, we denote $\rho(\mathcal{B}_i) = \{\rho(B) : B \in \mathcal{B}_i\}$, i.e. the set of roots in $\mathcal{B}_i$.

Using dynamic programming, we construct a table M whose rows are the roots in $\rho(\mathcal{B}_1)$ and whose columns are the roots in $\rho(\mathcal{B}_2)$. For $u \in \rho(\mathcal{B}_1), v \in \rho(\mathcal{B}_2)$, we define $\mathcal{N}_u$ as the subnetwork of $\mathcal{N}_1$ rooted at u, and $\mathcal{N}_v$ as the subnetwork of $\mathcal{N}_2$ rooted at v. We then define $M[u, v]$ as the number of *leaves* in a MACRS-SIMPLE of $\mathcal{N}_u$ and $\mathcal{N}_v$. If u is a tree vertex, its children are denoted u_1 and u_2.

In $\mathcal{N}_1$, we denote the two component paths on the same non-trivial biconnected component by $\pi_l^1 = p_{l,1}^1 \ldots$ and $\pi_r^1 = p_{r,1}^1 \ldots$ and in $\mathcal{N}_2$ these paths will be denoted $\pi_l^2 = p_{l,1}^2 \ldots$ and $\pi_r^2 = p_{r,1}^2 \ldots$. For a vertex p_i on path $\pi = p_1 \ldots$, let $h(p_i)$ be the child vertex of p_i such that $h(p_i) \neq p_{i+1}$. In other words, the edge $(p_i, h(p_i))$ is a bridge pendant π leading to a different biconnected component where $h(p_i)$ is rooting a distinct subnetwork. See Fig. 3 for an illustration of the component paths and the described labelings for an example $\mathcal{N}_1$ network.

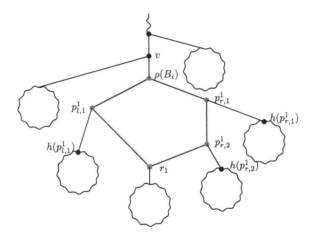

Fig. 3. In this figure, tree vertices are filled circles and reticulations are filled squares. A subnetwork is represented by a large open blob. Vertices in red are in the same non-trivial biconnected component. Yellow edges are path π_l^1 and green edges are path π_r^1. Tree vertex v is a trivial biconnected component itself such that $R(v) \neq \emptyset$.

We use Algorithm 4 to compute $M[u, v]$ for each $u \in \rho(\mathcal{B}_1), v \in \rho(\mathcal{B}_2)$ in postorder. The algorithm considers the possible cases for u and v. If one of them is a leaf, say u, then either we can reduce $\mathcal{N}_v$ to a leaf whose label intersects with u, or not (line 3). If u and v are trivial components, with children u_1, u_2 and v_1, v_2, respectively, an agreement subnetwork can either be obtained by "combining" agreement subnetworks between $\mathcal{N}_{u_1}, \mathcal{N}_{v_1}$ and between $\mathcal{N}_{u_2}, \mathcal{N}_{v_2}$, or "combining" agreement subnetworks between $\mathcal{N}_{u_1}, \mathcal{N}_{v_2}$ and between $\mathcal{N}_{u_2}, \mathcal{N}_{v_1}$. These cases are evaluated on line 6, with special cases when such combinations are not possible. When u is trivial and not v (or vice-versa), no such construction is possible and line 9 returns $-\infty$. When u and v are non-trivial, instead of mapping children of u and v as in the trivial component case, we map the child

---

Algorithm 4.

Input: Two networks $\mathcal{N}_1$, $\mathcal{N}_2$, vertices $u \in \rho(\mathcal{B}_1), v \in \rho(\mathcal{B}_2)$
Output: $M[u, v]$

1: **if** both u and v are trivial components **then**
2:     **if** u or v is a leaf **then**
3:

$$M[u, v] = \begin{cases} 1 & \text{if } X(u) \cap X(v) \neq \emptyset \text{ and } R(u) \cup R(v) = \emptyset \\ -\infty & \text{otherwise} \end{cases}$$

4:     **else**
5:         for each $i \in \{1, 2\}, j \in \{1, 2\}$, define $X_{ij} = X(u_i) \cap X(v_j)$
6:         $M[u, v] = max(M_1, M_2)$ where

$$M_1 = \begin{cases} 1 & \text{if } X_{11} \neq \emptyset \text{ and } X_{22} = \emptyset \text{ and } R(u) \cup R(v) = \emptyset \\ 1 & \text{if } X_{11} = \emptyset \text{ and } X_{22} \neq \emptyset \text{ and } R(u) \cup R(v) = \emptyset \\ M[u_1, v_1] + M[u_2, v_2] & \text{if } X_{11} \neq \emptyset \text{ and } X_{22} \neq \emptyset \\ -\infty & \text{otherwise} \end{cases}$$

$$M_2 = \begin{cases} 1 & \text{if } X_{12} \neq \emptyset \text{ and } X_{21} = \emptyset \text{ and } R(u) \cup R(v) = \emptyset \\ 1 & \text{if } X_{12} = \emptyset \text{ and } X_{21} \neq \emptyset \text{ and } R(u) \cup R(v) = \emptyset \\ M[u_1, v_2] + M[u_2, v_1] & \text{if } X_{12} \neq \emptyset \text{ and } X_{21} \neq \emptyset \\ -\infty & \text{otherwise} \end{cases}$$

7:     **end if**
8: **else if** u is a trivial biconnected component and v is in a non-trivial biconnected component (or vice versa) **then**
9:     $M[u, v] = -\infty$
10: **else** u and v are in non-trivial components with reticulations r_1, r_2 respectively and component paths $\pi_l^1, \pi_r^1, \pi_l^2, \pi_r^2$
11:     $M_1 = -\infty$
12:     $M_2 = -\infty$
13:     **if** $|\pi_l^1| = |\pi_l^2|$ and $|\pi_r^1| = |\pi_r^2|$ **then**
14:

$$M_1 = M[r_1, r_2] + \sum_{i=1}^{|\pi_l^1|} M[h(p_{l,i}^1), h(p_{l,i}^2)] + \sum_{i=1}^{|\pi_r^1|} M[h(p_{r,i}^1), h(p_{r,i}^2)]$$

15:     **end if**
16:     **if** $|\pi_l^1| = |\pi_r^2|$ and $|\pi_r^1| = |\pi_l^2|$ **then**
17:

$$M_2 = M[r_1, r_2] + \sum_{i=1}^{|\pi_l^1|} M[h(p_{l,i}^1), h(p_{r,i}^2)] + \sum_{i=1}^{|\pi_r^1|} M[h(p_{r,i}^1), h(p_{l,i}^2)]$$

18:     **end if**
19:     $M[u, v] = max(M_1, M_2)$
20: **end if**

---

paths under u and v that lead to their descending reticulation (there may be two ways to map these paths). The smaller subnetworks to be combined are those "dangling" outside of these paths, and the possibilities are calculated on lines 13–19.

We seek the result $M[\rho(\mathcal{N}_1), \rho(\mathcal{N}_2)] + |R(\mathcal{N}_1)|$ as M records only the number of leaves in an MACRS-SIMPLE of $\mathcal{N}_1$ and $\mathcal{N}_2$. From this information we can calculate more about the general size of the network because they are binary, $|V(\mathcal{N})| = 2|L(\mathcal{N})| + 2|R(\mathcal{N})| - 1$. Luckily, the number of reticulations in the solution is known ahead of time since it must have the same number of reticulations as each of the inputs. Note that we can also reconstruct the network that corresponds to the optimal size of the MACRS-SIMPLE of $\mathcal{N}_1$ and $\mathcal{N}_2$ by performing a traceback in the dynamic programming table.

Theorem 4. *Algorithm 4 runs in time $O(|V(\mathcal{N}_1)||V(\mathcal{N}_2)|(|V(\mathcal{N}_1)| + |V(\mathcal{N}_2)|))$.*

Proof. The algorithm fills a table M, a table of maximum size $|V(\mathcal{N}_1)||V(\mathcal{N}_2)|$, thus if we can show each table entry is calculated in at most linear $(|V(\mathcal{N}_1)| + |V(\mathcal{N}_2)|)$ time, then the algorithm is cubic as claimed. □

The preprocessing step to determine and label biconnected components is linear as it requires a modified depth-first search [7]. Then, the calculations being performed for lines 1 through 9 consist of finding and checking the labelled components (linear), and checking up to 12 set intersections (linear) of a vertex's descendants leaves (linear to find). Lines 10 and on perform a linear number of table lookups/calls. The paths themselves are also linear to find as they are simply the paths that leave each child of the rooting vertex of the biconnected component and end on the next reticulation, the length of which can also be calculated on a single pass. Thus the claim holds.

Theorem 5. *The entry $M[\rho(\mathcal{N}_1), \rho(\mathcal{N}_2)]$ correctly contains $|L(\mathcal{N}^*)|$ for $\mathcal{N}^*$ a MACRS-SIMPLE of $\mathcal{N}_1$ and $\mathcal{N}_2$ if one exists, and $-\infty$ otherwise.*

4.3   Complexity of Algorithm 1

Theorem 6. *Let $\mathcal{N}_1, \mathcal{N}_2$ be two networks, let $n = \max(|V(\mathcal{N}_1)|, |V(\mathcal{N}_2)|)$, and $r = |R(\mathcal{N}_1)| + |R(\mathcal{N}_2)|$. Then the MACRS problem can be solved in time $O(3^r n^3)$.*

Proof. By Theorem 3, Algorithm 2 can enumerate all reticulation-trimmed subnetworks of $\mathcal{N}_1$ and $\mathcal{N}_2$ in total time $O(3^{|R(\mathcal{N}_1)|}n + 3^{|R(\mathcal{N}_2)|}n) = O(3^r n)$. The number of pairs of such networks for which we compute a MACRS-SIMPLE is $O(3^r)$, each of which can be handled in time $O(n^3)$ by Theorem 4. The total running time is thus $O(3^r n + 3^r n^3) = O(3^r n^3)$. □

5   Conclusion and Discussion

In this paper, we presented the first exact algorithm to find an MACRS of two rooted binary level-1 networks. The proposed approach starts by enumerating all reticulation-trimmed subnetworks for both input networks, and then compares all the possible pairs produced for each input network using a dynamic programming algorithm for the MACRS-SIMPLE problem. The enumeration step presented here is currently exponential in the sum of reticulation numbers of both input networks, and the MACRS-SIMPLE algorithm takes cubic time in the maximum number of vertices contained in the input networks.

In addition to the benefit of being able to extract a common subnetwork structure of maximum size from two orchard networks, the proposed algorithm permits to find a measure of the amount of differences between them. As shown in our previous work [19], there is a direct correspondence between finding an MACRS (more specifically, its size) and calculating one of the three equivalent distances presented in that work. As such, the algorithm presented here provides a first method to calculate exactly these distances. This can be used in the future to compare this distance with other distances (such as the mixed distance) or to evaluate the accuracy of different heuristic approaches.

Future Extensions

There is an obvious optimization to the approach presented in this work on the enumeration of the reticulation-trimmed subnetworks. Since the MACRS-SIMPLE algorithm by definition does not remove reticulations, comparing two input reticulation-trimmed subnetworks that do not share the same reticulation number or topology (in the sense that no mapping of the components containing reticulations can be made) will result in no solution. An obvious improvement to the enumeration step is to compare the topological relationships of the reticulations in both input networks (which, in the case of level-1 networks, can be modelled by trees), find the largest common reticulation topology between them, and start enumerating from there by gradually removing all possible reticulations. While this strategy does not achieve any additional formal bounding, it may reduce greatly the number of reticulation-trimmed subnetwork pairs to consider on many real inputs (potentially down to a linear number of pairs).

Another interesting avenue of work is to generalize our algorithm to higher level networks. A possible strategy would be to extend the MACRS-SIMPLE dynamic programming to consider, for each pair of biconnected components, all possible isomorphisms, find the maximum value and then summing to it the values of the exterior nodes that are matched in the isomorphism.

Attaching leaves to a non-orchard network was used previously to extend an approach to solve the minimum hybridization problem on any rooted phylogenetic network [20]. Exploring if and how a similar idea could be employed to generalize our proposed algorithm to non-orchard networks should be considered.

Finally, as mentioned earlier, the complexity of our method is exponential in the sum of the number of reticulations in both input networks because of the enumeration step. Ideally, we could find an approach for which the complexity would depend only on the level of the two input networks, which we leave as an open problem.

References

1. Allen-Savietta, C.: Estimating Phylogenetic Networks from Concatenated Sequence Alignments. The University of Wisconsin-Madison (2020)
2. Baroni, M., Semple, C., Steel, M.: A framework for representing reticulate evolution. Ann. Comb. **8**, 391–408 (2005)
3. Bernardini, G., van Iersel, L., Julien, E., Stougie, L.: Reconstructing phylogenetic networks via cherry picking and machine learning. In: 2nd International Workshop on Algorithms in Bioinformatics, WABI 2022 (2022)
4. Bordewich, M., Semple, C.: Determining phylogenetic networks from inter-taxa distances. J. Math. Biol. **73**(2), 283–303 (2016)
5. Cardona, G., Llabrés, M., Rosselló, F., Valiente, G.: Metrics for phylogenetic networks I: generalizations of the Robinson-Foulds metric. IEEE/ACM Trans. Comput. Biol. Bioinf. **6**(1), 46–61 (2008)
6. Erdős, P.L., Semple, C., Steel, M.: A class of phylogenetic networks reconstructable from ancestral profiles. Math. Biosci. **313**, 33–40 (2019)
7. Hopcroft, J.E., Tarjan, R.E.: Dividing a graph into triconnected components. SIAM J. Comput. **2**(3), 135–158 (1973)
8. Huber, K.T., Linz, S., Moulton, V.: The rigid hybrid number for two phylogenetic trees. J. Math. Biol. **82**(5), 1–29 (2021). https://doi.org/10.1007/s00285-021-01594-2
9. Huber, K.T., Linz, S., Moulton, V.: Cherry picking in forests: a new characterization for the unrooted hybrid number of two phylogenetic trees. arXiv preprint arXiv:2212.08145 (2022)
10. Humphries, P.J., Linz, S., Semple, C.: Cherry picking: a characterization of the temporal hybridization number for a set of phylogenies. Bull. Math. Biol. **75**(10), 1879–1890 (2013)
11. Huson, D.H., Bryant, D.: Application of phylogenetic networks in evolutionary studies. Mol. Biol. Evol. **23**(2), 254–267 (2005). https://doi.org/10.1093/molbev/msj030
12. Huson, D.H., Rupp, R., Scornavacca, C.: Phylogenetic Networks: Concepts, Algorithms and Applications. Cambridge University Press (2010)
13. van Iersel, L., Janssen, R., Jones, M., Murakami, Y.: Orchard networks are trees with additional horizontal arcs. Bull. Math. Biol. **84**(8), 76 (2022)
14. van Iersel, L., Janssen, R., Jones, M., Murakami, Y., Zeh, N.: A unifying characterization of tree-based networks and orchard networks using cherry covers. Adv. Appl. Math. **129**, 102222 (2021)
15. Janssen, R., Jones, M., Murakami, Y.: Combining networks using cherry picking sequences. In: Martín-Vide, C., Vega-Rodríguez, M.A., Wheeler, T. (eds.) AlCoB 2020. LNCS, vol. 12099, pp. 77–92. Springer, Cham (2020). https://doi.org/10.1007/978-3-030-42266-0_7
16. Janssen, R., Murakami, Y.: Linear time algorithm for tree-child network containment. In: Martín-Vide, C., Vega-Rodríguez, M.A., Wheeler, T. (eds.) AlCoB 2020. LNCS, vol. 12099, pp. 93–107. Springer, Cham (2020). https://doi.org/10.1007/978-3-030-42266-0_8
17. Janssen, R., Murakami, Y.: On cherry-picking and network containment. Theoret. Comput. Sci. **856**, 121–150 (2021)
18. Kong, S., Pons, J.C., Kubatko, L., Wicke, K.: Classes of explicit phylogenetic networks and their biological and mathematical significance. J. Math. Biol. **84**(6), 47 (2022)

19. Landry, K., Teodocio, A., Lafond, M., Tremblay-Savard, O.: Defining phyloge-
    netic network distances using cherry operations. IEEE/ACM Trans. Comput. Biol.
    Bioinf. **20**, 1654–1666 (2022)
20. Linz, S., Semple, C.: Attaching leaves and picking cherries to characterise the
    hybridisation number for a set of phylogenies. Adv. Appl. Math. **105**, 102–129
    (2019)
21. Lu, B., Zhang, L., Leong, H.W.: A program to compute the soft Robinson-Foulds
    distance between phylogenetic networks. BMC Genomics **18**, 1–10 (2017)
22. Lutteropp, S., Scornavacca, C., Kozlov, A.M., Morel, B., Stamatakis, A.: NetRAX:
    accurate and fast maximum likelihood phylogenetic network inference. Bioinfor-
    matics **38**(15), 3725–3733 (2022)
23. Nguyen, Q., Roos, T.: Likelihood-based inference of phylogenetic networks from
    sequence data by PhyloDAG. In: Dediu, A.-H., Hernández-Quiroz, F., Martín-Vide,
    C., Rosenblueth, D.A. (eds.) AlCoB 2015. LNCS, vol. 9199, pp. 126–140. Springer,
    Cham (2015). https://doi.org/10.1007/978-3-319-21233-3_10
24. Park, H.J., Jin, G., Nakhleh, L.: Bootstrap-based support of HGT inferred by
    maximum parsimony. BMC Evol. Biol. **10**(1), 1–11 (2010)
25. Solís-Lemus, C., Bastide, P., Ané, C.: Phylonetworks: a package for phylogenetic
    networks. Mol. Biol. Evol. **34**(12), 3292–3298 (2017)
26. Tan, M., et al.: QS-Net: reconstructing phylogenetic networks based on quartet
    and sextet. Front. Genet. **10**, 607 (2019)
27. Van Iersel, L., Janssen, R., Jones, M., Murakami, Y., Zeh, N.: Polynomial-time
    algorithms for phylogenetic inference problems involving duplication and reticula-
    tion. IEEE/ACM Trans. Comput. Biol. Bioinf. **17**(1), 14–26 (2019)
28. Van Iersel, L., Jones, M., Weller, M.: Embedding phylogenetic trees in networks of
    low treewidth. arXiv preprint arXiv:2207.00574 (2022)
29. Wen, D., Yu, Y., Zhu, J., Nakhleh, L.: Inferring phylogenetic networks using Phy-
    loNet. Syst. Biol. **67**(4), 735–740 (2018)

CONSULT-II: Taxonomic Identification Using Locality Sensitive Hashing

Ali Osman Berk Şapcı[1], Eleonora Rachtman[1], and Siavash Mirarab[1,2(✉)] ⓘ

[1] Bioinformatics and Systems Biology Graduate Program, University of California, San Diego, CA 92093, USA
{asapci,erachtman,smirarab}@ucsd.edu
[2] Electrical and Computer Engineering, University of California, San Diego, CA 92093, USA

Abstract. Metagenomics is widely used to study the microbiome using environmental samples, and taxonomic classification of reads is a precursor to many analyses of such data. Taxonomic classification requires comparing sample reads against a reference dataset of known organisms. Crucially, the genomes represented in a sample may be phylogenetically distant from their closest match in the reference set. Thus, simply mapping reads to genomes is insufficient; we need to find inexact matches to species with substantial distance. While k-mer-based methods, such as Kraken, have proved popular, they have limited ability to match against distant taxa. In this paper, we use locality sensitive hashing to design a k-mer-based method that can match reads to genomes with higher distance than existing methods. We build on an earlier contamination detection method, CONSULT, to add taxonomic classification abilities. We show in a series of experiments that our method, CONSULT-II, has higher recall than alternatives when precision is about the same. Its results can also be summarized to obtain a taxonomic profile, which we show outperforms leading methods with respect to some measurement criteria. CONSULT-II is available at https://github.com/bo1929/CONSULT-II.

Keywords: Metagenomics · Taxonomic classification · Abundace profiling

1 Introduction

Modern microbiome studies rely on profiling microbial communities using environmental shotgun sequencing, producing genome-wide data [13]. The resulting *metagenomic* datasets include millions of short DNA reads from a mixed set of unknown microorganisms, which then need to be identified taxonomically before most downstream analyses can proceed [30]. These identifications are with respect to reference datasets and seek to detect which, if any, of the taxonomic units represented in the reference set could have generated each read. The goal is to identify reads at the lowest taxonomic level possible, but precision is hampered by the sparsity of the reference set and the ambiguity of matches. The per-read

K. Jahn and T. Vinař (Eds.): RECOMB-CG 2023, LNBI 13883, pp. 196–214, 2023.
https://doi.org/10.1007/978-3-031-36911-7_13

identifications are often summarized to obtain a taxonomic profile, showing relative abundances of various taxonomic groups. Metagenomics promises a more precise identification compared to the more traditional environmental sequencing of 16S or 18S rRNA amplicons [12]. However, identifying reads sampled across the genome is challenging due to the large size of the genome on one hand and the relative sparsity of genome-wide reference datasets (a quickly diminishing problem) on the other. While taxonomic identification/profiling tools have matured over years [5], repeated benchmarking by several groups have demonstrated major gaps in the accuracy of existing methods [23,26,40,48].

All taxonomic identification methods, whether they use k-mers [1,15,20,33, 46], genome-wide alignment [49], marker-based alignment [18,28,41,44] or phylogenetics [2,42,43,45], are fundamentally looking for matches between short reads in the sample and a reference set. Much of the microbial diversity of the earth is not represented with close representatives in the reference datasets [10,11]. For samples coming from a poorly known microbial habitat (e.g., seawater or soil, etc.), many constituents are likely to be missing from the reference set [34]. Thus, most methods recognize that we cannot restrict results to exact matches and seek to find inexact matches.

Nevertheless, classification methods continue to show reduced accuracy for the so-called *novel* sequences (i.e., sequences without a close match in the reference set) [17,24,29,34,36]. For example, domain-level classification using Kraken-II [46], one of the leading profiling methods, degrades dramatically as the Hamming distance to the closest match in the reference set increases beyond 10% [36]. As an example from real data, Pachiadaki et al. [34] found that phylum-level classification of ocean microbiome fails with high frequency. Despite the constant increase in reference density, these databases include a fraction of the fully sequenced prokaryotic genomes – itself a fraction of the estimated 10^{12} microbial species [19]. In addition to better reference sets, we need more *sensitive* methods that can identify novel sequences with respect to distant reference genomes.

Perhaps the most principled approach for identifying novel sequences is sequence alignment followed by phylogenetic placement [22]. For example, the phylogenetic method TIPP [31] showed the best accuracy in an independent benchmarking that included high novelty [26]. However, this same benchmarking also showed a high computational cost for TIPP. As the size of reference sets grows, the alignment followed by the phylogenetic placement method will further fall behind other methods in terms of running time. Thus, methods that simply match k-mers to the reference datasets are used more often than phylogenetic methods in practice. The best k-mer-based methods tend to inhabit a sweet spot of high speed, good accuracy, and ease of use [23,40,48]. However, k-mer-based methods can be particularly sensitive to the completeness of reference sets if they allow only exact matches. In this paper, we seek to make k-mer matching more sensitive than existing methods so that we can match with more distant references without compromising speed and ease of use.

The k-mer-based methods that use long k and look for presence/absence can accommodate novel sequences by allowing inexact k-mer matches. Kraken-II accomplishes this by masking some positions in a k-mer (by default, 7 out of 31). Recently, focusing on the problem of contamination removal instead of taxonomic classification, Rachtman et al. [35] showed that novel reads (e.g., those with 10–15% distance to the closest match) can be identified with higher accuracy by making inexact matches a central feature of the search. The proposed method, CONSULT, uses locality sensitive hashing (LSH) to partition the k-mers in the reference set (multiple ways) such that for a given k-mer, the reference k-mers with distance up to a certain threshold (e.g., 3 or 4) are with high probability within a constant-sized easily-indexed subset. By explicitly allowing mismatches, CONSULT allows matching against more phylogenetically distant references. Unlike the masking technique of Kraken-II, CONSULT outputs the Hamming distance of the query to the matched k-mers. Thus, the core of CONSULT is to find low hamming distance matches efficiently, using LSH. The idea of using LSH for finding similar DNA sequences efficiently was not novel to CONSULT [4,9,21, 25,38]. For example, Brown and Truszkowski [8] addressed the related problem of phylogenetic placement by using LSH to find a restricted area of the tree that is then examined more carefully to find the best placement. What distinguishes CONSULT from these past works is its focus on including a large number of entire genomes in the reference set.

Designed for contamination detection, CONSULT has no ability to detect *which* reference species are present in a sample – only that *some* reference species match a k-mer. To make the tool able to perform taxonomic identification is not trivial; in principle, it requires remembering what reference genome(s) include each reference k-mer. As is, CONSULT is memory-hungry, using close to 120 Gb to store its standard microbial dataset with 8 billion k-mers. Adding taxonomic information makes the situation only worse. It is impossible to keep track of *every* genome that included a k-mer given the memory available in present-day machines. Thus, we need heuristic strategies to keep a more limited set of taxonomic information per reference k-mer.

In this paper, we introduce CONSULT-II for extending the algorithmic ideas from CONSULT to enable taxonomic identification. We design probabilistic heuristics that keep only one taxonomic (or phylogenetic) ID per k-mer for the reference set, allowing the construction of databases that fit in the memory of available machines. We then show how a potentially novel query read can be classified using information about the distance of its k-mers to its closest reference matches. Finally, we show how the hashing parameters can be automatically set to allow adjustment to variable size references (a feature missing from CON-SULT). We evaluate CONSULT-II against a large reference set in simulations studies and show much-improved sensitivity with little loss of precision.

2 Algorithm

2.1 Background: Review of CONSULT

For a given set of reference k-mers and a query k-mer, CONSULT seeks to answer this question: Is there any reference k-mer with Hamming distances less than some threshold p to the query k-mer? It uses $k = 32$ and allows a large (e.g., 2^{33}) set of reference k-mers; p is set to 3 by default, but can change.

CONSULT represents a given set of reference k-mers using two main data structures: an array $\mathcal{K}$ that encodes each 32-mer as a 64-bit number (by default, $|\mathcal{K}| = 2^{33}$), and a set of l (default: 2) fixed-sized hash tables $\mathbf{H}^1 \ldots \mathbf{H}^l$ with 4-byte pointers to $\mathcal{K}$ (and extra bits when $|\mathcal{K}| > 2^{32}$). Each hash table is a simple $2^{2h} \times b$ matrix (default: $h = 15$ and $b = 7$) where each row corresponds to a hash index and the columns are the limited number of k-mers with that index kept in the database. For any given 32-mer, we can quickly compute its hash index as follows. For each hash table $\mathbf{H}^i$, we preselect h positions of the 32-mer as its hash. Computing the hash index for a query k-mer simply requires extracting the corresponding bits from the 64-bit encoded k-mer, which is facilitated by the specific 64-bit representation of 32-mers used. Once the l hashes are computed, the pointers from all $\leq b \times l$ entries in the corresponding $\mathbf{H}^j$ tables are followed to the encodings in the $\mathcal{K}$ array; for each such encoding, CONSULT computes the Hamming distance explicitly and returns a match if the distance is below p. Thus, CONSULT uses LSH only to limit the number of comparisons; it has no false positive results but can have false negatives.

Our hashing scheme is the classic example of an LSH: two k-mers at Hamming distance d have the same hash with probability $(1 - d/k)^h$. Given two k-mers, the probability that *at least one* of the hash functions is the same for both k-mers is:

$$\rho(d) = 1 - (1 - (1 - \frac{d}{k})^h)^l \tag{1}$$

It can be easily confirmed that $\rho(d)$ is close to 1 for $d \leq p$ and drops quickly to small values for $d \gg p$ for some small p (e.g., $p = 3$) for several choices of l and h, an issue we will return to. Furthermore, classification is done at the read level, and each read includes $L - k + 1$ k-mer-s where L is the length of the read. We do not need every k-mer of the read to match to be able to classify a read. When the probability of a mismatch between a read and a reference species is d/k, the expected number of matching k-mers is $(L - k + 1)\rho(d)$, which can be a large value for realistic settings (Fig. 1a). Note that with the default settings of $k = 32, h = 15, l = 2$, for a read at distance 25% distance from the closest reference, we still expect 3.2 k-mer matches and can potentially classify it.

2.2 Overview of Changes

To extend CONSULT to enable taxonomic classification, several challenges needed to be addressed. *i*) CONSULT was designed for a fixed reference library size. As a result, all the hashing settings were fixed for a library of roughly 2^{33}

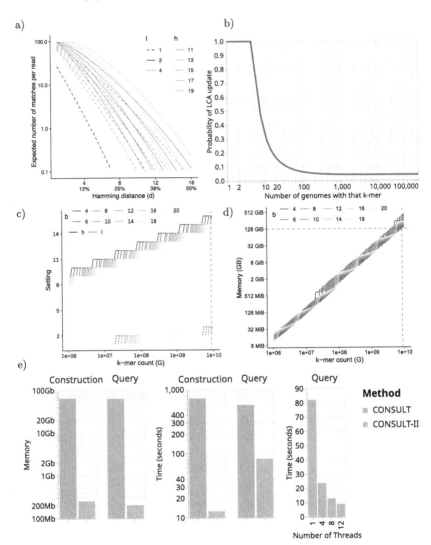

Fig. 1. Method design. a) The expected number of 32-mers matched by the LSH approach for a read of length $L = 150$ as the normalized distance d/k of the read to the closest match varies is $(L - k + 1)\rho(d)$, shown here. Lines show different settings of l and h for an infinite b. The black line corresponds to $k = 35$, $h = 35 - 7$, and $l = 1$, which is similar to the default Kraken-II settings. b) The probability of updating the LCA per each new k-mer goes down with how often that k-mer is observed in genomes (m); specifically, we use $\min\{^2/\max\{N + 2 - 5, 2\} + 1/5^2, 1\}$ as shown. c) The setting of h and l to accommodate G k-mers and achieve performance $\rho(3) = 0.45$ for various choices of b. d) Fixing the performance, and selecting the required h and l according to part (c) results in substantially different memory requirements for various b; note the log-scale y-axis. e) Comparison of CONSULT and CONSULT-II in terms of computational resources. CONSULT-II achieves the same level of performance with CONSULT but with much lower computational resource requirements. Reported memory usages and runtimes are measured in a real dataset, consisting of six E. coli genomes in the reference set (12 million 32-mers), and four E. coli genomes in the query set (1 million 150bp reads).

32-mers. To make the method more usable, we it needs to adjust to the size of the input library. *ii*) At the time of building the reference library, we need to keep track of which k-mers belong to which set of species. Since keeping track of this mapping in a fine-grained way will lead to an explosion of memory, we need heuristics that keep *some* taxonomic information but keep the memory manageable. *iii*) At the time of querying a read, we need some way of summarizing all the inexact matches from all the k-mers of a read into one final taxonomic classification. Next, we delve into the details of our solutions for each of these challenges.

2.3   Automatic Setting of Hash Table Parameters

CONSULT is highly sensitive to parameter values, both in terms of computational resources and accuracy. Rachtman et al. [35] determines a robust set of default values for its parameters but fails to provide a systematic way of choosing the parameter values for input sets with varying sizes. As a result, with its default parameters, CONSULT consumes more resources, i.e., memory and runtime, than it actually needs to maintain a certain level of performance. Here, we extend CONSULT-II by developing a heuristic to determine parameter values as a function of the number of k-mers in the reference set: $G = 2^g$. Thanks to this heuristic, CONSULT-II requires less memory for prevalent cases with $2^g < 2^{33}$ k-mers while preserving the same performance as CONSULT.

We next compute the needed h and l to accommodate G k-mers and maintain the same performance for a fixed b. First, note that the total number of k-mers saved in each hash table is $2^{2h}b$. Thus, we need $h = 1/2 \log_2(G/b)$ to accommodate G k-mers in all l tables. We desire to maintain a certain level of performance by ensuring $\rho(p) = \alpha$. Solving for l in (1) yields $l = {}^{\log(1-\alpha)}/{}_{\log\left(1-(1-p/k)^h\right)}$. Thus, for a given b, we can select h and l to accommodate all k-mers and fix the performance (Fig. 1c). Remarkably, l only ranges between 1 and 3 for even very large G. Then, to choose b, we simply scan a reasonable range to find the value that minimizes the total memory consumption of CONSULT-II, which is $4l2^{2h}b + 8G$. Adding some bounds and appropriate rounding, we obtain:

$$h(b) = \max\left\{\lceil 1/2 \log_2(G/b)\rceil, 4\right\} \tag{2}$$

$$l(b) = \max\left\{\text{round}\left({}^{\log(1-\alpha)}/{}_{\log\left(1-(1-p/k)^h\right)}\right), 2\right\} \tag{3}$$

$$\hat{b} = \underset{4 \leq b \leq 20}{\arg\min}\left(4 \cdot l(b)2^{2h(b)}b + 8 \cdot 2^g\right). \tag{4}$$

Thus, we set $b = \hat{b}, l = l(\hat{b})$, and $h = h(\hat{b})$. The user can select α and p. Note that $\rho(3) = 0.4$ with default CONSULT. In CONSULT-II, by default, we slightly increase the level of performance to $\rho(3) = 0.45$, which corresponds to $\alpha = 0.45$ for $p = 3$. With these settings, the right choice of b as indicated above can lead to substantial memory saving, all else equal (Fig. 1d). Note that the best choice of b depends on G in very non-linear ways due to the rounding necessary in the settings. Finally, note that all the calculations above are assuming every k-mer will appear in all l hash tables. In practice, this is neither necessary nor possible when available memory is limited and k-mers are not strictly random. To account for the fact that some k-mers may appear in some of the tables, we can slightly decrease G before plugging it in (by default, we use $7/8G$).

Figure 1e demonstrates the dramatic improvement in terms of computational resources obtained using this heuristic. Here, the reported memory usages and runtimes are measured in a dataset consisting of six E. coli genomes in the reference set (12 million 32-mers), and four E. coli genomes in the query set (1 million 150bp reads). Using the default CONSULT setting ($h = 15$, $b = 7$, $l = 2$) results in a library that exceeds 60 Gb. Our heuristic determines the optimal parameter configuration as $h = 10$, $b = 12$, and $l = 2$ and achieves the same level of accuracy with less than 500 Mb. Having a smaller library also leads to shorter runtimes, both for library construction and querying. Library construction takes $\approx 60\times$ shorter thanks to CONSULT-II's heuristic. At the query time, CONSULT-II is $\approx 7\times$ faster given a single thread. CONSULT-II is also efficiently parallelized and obtains close to linear speed-ups up to 8 threads.

2.4   Library Construction: Saving Taxonomic IDs per k-mer

Given the density of modern reference datasets, each k-mer can appear in several (tens or even hundreds of) reference species. Assuming tens of thousands of genomes in the reference genome, keeping each taxonomic ID requires 2 bytes. Keeping one ID per k-mer would require 16 Gb for our standard libraries with 2^{33} k-mers and is perhaps doable. However, keeping all or even a smaller subset of taxonomic IDs would quickly expand the size of the memory needed beyond what the standard machines offer. We solved the problem by keeping a single ID for a "probable" LCA of all the genomes that include the k-mer, where probable LCA is defined as follows.

We pre-process all k-mers to compute the number of genomes that includes each k-mer $x_i \in \mathcal{K}$ (call this quantity N_i). For each $x_i \in \mathcal{K}$, we assign a 2-byte taxon ID, denoted by t_i. We process through each reference genome r, and for each k-mer, if it is found in the database (i.e., $x_i \in \mathcal{K}$), we simply set or update t_i to be the LCA of the current t_i (if it exists) and the species r with the probability

$$p_u(N_i) = \min \left\{ \frac{w}{\max\{N_i + w - s, w\}} + \frac{1}{s^2}, 1 \right\} \qquad (5)$$

where w and s parameterize the rate of decrease and the offset of the probability function p_u, respectively. We set $s = 5$ and $w = 2$ as the default values, and the corresponding p_u is shown in Fig. 1b. Note that the order of processing the reference genomes has no systematic effect on the final taxonomic ID of a k-mer. Every k-mer, including the first encountered, will be ignored with the same probability regardless of their order. Also, k-mers appearing in more than s genomes have a very small, but non-zero, probability of not having a taxonomic ID at all. The goal of this probabilistic assignment is to avoid pushing up taxonomic identifications due to errors in the reference libraries. Imagine a

k-mer that is found exclusively in 20 species of a particular genus, but due to incorrect assembly (or real HGT), is also found in one species of a completely different phylum. Simply using the LCA would push classifications up to the domain level. Using this approach, in a situation like the one described, with 85% probability, we will *not* update the LCA beyond the genus. Note that our particular choice of the probability function is a heuristic with no theory behind it; we chose it to ensure k-mers are labeled by the LCA as long as they appear in a handful (hence $s = 5$) of genomes and smoothly become more probable to be labeled by a node lower than the strict LCA as N_i grows. The $1/s^2$ term ensures that each k-mer has a non-zero probability of having a taxonomic ID associated with it, even if it is very common.

2.5   Query and Taxonomic Identification

Querying a read against the library described above results in a list of matched k-mers, their saved taxonomic IDs, and distances between these k-mers and their closest matches with the same hash index, summarizing the similarity between the queried read and various taxonomic IDs. Using this rich information, our goal is to assign a taxonomic label to the query read. Considering each match as a vote to the corresponding taxonomic ID weighted by its distance, we developed a voting-based approach utilizing the taxonomic tree.

Let $\mathcal{T}$ be the set of all taxonomic IDs, and let $\mathcal{R}$ be the set of k-mers of the query read. Note that, a k-mer $x \in \mathcal{R}$ might have matches with multiple k-mers with varying distances, or might not have a match at all. We compare x with all k-mers with the same hash index and choose the one with the minimum distance d; if multiple k-mers have distance d, we arbitrarily choose one. Let t be the taxonomic ID of this chosen reference k-mer. We define the vote of x for the taxonomic ID t using the equation

$$v_t(x) = \left(1 - \frac{d}{k}\right)^k \mathbb{1}\{d \leq d_{\max}\} \tag{6}$$

where $x \in \mathcal{R}$ is a k-mer, d is the Hamming distance between x and its closest k-mer in the reference, and $\mathbb{1}\{d \leq d_{\max}\}$ evaluates if x is a match with Hamming distance $d \leq d_{\max}$ (default is round($3p/2$), in our analysis $d_{\max} = 5$). The voting function (6) drops close to exponentially with distance d. Note that, when searching a k-mer, matches with distance above p are also found with non-zero probability; the LSH guarantees that k-mers with distance $< p$ to the reference are matched with high probability, but more distant matches can also be found (see the smooth curves on Fig. 1a). Thus, allowing $d_{\max}$ to be different from p gives us more flexibility to control sensitivity.

After computing the individual votes, we further incorporate the hierarchical relationship between taxonomic IDs, and recursively sum up individual votes contributed by each k-mer in a bottom-up manner using the taxonomic tree as follows to derive a final vote value for each taxonomic ID:

$$\bar{v}(t) = \sum_{x \in \mathcal{R}} v_t(x) + \sum_{t' \in C(t)} \bar{v}(t') \tag{7}$$

where $C(t)$ is the set of children of the taxonomic ID t in the taxonomic tree.

By design, the votes $\bar{v}(t)$ increase as we go up in the taxonomic tree, and reach their maximum at the root. To balance specificity and sensitivity, we require a majority vote. Let $\tau = 0.5 \max_{t \in T} \bar{v}(t)$. Then, the taxonomic ID $\hat{t}$ which belongs to the lowest taxonomic level satisfying the condition $\bar{v}(\hat{t}) > \tau$ is assigned to the queried read with k-mer set $\mathcal{R}$. This τ threshold has a special property that only a single taxonomic ID t can exceed τ at a given taxonomic level. Therefore, the taxonomic ID predicted by the described classification scheme is unique, and the interpretation can be described as follows: the classifier tries to be as specific as possible but also requires a certain level of confidence, and hence immediately stops considering upper taxonomic levels when the vote value is large enough. Additionally, we require $\bar{v}(\hat{t})$ to be greater than some small threshold, which we explore in our experimental results.

We also use computed vote values to derive taxonomic abundance profiles. We consider each taxonomic level separately and simply derive a profile vector based on the vote values $\bar{v}(t)$ for $t \in T_l$ for taxonomic level l, e.g., genus. To compute this abundance profile, one needs to normalize the vote values at a given taxonomic level to obtain a profile of values summing up to 1. While there are many alternatives to consider, we opted to use the following;

$$p_t^l = \frac{\sqrt{\bar{v}(t)}}{\sum_{t' \in T_l} \sqrt{\bar{v}(t')}} \tag{8}$$

where p_t^l is the profile value of taxonomic ID t from level l. Then, the abundance profile for taxonomic level l is given by $p^l = \left[p_t^l \right]_{t \in T_l}$.

3   Experimental Setup

We compared performance of CONSULT-II with two methods: Kraken-II [46, 47] and CLARK [33]. Based on benchmarking studies [23,27,39,48], these are some of the leading metagenomic identification tools. Kraken-II maps each query read to the lowest common ancestor (LCA) of all genomes that contain k-mers identified in a query. CLARK is a supervised sequence classification method that builds a database of discriminative k-mers of reference genomes and relies on exact k-mer matching.

3.1   Experiments

Reference Library Construction. To benchmark software, we constructed reference libraries using the Tree of Life (ToL) [50] microbial genomic dataset. ToL is composed of 10,575 microbial species and a reference phylogeny. Five genomes had IDs that did not exist in NCBI and were excluded from this set. Following previous analyses [35], 100 archeal genomes were left out and the remaining 10,460 genomes were used as the reference. We used this set for all methods in all experiments. We constructed the Kraken-II reference library for

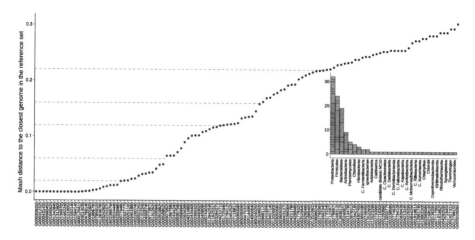

Fig. 2. The distribution of the 120 queries selected. For each query, we show their distance to the closest species in the reference database, as computed by Mash [32]. Dashed lines show the boundaries between bins of queries by distance. The inset bar graph shows that selected queries come from 29 distinct phyla (bars), some of which are the relatively novel *Candidatus* phyla (abbrev. *C.*). Several of the phyla include multiple orders (bar segments, arbitrarily colored) for a total of 54 orders.

genomes that belonged to ToL using the corresponding ToL taxonomy. Kraken reference libraries were built without masking low-complexity sequences, but using default settings otherwise. We note that prior studies [37] found default settings to be preferable for query identification. We constructed the CLARK database for ToL genomes using standard parameters (e.g., $k = 31$, default classification mode, species rank for classification) and ToL taxonomy.

Experiment 1: Controlled Novelty. We created a set of queries consisting of 120 bacterial genomes among genomes added to RefSeq after ToL (i.e., absent from ToL) to generate queries with controlled distances to the reference genome. These queries are selected carefully to span a range of different distances to the closest species in the reference dataset measured by Mash [32] (Fig. 2). We bin the queries into seven groups based on these distances (cutting at $0.001, 0.02, 0.06, 0.12, 0.16, 0.22$), making sure each bin has at least 11 queries. Queries span 29 phyla (Fig. 2). Most queries are from distinct genera (102 genera across 120 queries) and only one species contributes two queries. For each of these genomes, we generated 150bp synthetic reads at higher coverage using ART [14] and then subsampled reads for each query down to 66667 reads ($\approx 2\times$ coverage for bacterial genome) using seqtk [16].

To measure the accuracy, we compare the taxonomic classification of each method against the known NCBI taxonomy. For each read, we classify it as follows for *each* classification level l. If the reference library has at least one genome matching the query at a level l, we call it a *positive*. Then, the result of a method is called a TP if it finds the correct label, FP if it finds an incorrect label, and FN if it does not classify at level l. When the reference library does not have any genomes from the level l, we call it a *negative*. Then, the result of a method is a TN if it does not classify at level l, FP if it classifies it which is necessarily a false classification. We show precision= $TP/(TP + FP)$, recall= $TP/(TP + FN)$. We also report a measure that combines sensitivity and specificity: F1= $2TP/(2TP + FP + FN)$. We ignore the query read at level l when its true taxonomic rank given by NCBI is 0 (which indicates a missing rank).

Experiment 2: Abundance Profiling. We also evaluated the ability of CONSULT-II to perform taxonomic abundance profiling. We compared the performance of CONSULT-II with Bracken [20] and CLARK [33]. Bracken is a Bayesian process on top of the Kraken-II results that uses the taxonomic identification results to arrive at a profile. For evaluation, we use the critical assessment of metagenome interpretation (CAMI) [39] dataset. CAMI-I high complexity data set is a metagenomic time series profile simulated for the first CAMI benchmarking challenge. The profile included five microbial samples that were created to mimic changing abundance distribution of the underlying microbial community. For our analysis, we arbitrarily selected the first sample, subsampled it down to 10^7 reads, and computed abundance profiles using default settings for Bracken and CLARK. We compared the performance of different tools using metrics computed by the open-community profiling assessment tool (OPAL) [27]. In particular, we focus on the two metrics the original publication focused on: the Bray-Curtis dissimilarity between the estimated profile and the true profile and the Shannon equitability measure of alpha diversity.

4   Results

4.1   Controlled Novelty Experiment

Impact of the Total Vote. We start by examining the impact of removing classifications that despite reaching the majority vote have a low total $\bar{v}\left(\hat{t}\right)$. Results confirm that the total vote is a major determinant of whether a classification is correct or not (Fig. 3a). In particular, more than 25% of the FP predictions have $\bar{v}\left(\hat{t}\right) < 0.01$, corresponding to up to two votes from k-mers of distance 5; around 50% of FPs have $\bar{v}\left(\hat{t}\right) < 0.03$ (e.g., less than two matches of distance 4). In each case, however, removing the FPs comes at the expense of also removing some TPs. Thus, the total votes give us a way to trade off precision and recall. In terms of the F1 measure that combines both aspects, the three thresholds are quite similar (Fig. 4a), despite having very different precision and recall. There is a small but consistent improvement in F1 when fewer classifications are discarded. Nevertheless, we discard classifications with less than 0.03 vote when comparing to other methods to ensure precision levels are similar, making comparisons easier.

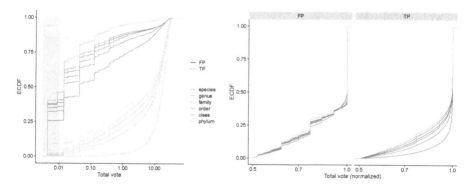

Fig. 3. Predictive power of votes. a) The Empirical cumulative distribution function (ECDF) of votes for all CONSULT-II classifications that were correct (TP) or incorrect (FP). The shaded areas note 0.01 and 0.03 thresholds used below to filter out spurious classifications. Note that the x-axis is log-scale. b) Similar distributions for the normalized votes (share of the total vote).

Examining the distribution of the normalized votes $\dfrac{\bar{v}(t)}{\max_{t \in \mathcal{T}} \bar{v}(t)}$ shows that more precise classifications often have a higher share of the vote (Fig. 3b). While every classification by construction has at least 50% of the vote, around 1/4 of phylum level classification have less than 80% of the votes, compared with around 10% of species-level classifications. The normalized votes do not effectively distinguish FP and TP results, and thus, are not used for any filtering.

Comparison to Other Methods. CONSULT-II has better F1 than CLARK-S and Kraken-II, regardless of the voting threshold used (Fig. 4a) on the bacterial query set. As expected, as queries become more novel, accuracy drops across all levels for all methods. However, there are significant differences between the methods. For queries with almost identical matches in the reference, all methods are about equal at the species level, and CONSULT-II has a small advantage at the higher levels. For queries that are even slightly novel (in the [0.001, 0.02] distance range), CONSULT-II starts to outperform Kraken-II and CLARK across all levels. For yet more novel queries, CONSULT-II has much better F1 across all other levels except for the species level. F1 scores reduce for all methods as queries become more novel. The gap between CONSULT-II and alternatives is especially high for the more novel queries (e.g., ≥0.6 distance to the closest) and is more wide at phylum and class levels. Between CLARK and Kraken-II, it appears that Kraken-II has a slight advantage in all levels, except perhaps at the species level (with some exceptions in the [0.001, 0.02] bin).

Comparing the precision and recall of the methods shows that the advantage of CONSULT-II is due to universally higher recall levels (Fig. 4b). In contrast, the precision is comparable in most cases, is slightly higher for CONSULT-II in some cases (e.g., most bins in the family level), and is slightly worse in other cases (e.g., most bins in the phylum level). At the species level, all methods have low

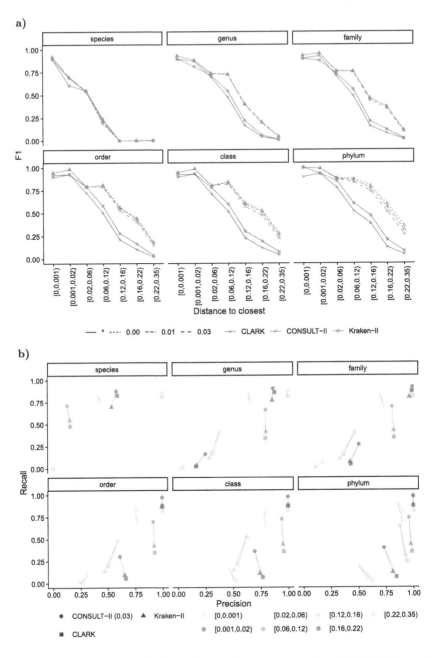

Fig. 4. Results on the controlled novelty experiment, comparing CONSULT-II against Kraken-II and CLARK. The set of 120 queries is binned into 7 groups based on their Mash distance to their closest neighbor in the reference set. Results summarize classifications across 66667 reads per query. a) F1 $\left(2TP/(2TP + FP + FN)\right)$ score across taxonomic groups and divided into query bins. b) At different taxonomic levels, recall $\left(TP/(TP + FN)\right)$ versus precision $\left(TP/(TP + FP)\right)$ for each bin of queries.

precision, except for the least novel queries. At the higher levels, CONSULT-II continues to have a better recall and the precision of all methods tends to improve (e.g., notice the [0.06, 12] bin). In most cases, this better recall comes at little or no expense to precision. Note that the better recall of CONSULT-II over other methods is despite using the more conservative version; requiring fewer total votes can further increase the recall at the expense of precision.

4.2   CAMI Profiling Results

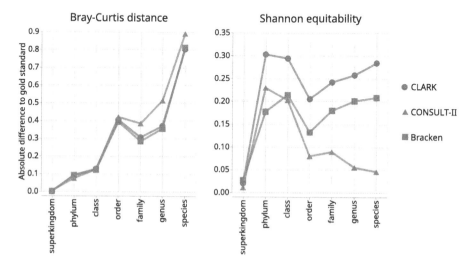

Fig. 5. Abundance estimates of CONSULT-II in comparison to other tools at different taxonomic ranks. a) Bray-Curtis dissimilarity as a measure of beta diversity. Beta diversity indicates the similarity between estimated and gold standard profiles. The smaller the value, the more similar distributions of abundance profiles are. b) Shannon equitability as a measure of alpha diversity. The closer the Shannon equitability of the predicted profile to the gold standard, the better it reflects the actual alpha diversity in the gold standard in terms of the evenness of the taxa abundances. Here, we show the absolute difference of Shannon alpha diversity to the gold standard.

Focusing on the taxonomic profiling, CONSULT-II outperforms Bracken and CLARK in its ability to predict taxon abundances according to some measures but not others (Fig. 5). In terms of the Shannon equitability, which is a measure of the variety and distribution of taxa present in a sample, CONSULT-II outperforms other methods in all taxonomic levels except for phylum (Fig. 5b). The improvements are especially large for the species and genus level. For example, the Shannon equitability of CONSULT-II is only 0.05 away from the correct value (i.e., gold standard) whereas Bracken and CLARK are 0.2 and 0.28 away. This improved estimate of the alpha diversity can be explained by the improved classification ability of CONSULT-II at lower taxonomic levels.

In terms of the Bray-Curtis dissimilarity, CONSULT-II produces comparable results to other methods at phylum, class, and order levels. In phylum and class levels, all methods have relatively small errors, and CONSULT-II slightly outperforms others at the phylum level. Moving to lower taxonomic levels, the error goes up rapidly, and CONSULT-II becomes worse than alternatives at family and genus levels (Fig. 5a). At the species level, all methods have extremely high error and CONSULT-II is slightly worse than others. Given the sensitivity of Bray-Curtis to FP classification, these higher errors at family and genus level may be due to less precision of CONSULT-II.

5   Discussion

We introduced an LSH-based method for the taxonomic identification of individual reads and extended it to allow taxonomic profiling. The main idea behind our method, CONSULT-II, is that by using LSH, we can find a limited number of k-mers to compare to each k-mer in the read; we choose LSH settings such that with high enough probability, we will find the reference k-mers that match the query k-mer at low enough distances. Around this central idea, many heuristics had to be developed to allow taxonomic classification and profiling. Our heuristics have no theoretical proof, but performed well empirically on the datasets we tested.

This presented work is only a first step and can be extended in several ways. First and foremost, we did not explore alternative heuristics nor did we explore alternative parameter settings for our heuristics. For example, our equations for LCA update (5), vote-versus-distance (6), vote aggregation (7), and vote-based profiling (8) were all based on our intuitions with no theory nor empirical explorations. Such explorations can be done in the future, provided that evaluations are performed on a different dataset. Even better would be developing a theoretical framework that leads to the proper choice of these functions under certain assumptions. Connecting taxonomic profiling to distance-based phylogenetic placement [3] may provide a framework to achieve this goal. Such a framework may allow us to go beyond taxonomic identification and provide alignment-free phylogenetic placement of reads, as others have attempted [6].

Beyond taxonomic or phylogenetic identification, we note that our method provides a very information-rich representation of a read. It can give us the number of k-mers at each distance from each reference taxon. Such representations may be able to help us eliminate the need for taxonomic profiling in downstream analyses. Defining taxonomic units for microbes is notoriously difficult [7], and taxonomies rarely agree with each other or with phylogenies. Thus, a rich representation of reads may enable statistical associations without the need to build arbitrary units such as taxon labels or even OTUs.

References

1. Ames, S.K., Hysom, D.A., Gardner, S.N., Lloyd, G.S., Gokhale, M.B., Allen, J.E.: Scalable metagenomic taxonomy classification using a reference genome database. Bioinformatics **29**(18), 2253–2260 (2013). ISSN 1367-4811 (Electronic). https://doi.org/10.1093/bioinformatics/btt389

2. Asnicar, F., et al.: Precise phylogenetic analysis of microbial isolates and genomes from metagenomes using PhyloPhlAn 3.0. Nat. Commun. **11**(1), 2500 (2020). ISSN 2041-1723. https://doi.org/10.1038/s41467-020-16366-7

3. Balaban, M., Sarmashghi, S., Mirarab, S.: APPLES: scalable distance-based phylogenetic placement with or without alignments. Syst. Biol. **69**(3), 566–578 (2020). ISSN 1063-5157. https://doi.org/10.1093/sysbio/syz063

4. Berlin, K., Koren, S., Chin, C.S., Drake, J.P., Landolin, J.M., Phillippy, A.M.: Assembling large genomes with single-molecule sequencing and locality-sensitive hashing. Nat. Biotechnol. **33**(6), 623–630 (2015), ISSN 1546–1696 (Electronic). https://doi.org/10.1038/nbt.3238

5. Bharti, R., Grimm, D.G.: Current challenges and best-practice protocols for microbiome analysis. Briefings Bioinf. **22**(1), 178–193 (2021). ISSN 1477-4054. https://doi.org/10.1093/bib/bbz155

6. Blanke, M., Morgenstern, B.: Phylogenetic placement of short reads without sequence alignment. bioRxiv, October 2020

7. Brenner, D.J., Staley, J.T., Krieg, N.R.: Classification of procaryotic organisms and the concept of bacterial speciation. In: Bergey's Manual of Systematics of Archaea and Bacteria, pp. 1–9. Wiley, Chichester, UK, September 2015. https://doi.org/10.1002/9781118960608.bm00006

8. Brown, D., Truszkowski, J.: LSHPlace: fast phylogenetic placement using locality-sensitive hashing. In: Pacific Symposium on Biocomputing, pp. 310–319, November 2013. ISBN 978-981-4596-36-7. ISSN 2335-6936

9. Buhler, J.: Efficient large-scale sequence comparison by locality-sensitive hashing. Bioinformatics **17**(5), 419–428 (2001). ISSN 1367-4803. https://doi.org/10.1093/bioinformatics/17.5.419

10. Choi, J., et al.: Strategies to improve reference databases for soil microbiomes. ISME J. **11**(4), 829–834 (2017). ISSN 1751-7362. https://doi.org/10.1038/ismej.2016.168

11. Dress, A.W., et al.: Noisy: identification of problematic columns in multiple sequence alignments. Algorithms Mol. Biol. **3**(1), 7 (2008). ISSN 1748-7188. https://doi.org/10.1186/1748-7188-3-7

12. Gill, S.R., et al.: Metagenomic analysis of the human distal gut microbiome. Science **312**(5778), 1355–9 (2006). ISSN 1095-9203. https://doi.org/10.1126/science.1124234

13. Handelsman, J.: Metagenomics: application of genomics to uncultured microorganisms. Microbiol. Mol. Biol. Rev. **68**(4), 669–85 (2004). ISSN 1092-2172. https://doi.org/10.1128/MMBR.68.4.669-685.2004

14. Huang, W., Li, L., Myers, J.R., Marth, G.T.: ART: a next-generation sequencing read simulator. Bioinformatics **28**(4), 593–594 (2012). ISSN 1367-4803. https://doi.org/10.1093/bioinformatics/btr708

15. Lau, A.K., Dörrer, S., Leimeister, C.A., Bleidorn, C., Morgenstern, B.: ReadSpaM: assembly-free and alignment-free comparison of bacterial genomes with low sequencing coverage. BMC Bioinf. **20**(S20), 638 (2019). ISSN 1471-2105. https://doi.org/10.1186/s12859-019-3205-7

16. Li, H.: Seqtk, toolkit for processing sequences in FASTA/q formats (2018). https://github.com/lh3/seqtk

17. Liang, Q., Bible, P.W., Liu, Y., Zou, B., Wei, L.: DeepMicrobes: taxonomic classification for metagenomics with deep learning. NAR Genomics Bioinf. **2**(1) (2020). ISSN 2631-9268. https://doi.org/10.1093/nargab/lqaa009

18. Liu, B., Gibbons, T., Ghodsi, M., Pop, M.: MetaPhyler: taxonomic profiling for metagenomic sequences. In: 2010 IEEE International Conference on Bioinformatics and Biomedicine (BIBM), pp. 95–100. IEEE (2011). ISBN 978-1-4244-8305-1

19. Locey, K.J., Lennon, J.T.: Scaling laws predict global microbial diversity. Proc. Nat. Acad. Sci. **113**(21), 5970–5975 (2016). ISSN 0027-8424. https://doi.org/10.1073/pnas.1521291113

20. Lu, J., Breitwieser, F.P., Thielen, P., Salzberg, S.L.: Bracken: estimating species abundance in metagenomics data. PeerJ Comput. Sci. **3**, e104 (2017). ISSN 2376-5992. https://doi.org/10.7717/peerj-cs.104

21. Luo, Y., Yu, Y.W., Zeng, J., Berger, B., Peng, J.: Metagenomic binning through low-density hashing. Bioinformatics **35**(2), 219–226 (2019). ISSN 1367-4803. https://doi.org/10.1093/bioinformatics/bty611

22. Matsen, F.A.: Phylogenetics and the human microbiome. Syst. Biol. **64**(1), e26–e41 (2015). ISSN 1076-836X. arXiv:1407.1794. https://doi.org/10.1093/sysbio/syu053

23. McIntyre, A.B.R., et al.: Comprehensive benchmarking and ensemble approaches for metagenomic classifiers. Genome Biol. **18**(1), 182 (2017). ISSN 1474-760X. https://doi.org/10.1186/s13059-017-1299-7

24. von Meijenfeldt, F.A.B., Arkhipova, K., Cambuy, D.D., Coutinho, F.H., Dutilh, B.E.: Robust taxonomic classification of uncharted microbial sequences and bins with CAT and BAT. Genome Biol. **20**(1), 217 (2019). ISSN 1474-760X. https://doi.org/10.1186/s13059-019-1817-x

25. Metsky, H.C., et al.: Capturing sequence diversity in metagenomes with comprehensive and scalable probe design. Nat. Biotechnol. **37**(2), 160–168 (2019). ISSN 1087-0156. https://doi.org/10.1038/s41587-018-0006-x

26. Meyer, F., Bremges, A., Belmann, P., Janssen, S., McHardy, A.C., Koslicki, D.: Assessing taxonomic metagenome profilers with OPAL. Genome Biol. (2019). ISSN 1474-760X. https://doi.org/10.1186/s13059-019-1646-y

27. Meyer, F., Bremges, A., Belmann, P., Janssen, S., McHardy, A.C., Koslicki, D.: Assessing taxonomic metagenome profilers with OPAL. Genome Biol. **20**(1), 51 (2019). ISSN 1474-760X. https://doi.org/10.1186/s13059-019-1646-y

28. Milanese, A., et al.: Microbial abundance, activity and population genomic profiling with mOTUs2. Nat. Commun. **10**(1), 1014 (2019). ISSN 2041-1723. https://doi.org/10.1038/s41467-019-08844-4

29. Nasko, D.J., Koren, S., Phillippy, A.M., Treangen, T.J.: RefSeq database growth influences the accuracy of k-mer-based lowest common ancestor species identification. Genome Biol. **19**(1), 165 (2018). ISSN 1474-760X. https://doi.org/10.1186/s13059-018-1554-6

30. National Research Council (US). Committee on Metagenomics, Functional Applications, National Academies Press (US): The New Science of Metagenomics. National Academies Press, Washington, D.C., May 2007. ISBN 978-0-309-10676-4. https://doi.org/10.17226/11902

31. Nguyen, N., Mirarab, S., Liu, B., Pop, M., Warnow, T.: TIPP: taxonomic identification and phylogenetic profiling. Bioinformatics **30**(24), 3548–3555 (2014), ISSN 1460-2059. https://doi.org/10.1093/bioinformatics/btu721

32. Ondov, B.D., et al.: Mash: fast genome and metagenome distance estimation using MinHash. Genome Biol. **17**(1), 132 (2016). ISSN 1474-760X. https://doi.org/10.1186/s13059-016-0997-x

33. Ounit, R., Wanamaker, S., Close, T.J., Lonardi, S.: CLARK: fast and accurate classification of metagenomic and genomic sequences using discriminative k-mers. BMC Genomics **16**(1), 236 (2015). ISSN 1471-2164. https://doi.org/10.1186/s12864-015-1419-2

34. Pachiadaki, M.G., et al.: Charting the complexity of the marine microbiome through single-cell genomics. Cell **179**(7), 1623–1635.e11 (2019). ISSN 0092-8674. https://doi.org/10.1016/j.cell.2019.11.017

35. Rachtman, E., Bafna, V., Mirarab, S.: CONSULT: accurate contamination removal using locality-sensitive hashing. NAR Genomics Bioinf. **3**(3) (2011). ISSN 2631-9268. https://doi.org/10.1093/nargab/lqab071

36. Rachtman, E., Balaban, M., Bafna, V., Mirarab, S.: The impact of contaminants on the accuracy of genome skimming and the effectiveness of exclusion read filters. Mol. Ecol. Resour. **20**(3), 649–661 (2020). ISSN 1755-098X. https://doi.org/10.1111/1755-0998.13135

37. Rachtman, E., Balaban, M., Bafna, V., Mirarab, S.: The impact of contaminants on the accuracy of genome skimming and the effectiveness of exclusion read filters. Mol. Ecol. Resour. (2020). ISSN 1755-0998 (Electronic). https://doi.org/10.1111/1755-0998.13135

38. Rasheed, Z., Rangwala, H., Barbará, D.: 16S rRNA metagenome clustering and diversity estimation using locality sensitive hashing. BMC Syst. Biol. **7**(Suppl. 4), S11 (2013). ISSN 1752–0509. https://doi.org/10.1186/1752-0509-7-S4-S11

39. Sczyrba, A., et al.: Critical assessment of metagenome interpretation-a benchmark of metagenomics software. Nat. Meth. **14**(11), 1063–1071 (2017). ISSN 1548-7105. https://doi.org/10.1038/nmeth.4458

40. Sczyrba, A., et al.: Critical assessment of metagenome interpretation-a benchmark of metagenomics software. Nat. Meth. **14**(11), 1063–1071 (2017). ISSN 1548-7091. https://doi.org/10.1038/nmeth.4458

41. Segata, N., Waldron, L., Ballarini, A., Narasimhan, V., Jousson, O., Huttenhower, C.: Metagenomic microbial community profiling using unique clade-specific marker genes. Nat. Meth. **9**(8), 811–814 (2012). ISSN 1548-7091. https://doi.org/10.1038/nmeth.2066

42. Shah, N., Molloy, E.K., Pop, M., Warnow, T.: TIPP2: metagenomic taxonomic profiling using phylogenetic markers. Bioinformatics **37**(13), 1839–1845 (2021). ISSN 1367-4803. https://doi.org/10.1093/bioinformatics/btab023

43. Stark, M., Berger, S.A., Stamatakis, A., von Mering, C.: MLTreeMap-accurate Maximum Likelihood placement of environmental DNA sequences into taxonomic and functional reference phylogenies. BMC Genomics **11**(1), 461 (2010). ISSN 1471-2164. https://doi.org/10.1186/1471-2164-11-461

44. Sunagawa, S., et al.: Metagenomic species profiling using universal phylogenetic marker genes. Nat. Meth. **10**(12), 1196–1199 (2013). ISSN 1548-7091. https://doi.org/10.1038/nmeth.2693

45. Truong, D.T., et al.: MetaPhlAn2 for enhanced metagenomic taxonomic profiling. Nat. Meth. **12**(10), 902–903 (2015). ISSN 1548-7091. https://doi.org/10.1038/nmeth.3589

46. Wood, D.E., Lu, J., Langmead, B.: Improved metagenomic analysis with Kraken 2. Genome Biol. **20**(1), 257 (2019). ISSN 1474-760X. https://doi.org/10.1186/s13059-019-1891-0

47. Wood, D.E., Salzberg, S.L.: Kraken: ultrafast metagenomic sequence classification using exact alignments. Genome Biol. **15**(3) (2014). ISSN 1474-760X. https://doi.org/10.1186/gb-2014-15-3-r46

48. Ye, S.H., Siddle, K.J., Park, D.J., Sabeti, P.C.: Benchmarking metagenomics tools for taxonomic classification. Cell **178**(4), 779–794 (2019). ISSN 1097-4172 (Electronic). https://doi.org/10.1016/j.cell.2019.07.010

49. Zhu, Q., et al.: Phylogeny-aware analysis of metagenome community ecology based on matched reference genomes while bypassing taxonomy. mSystems **7**(2), e0016722 (2022). ISSN 2379-5077. https://doi.org/10.1128/msystems.00167-22

50. Zhu, Q., et al.: Phylogenomics of 10,575 genomes reveals evolutionary proximity between domains Bacteria and Archaea. Nat. Commun. **10**(1), 5477 (2019). ISSN 2041-1723. https://doi.org/10.1038/s41467-019-13443-4

MAGE: Strain Level Profiling
of Metagenome Samples

Vidushi Walia, V. G. Saipradeep, Rajgopal Srinivasan,
and Naveen Sivadasan$^{(\boxtimes)}$

TCS Research, Hyderabad, India
{vidushi.walia,saipradeep.v,rajgopal.srinivasan,
naveen.sivadasan}@tcs.com

Abstract. Metagenomic profiling from sequencing data aims to disentangle a microbial sample at lower ranks of taxonomy, such as species and strains. Deep taxonomic profiling involving accurate estimation of strain level abundances aids in precise quantification of the microbial composition, which plays a crucial role in various downstream analyses. Existing tools primarily focus on strain/subspecies identification and limit abundance estimation to the species level. Abundance quantification of the identified strains is challenging and remains largely unaddressed by the existing approaches. We propose a novel algorithm MAGE (Microbial Abundance GaugE), for accurately identifying constituent strains and quantifying strain level relative abundances. For accurate profiling, MAGE uses read mapping information and performs a novel local search-based profiling guided by a constrained optimization based on maximum likelihood estimation. Unlike the existing approaches that often rely on strain-specific markers and homology information for deep profiling, MAGE works solely with read mapping information, which is the set of target strains from the reference collection for each mapped read. As part of MAGE, we provide an alignment-free and kmer-based read mapper that uses a compact and comprehensive index constructed using FM-index and R-index. We use a variety of evaluation metrics for validating abundances estimation quality. We performed several experiments using a variety of datasets, and MAGE exhibited superior performance compared to the existing tools on a wide range of performance metrics. (Supplementary material available at https://doi.org/10.5281/ zenodo.7746145.)

Keywords: Metagenomics · Microbial strain profiling · Abundance estimation

1 Introduction

The dynamics of a microbial ecosystem is governed by the diversity and abundance of microorganisms present in the microbial environment. Metagenomics research empowers the detailed characterization of microbial population in the microbial environment or host-associated microbial communities. The detailed

© The Author(s), under exclusive license to Springer Nature Switzerland AG 2023
K. Jahn and T. Vinař (Eds.): RECOMB-CG 2023, LNBI 13883, pp. 215–231, 2023.
https://doi.org/10.1007/978-3-031-36911-7_14

analysis of microbial communities in metagenomic samples is tremendously useful in clinical study, disease study, study of environmental habitats, characterization of an industrial product, understanding the host-pathogen interactions and many more. Advancements in Next Generation Sequencing (NGS) technologies have significantly facilitated culture independent analysis of the microbiome, which has increased the scope of metagenomic studies [2,12].

An essential prerequisite for any metagenomic analysis is to disentangle the microbial sample at lower ranks of taxonomy such as species/strain with precise measurements of their abundances. Within the same species individual strains can elicit distinct functions and metabolic responses [1,3], thus governing the dynamics of the microbial environment. This establishes strains as the functional unit of microbial taxonomy and necessitates identification and quantification of metagenomic samples down to the level of strains. The dynamics of metagenomic samples depends not only on the taxonomic composition but also on the taxa abundance. The quantity of various taxa present in the metagenomics sample can significantly affect the properties of microbial environment [22].

Characterizing or profiling of metagenomic samples at the level of strains becomes even more complicated due to the high similarity of the genome sequences present in the samples. Selective amplification of highly conserved regions is widely used for microbial profiling. These conserved regions include housekeeping genes such as 16s and 18s rRNA, ITS. 16srRNA has been extensively used by the microbial community. However, it is not only incapable of capturing the viruses, plasmids and eukaryotes present in the microbial community, but also has a low discriminating power for the classification at species and strain level [19]. Other conserved gene regions or the repetitive regions which have high evolutionary rate offer improved classification at species and strain level but suffer from underdeveloped reference collections [19] and low accuracy in estimating the abundances at strain-level. Whole genome sequencing approaches on the other hand are promising and address most of the drawbacks of amplicon-based approaches but suffer from the problems of high cost and shorter read length which results in ambiguous mapping. There are several strategies to accomplish metagenomic profiling which includes alignment/mapping of reads to complete reference sequences or to a well curated database of unique taxonomic markers, k-mer composition-based approaches and assembly-based approaches. The approaches based on unique taxonomic markers work well in discriminating strains in microbial samples but lacks accuracy in estimating their abundance. These approaches do not allow for easy updates in the reference database as they rely on complex and compute intensive methods to build reference collection of signature markers for every level of taxonomy.

There is a plethora of microbial abundance estimation tools [19,29] that address the problem of species level profiling and strain level identification. However, problem of strain level abundance estimation remains challenging and largely unaddressed. Tools like MetaPhlAn_strainer [4,31,31] provide strain tracking feature along with abundance estimation at strain level. StrainPhlAn2 [4] solves a different problem of SNP profiling of strains present in a metagenomic sample. PanPhlAn [28] and StrainPhlAn2 both work at the strain-level resolution. StrainPhlAn2 provides the SNP profile of the detected strains and PanPhlAn provides the gene specific information for the strains present in the sample. Other approaches like GOTTCHA [10] (a gene-independent, signature-based approach) and Centrifuge [14] (a high speed metagenomic classifier), provide more detailed taxonomic profiles which includes abundance estimates at the strain-level.

Kraken2 [32,33] solves the problem of sub-species/strain level identification and relies on Bracken [18] for efficient estimation of species-level microbial abundance. Kraken2 also reports the number of reads assigned to a taxa in a metagenomic sample, this information can be further utilized to estimate the abundance at various levels of the taxonomy. Other approaches like StrainGE [8], Strain-FLAIR [7] provide strain level resolution. However, these approaches work with a very limited reference collections and thus are not suitable as general purpose profiling tools. For example, StrainGE [8] provides strain-level profiling but limits the analyses to within species strains in bacterial population and focuses on strain-aware variant calling. Strainseeker [27] is a k-mer based approach for clade detection of bacterial isolates. Strainseeker requires a guide tree of the reference strains to be supplied by the user. Based on the hierarchy of k-mers that are shared by the strains under each clade, it performs a clade search. However, the estimation process is sensitive to the user provided guide tree and the k-mer length [27]. These choices can lead to Type 1,2 and 3 errors [27]. Further, k-mers that are shared by distant strains lack representation in the index.

We develop a novel algorithm MAGE (Microbial Abundance GaugE), for accurately identifying constituent strains and quantifying strain level relative abundances. For accurate and efficient profiling, MAGE uses read mapping information and performs a novel local search-based profiling guided by a constrained optimization based on maximum likelihood estimation. Unlike the existing approaches that often rely on taxonomic markers and homology information for deep profiling, MAGE works solely with read mapping information, which is the set of target strains from the reference collection for each mapped read. As part of MAGE, we provide an alignment-free and k-mer-based read mapper (MAGE mapper) that uses a compact and comprehensive index constructed using FM-index [9] and R-index [11]. In MAGE mapper, the reads are mapped to strains solely based on the k-mer composition of the reads without requiring any gapped alignment. We use a variety of evaluation metrics for validating abundances estimation quality. We benchmark MAGE against state-of-the-art methods using a variety of low complexity and high complexity datasets. MAGE showed superior performance compared to the state-of-the-art on a wide range of performance measures.

2   Methods

In this section, we discuss details of MAGE abundance estimation and MAGE mapper. First, we discuss how MAGE estimates strain level abundances given read mapping information. In the later sections, we present an alignment free and k-mer based mapper to generate read mapping information.

2.1   Abundance Estimation and MLE

Let $S = \{s_1, \ldots, s_N\}$ denote a reference collection of N strains (sequences) spanning the various members of the microbial community such as bacteria, virus, fungi etc. We will use the terms sequence and strain interchangeably. For strain level abundance estimation, MAGE uses read mapping information, which is the set of target strains from the reference collection S for each mapped read. Let $\mathcal{R}$ denote the set of mapped reads. For a read $r \in \mathcal{R}$, let $Q(r) \subseteq S$ denote the subset of strains from S to which read r maps. The read mapping information is simply $\{Q(r)\}$. In the following we first discuss how MAGE estimates strain level abundances given $\{Q(r)\}$.

Before discussing the details of the approach used by MAGE, we first discuss the standard MLE (Maximum Likelihood Estimation) formulation similar to [5,26] for abundance estimation. Each read r in $\mathcal{R}$ is assumed to be sampled independently and randomly from its source strain. The source strain is chosen with probability a_i, which is the relative abundance of s_i in the sample. We make the simplifying assumption of uniform coverage, which implies that a read is generated from a location sampled uniformly at random from its source strain. Let ℓ'_j denote the effective length of strain s_j [5,26]. Under the uniform coverage assumption, we approximate ℓ'_j as $\ell_j - \hat{r}$ where $\hat{r}$ is the average read length. More accurate estimates for ℓ'_j can also be used [26]. For a given sample, the relative abundances of all the strains is denoted by the vector $\mathbf{a}$ where the ith component denotes the relative abundance of the ith reference strain. Clearly, $\sum a_i = 1$. The likelihood for $\mathcal{R}$ is given by

$$\Pr(\mathcal{R} \mid \mathbf{a}) = \prod_{r \in \mathcal{R}} \Pr(r \mid \mathbf{a}) = \prod_{r \in \mathcal{R}} \sum_{s_j \in Q(r)} (x_j/\ell'_j) \tag{1}$$

where x_j and a_j are related as $x_j = (a_j \ell'_j) / \sum_{i=1}^m (a_i \ell'_i)$. Here, x_j is simply the length adjusted relative abundance of strain s_j. Clearly, $\sum x_i = 1$. For several reads, their corresponding sequence sets $Q(r)$ will be identical. In other words, the mapped sequence sets of the reads define a partition of $\mathcal{R} = \mathcal{C}_1 \cup \cdots \cup \mathcal{C}_R$ where $\mathcal{C}_i$ denotes the subset of reads that are mapped to the same set of sequences denoted as Q_i. Therefore, the log likelihood with respect to the strain collection S is given by

$$\mathcal{L}(S) = \sum_{i=1}^R n_i \log \left(\sum_{s_j \in Q_i} (x_j/\ell'_j) \right) \tag{2}$$

where n_i is the cardinality of $\mathcal{C}_i$. We solve for x_js that maximizes $\mathcal{L}$. From x_j, we solve for a_j using their relation. To reduce estimation noise, estimated abundance values below a small predefined threshold can be made zero.

The standard MLE approach usually works well for species level abundance estimation. However, MLE suffers from the problem of distributing abundance values across several related strains that are present in the reference leading to reduced specificity and strain level profiling quality. This is because, inter-strain sequence similarity within a species is much higher than inter-species similarity.

2.2 Constrained Optimization and Strain Level Coverage

To perform accurate strain level profiling, MAGE performs a local search based profiling guided by the following constrained optimization problem based on the MLE formulation discussed earlier. EM based solution of the above MLE formulation also yields an estimate of the latent matrix $M_{R \times N}$, where $M(i, j)$ is the number of reads from the read partition $\mathcal{C}_i$ that have originated from strain s_j. We recall that all reads in $\mathcal{C}_i$ map to the same subset of strains Q_i. If strain s_j does not belong to Q_i then clearly $M(i, j) = 0$. From an optimal MLE solution, estimate for $M(i, j)$, where strain s_j belongs to Q_i, is obtained as

$$M(i, j) = n_i \cdot (x_j/\ell'_j) / \sum_{s_k \in Q_i} (x_k/\ell'_k) \qquad (3)$$

where we recall that n_i is the total number of reads in the partition $\mathcal{C}_i$ and Q_j is the set of sequences to which each read in $\mathcal{C}_i$ maps to. Using matrix M, estimate for the total number of reads originating from strain s_j denoted by Y_j is given by

$$Y_j = \sum_{i=1}^{R} M(i, j) \qquad (4)$$

Also, Y_j and x_j are related as $x_j = Y_j/N_r$, where $N_r = \sum n_i = |\mathcal{R}|$ is the total number of reads in $\mathcal{R}$.

Using the estimate of Y_j, estimated coverage of strain s_i, denoted by $cov(j)$, is given by

$$cov(j) = Y_j \, \hat{r}/\ell'_j \qquad (5)$$

where we recall that $\hat{r}$ is the average read length and ℓ'_j is the effective length of strain s_j. From the above set of equations, we obtain that optimal estimates for x_j and $cov(j)$ are related as

$$cov(j) = x_j N_r \hat{r}/\ell'_j \qquad (6)$$

Coverage values $cov(j)$ for different strains can vary as it depends on both read coverage as well as the abundance of the strain in the sample. Using Eq. (6) and noting that N_r and $\hat{r}$ are constants, it is straightforward to verify that at an optimal MLE solution for (2), the objective value $\mathcal{L}^o$ is given by

$$\mathcal{L}^o(S) = \sum_{i=1}^{R} n_i \log \left(\sum_{s_j \in Q_i} cov(j) \right) - c \qquad (7)$$

where the $c = N_r \log(\hat{r} N_r)$ is a constant. From the above expression, we see that the final optimum value depends on the cumulative coverages of the strain sets $Q_1, \ldots, Q_R$ where the cumulative coverage of set Q_i is the sum total of the strain coverages of strains in Q_i.

Suppose we wish to perform a constrained MLE where S is restricted to a subset say $S' \subset S$. When S' is specified, the original read mapping, namely, $Q_1, \ldots, Q_R$, undergoes a projection to obtain the modified mapping information $Q'_1, \ldots, Q'_R$ where Q'_i is simply $Q'_i = Q_i \cap S'$. That is, only the strains in S' are retained in Q'_i. In the rest of the paper, when a strain subset S' is specified, we assume without explicitly stating that the associated mapping information Q'_is are the projections of Q_is obtained with respect to S'. It is easy to see that restricting the strain set to S' amounts to constraining remaining strains to have zero abundance. MLE with respect to S' has its optimum value again given by $\mathcal{L}^o(S')$.

In the next section, we discuss how MAGE performs strain set refinement using a local search, guided by strain coverages $cov(j)$ and $\mathcal{L}^o(S')$, in order to improve the specificity and profiling accuracy. We note that under strain set constraints, for a strain $s' \in S - S'$, the reads that are estimated to be from strain s' in the unconstrained MLE are redistributed to those strains in S' that occur along with s' in any of the original strain subset Q_i. This redistribution leads to an increase in the estimated coverages of the strains in S'. From $\mathcal{L}^o(S')$, we observe that the final optimum objective value depends not only on the individual coverages $cov(j)$ of the strains in S' but more importantly the cumulative coverages with respect to the associated Q'_is. Hence, strain set restriction should be guided in a manner that increases the cumulative coverages of associated Q'_is.

One additional aspect to be considered for strain set refinement is that when the strain set S is restricted to S', some of the reads in $\mathcal{R}$ can become unmapped because none of the mapped strains for this read are present in S'. This is a low probability event if S' contains all the strains present in the sample. Let p_ϵ denote the probability of observing such an unmapped read. From the independence assumption, the probability of observing t such reads is p_ϵ^t. As a consequence, in the likelihood expression given by Eq. (1), each unmapped read contributes a multiplicative term p_ϵ. Lastly, an additional sparsity inducing regularizer is introduced in the objective function to favour sparse solutions. Consequently, the optimum objective value $\mathcal{L}^o$ in Eq. (7) is extended as

$$\mathcal{L}^*(S') = \sum_{Q_i \neq \emptyset} n_i \log \left(\sum_{s_j \in Q_i} cov(j) \right) + \lambda N_\epsilon - \beta |S'| - c \qquad (8)$$

where S' is the restricted set, Q_is are the projected strain sets with respect to S', $\lambda = \log(p_\epsilon)$, β is the sparsity parameter, and N_ϵ is the number of unmapped reads with respect to S'. In the next section, we discuss how MAGE performs iterative strain set refinement using a local search guided by the above constrained objective function.

2.3   Local Search and Strain Set Refinement

Suppose we know the true set of strains S'. Then the abundance estimation can be done by simply performing the constrained MLE ($\mathcal{L}^o(S')$) as discussed in the previous section. However, in practice we do not have such knowledge. Estimating $\mathcal{L}^*(S')$ for all possible subsets is also prohibitively expensive as there are exponentially many subsets. Additionally, if we have prior information on lower bounds for strain level coverages ($cov(j)$), we could further prune the search space to those S' for which the estimated coverages satisfy the additional constraints. However, this information is also hard to obtain, because, as discussed earlier, coverages $cov(j)$ depends on the strain abundances and coverage induced by the mapped reads, which can deviate due to strain mutations. Further, the regularizers used in the above objective do not exploit any specific priors on the cardinality of the distinct strains in the given sample or its composition. Because of this, a set corresponding to an optimal solution for the above objective can still contain several false positives. MAGE therefore uses the above objective $\mathcal{L}^*$ as a surrogate for the true unknown objective. MAGE performs a local search based iterative strain set refinement guided by the surrogate objective $\mathcal{L}^*$ in order to arrive at its final strain set S' which is close to the (unknown) true set. In the following, we discuss the iterative strain set refinement performed by MAGE.

Starting with the initial MLE solution on the on the full strain set S, MAGE iteratively removes strains from among the candidates that have low estimated coverages. The details are discussed later. Removal of such strains can induce only minimal changes to $\mathcal{L}^o(S')$ in each iteration as there are only few reads that gets redistributed in each iteration. Therefore, performing refinement in this manner helps the local search and refinement not to deviate from the desired trajectory in the solution space that leads to the true final S^*.

To start with, all strains whose coverage is below a low threshold are removed at once before performing the iterative refinement. In each iteration of strain set refinement, greedily removing the strain with the lowest estimated coverage can lead to false negatives. This is because, some of the true strains can have low estimated coverages in the current solution S' due to the presence of several similar false positive strains in S'. MAGE therefore considers the bottom b strains $\{s'_i, \ldots, s'_b\}$ having the lowest estimated coverages $cov(j)$, for some fixed b, and computes the modified $\mathcal{L}^*(S' \backslash s'_j)$ values for each of these b candidates. Among these b candidates, we finally remove, from the current solution, the strain s' whose corresponding $\mathcal{L}^*(S' \backslash s')$ is the largest. The refined set after this iteration is given by $S' \backslash s'$. The coverage values of the strains in the refined set are updated using the outcome of MLE performed (using Eq. (5)) for estimating $\mathcal{L}^*(S' \backslash s')$.

To reduce the overall number of MLE computations, MAGE further approximates $\mathcal{L}^*(S' \backslash s'_j)$ for a low coverage strain s'_j as follows. The current sets Q_i are first projected with respect to $S \backslash s'_j$ as discussed earlier. Using the projected Q'_is and using Eq. (3), $M(i, j)$ values are directly recomputed. Using these updated $M(i, j)$ values, modified coverage values $cov'(j)$ for the strains in $S \backslash s'_j$ are computed using Eqs. (4) and (5). These modified coverage values and the modified count of unmapped reads N'_ϵ (as a consequence of removing s'_j) are used in Eq. (7) to approximate $\mathcal{L}^*(S\backslash s_j)$. The above approximation is performed for every t consecutive iterations, for a fixed t, followed by MLE and computation of $\mathcal{L}^*(S')$ without approximation in the subsequent iteration.

As discussed earlier, after each iteration, the coverages $cov(j)$ of the remaining strains in the solution set S' gradually increase. However, excessive refinement will lead to significant decline in the $\mathcal{L}^*(S')$ value because of removing true positives and thereby leading to several unmapped reads. MAGE identifies the final subset S' in the following manner. MAGE first performs the above strain filtering operation till no more strains are left. This produces a sequence of strains $s_1, \ldots, s_m$ in the order of their removal, where the strain s_m in the end of this sequence was removed in the last filtering iteration m. We expect that the true positive strains are present in abundance towards the end of this sequence. In other words, the rank of a strain in the above sequence in a sense indicates the likelihood of the strain to be present in the sample. Let the change of the objective function value at each of these m iterations be denoted by the corresponding sequence $\delta_1, \ldots, \delta_m$. Using this sequence of δs, MAGE attempts to identify an iteration $k \leq m$ such that the final strain set S' given by the sequence $s_k, \ldots, s_m$ has reduced number of false positives and false negatives. For computing k, MAGE performs change point detection on the sequence of δs where the change point indicates the iteration after which a sizeable number of true positives are filtered. The change point is detected using a simple heuristic: scan in the order $\delta_m, \ldots, \delta_1$ and consider the corresponding moving averages $\tilde{\delta}_m, \ldots, \tilde{\delta}_1$ where $\tilde{\delta}_j$ is the average of δ_j, δ_{j+1} and δ_{j+2}. Stop at k such that $\tilde{\delta}_k$ is more than a threshold Δ and $|\tilde{\delta}_{k+1}| \geq \gamma|\tilde{\delta}_k|$ for a threshold γ. For the final strain set S', the associated abundance values are computed using MLE as discussed earlier.

2.4   Read Mapping

As part of MAGE, we provide an alignment-free and k-mer-based read mapper to compute the read mapping information $\{Q(r)\}$ which forms the input for the abundance estimation discussed above. The read mapper works solely based on k-mer lookup on the reference collection. That is, for each k-mer in the read r, the target set of strains in the reference where the k-mer is present is identified. After querying all k-mers in the read, we finally obtain a candidate subset of strains C from the reference that received one or more k-mer hits. The final set of strains $Q(r)$ that qualify as target strains for the mapped read r is identified by performing a random selection from the candidate strain set C. The random selection is based on the cumulative k-mer hits that a strain in C has received. For this, MAGE maintains a probability distribution that encodes Pr(cumulative k-mer hits $= t$ for a read on its source strain). This distribution is estimated apriori using read simulations. Using this probability distribution, each candidate strain in C is included in $Q(r)$ based on the probability associated with the cumulative k-mer hits for the strain.

Reference Index: To support k-mer lookup required by the MAGE mapper, we also constructed an index of the reference collection. We remark that any index of the reference that supports k-mer search would suffice for the MAGE read mapper. However, as part of MAGE, we constructed a simple index by full text indexing of the reference using FM-index [9] and R-index [11]. R-index provides superior compression when the constituent strains share high similarity. However, for divergent set of strains, R-index exhibits poor compression compared to FM-index. We therefore used R-index to index the sub-collection of species that have several strains with high inter-strain similarity. For the remaining species, we used FM-index. In order to link the sub-collection indexes, we created an additional meta-index that directs the k-mer queries to the appropriate sub-collection indexes. The meta-index is a simple lookup that maps a k-mer to a binary vector where a bit is ON iff the k-mer is present in the associated sub-collection. Clearly, the above meta-index allows the reference to be chunked if required into several sub-collections. However, we remark that the meta-index is optional. For instance, no additional meta-index is required when the whole reference indexed as a single chunk. We used the RefSeq strain collection [24] as the reference in the current implementation and the associated index that we created used meta-index. For a given read, all the read k-mers are first searched only in the meta-index. Only those index chunks that received sufficiently many k-mer hits, parameterized by a threshold h, are subsequently searched for the candidate strains.

3   Results

We compare MAGE with state-of-the-art metagenome strain level profiling tools Centrifuge [14], GOTTCHA [10], Kraken2 [33] and MetaPhlAn_strainer [31].

For MetaPhlAn_strainer, we used the tool accessed from [21], the strain tracking feature of MetaPhlAn_strainer enables profiling the samples down to the resolution of strains. MetaPhlAn_strainer [31] database has extended set of markers which enables strain identification and strain tracking of metagenomic samples. StrainPhlAn2 and PanPhlAn do provide strain level relative abundances. We therefore chose MetaPhlAn_strainer over the other tools offered by bioBakery [20] in our benchmark. Centrifuge recommends uncompressed index for taxonomic profiling at the strain level [14]. Hence, we considered only those indexes that are constructed from uncompressed genomic sequences. For benchmarking MAGE's results with Centrifuge, we have considered two different indexes provided by Centrifuge, which we denote as Centrifuge1 and Centrifuge2 in the results. Centrifuge1 uses "p+h+v" index for read mapping. It includes reference genomes from Bacteria, Archae, Viruses and Human (2016 release). Centrifuge2 uses the massive "nt-database" (compressed index size 64 GB) index built using NCBI nucleotide non-redundant sequences (2018 release). These indexes are available for download from [6]. GOTTCHA also provides multiple indexes for metagenomic profiling. In the results GOTTCHA 1 and GOTTCHA 2 denote the results from the two indexes that we have considered in our benchmarking. GOTTCHA 1 uses bacterial index "GOTTCHA_BACTERIA_c4937_k24_u30_xHUMAN3x.strain" for profiling metagenome samples while GOTTCHA 2 uses the combined index "GOTTCHA_COMBINED_xHUMAN3x.strain" for profiling. Both GOTTCHA indexes selected for benchmarking offer strain-level resolution.

Kraken2 also offers two mini databases, namely MiniKraken version1 and MiniKraken version2, that are constructed by downsampling the minimizers in the standard Kraken2 database. The sole difference between the two MiniKraken databases is that MiniKraken version2 also includes k-mers from GRCh38 human genome. The results for these two different databases are denoted as Kraken V1 and Kraken V2 respectively. Since Kraken2 does not report the relative abundances directly, we estimate the strain level abundances as relative fraction of the read counts (reported by Kraken2) after normalizing by the corresponding genome length. Since all ground truth strains belong to RefSeq, we considered only the RefSeq strains reported by Kraken2.

The strain level outputs of MetaPhlAn_strainer, GOTTCHA 1&2, Kraken2 V1&V2 and Centrifuge 1&2 were mapped to the NCBI taxonomy to compare them with the ground truth. For all experiments, the MAGE parameters were fixed as 21 for k-mer length in level 2 index, 16 for k-mer length in level 1 index and 3 for parameter h used in read mapping. In the abundance estimation phase, the parameter values used were $b = 5, t = 5, \beta = 2000, \gamma = 2, \Delta = -10^6$.

We additionally included the standard MLE solution (described in the Sect. 2.1) in our benchmarking. For the standard MLE, the pruning threshold for relative abundance was set to 10^{-4}. Additional details regarding the various experiments are provided in the Supplementary material [34].

Reference Collections: MAGE index was constructed using reference genome sequences from NCBI's RefSeq (complete genomes) [24], that were available at the time of download (2018). In total, MAGE reference database consisted of ~24k strains spanning ~13k species. The species were from 3,559 genus belonging to Virus, Bacteria, Archae, Fungi and Protozoa. The final MAGE index was ~38 GB in size. The level 2 MAGE index consisted of 32 chunks of which 22 were indexed using R-index and the remaining 10 were indexed using FM-index.

The RefSeq strains were part of the reference collection used by each of the tool used in our benchmarking. MetaPhlAn_strainer uses reference genomes from Integrated Microbial Genomes (IMG) system to build its database of clade-specific marker genes. The IMG database contains sequence data from multiple sources including RefSeq.

The "nt-database" index of Centrifuge [14] used by Centrifuge2 contains prokaryotic sequences from NCBI BLAST's nucleotide database which contains all spliced non-redundant coding sequences from multiple databases including RefSeq. The "p+h+v" Centrifuge index used by Centrifuge1 consists of prokaryotic and human genome sequences downloaded from NCBI and contains sequences from Bacteria, Archae, Viruses and Human.

GOTTCHA provides a database of unique signatures for prokaryotic and viral genomes. These unique genome segments are scattered across multiple levels of taxonomy. MiniKraken indexes version1 and version2 used by Kraken2 V1 and V2 respectively also incorporate reference genomes from RefSeq.

Datasets: We experimented with multiple datasets, each containing strains from across the various kingdoms of prokaryotes. We used ART [13] simulator to generate Illumina paired-end reads of 150bp length and with 20X coverage, to generate the various synthetic read datasets.

We created three simulated read datasets, namely, HC1, HC2 and LC. Both HC1 and HC2 datasets are high complexity metagenomic samples, where multiple strains from same species or species with same genus are present in a sample, such that the constituent strains bear high similarity.

HC1 dataset consisted of 34 strains from 10 species spanning 8 genus. Out of the 34 strains, 20 strains were from Escherichia Coli and the remaining 14 strains were spread across 9 other species. Of the 10 species in HC1, 9 species were from bacteria and the remaining 1 was from a viral species ("Murine leukemia"). In HC1 sample, most of the constituent strains had similar relative abundances and a very few strains had higher abundances.

The HC2 dataset consisted of 60 strains from 25 species spanning 25 genus. HC2 dataset had 5 strains from Salmonella enterica, 4 strains from "Bordetella pertussis". The remaining 23 species had 2 to 3 strains each. Out of the 25 species, 24 species are from Bacteria and 1 was a viral species ("Bixzunavirus Bxz1"). In HC2 sample, the constituent strains had varying relative abundances.

LC is a low complexity dataset where most constituent strains do not bear very high similarity with one another. The LC dataset consisted of 50 strains from 47 species in total spanning 38 genus. Each of the 47 species contributed 1 to 2 strain to the LC sample. In LC sample, close to 20% of the strains had higher relative abundances and the remaining 80% of the strains had similar and comparatively lower relative abundances.

To mimic the real world metagenomic samples, we also constructed simulated read datasets where the reference strains were included in the sample after introducing mutations. This allows us to evaluate the ability of the benchmarked tools to correctly identify and quantify the target strains in the presence of mutations. In this study, we used three datasets, namely HC1-Mutated, HC2-Mutated and LC-Mutated, after introducing mutations to the strains in HC1, HC2 and LC respectively. We used wgsim [16,17] pipeline to create the read datasets with 20X coverage. Wgsim introduces SNPs and insertion/deletion (INDEL) mutations to the input genomes and simulates reads from the mutated genomes. We used default parameters to generate mutated read dataset with 0.1% mutation, of which 15% of the mutations were INDELS and the remaining 85% mutations were SNPs. For HC1, HC2 and LC, mutations were introduced in all the strains present in the sample.

To further benchmark the performance of MAGE on low abundance datasets, we considered two additional datasets. One was a real dataset with known strain composition available for download from NCBI's Sequence Read Archive (SRA) under bioproject PRJNA685748 (biosample SAMN17091845) [8]. The real metagenomic read sample consisted of approximately 99% human reads and 1%

E.coli reads coming from four different E.coli strains. The relative abundance of the four strains was unequally distributed as approximately 80:15:4.9:0.1. We also created a simulated low abundance dataset named LA1 using ART under the same setting discussed above. This dataset consisted of 5 phylogenetically different strains. Four strains out of 5 had a very high abundance while 1 out of 5 strains had a very low abundance. The relative abundance of the 5 strains was approximately 24.9:24.9:24.9:24.9:0.2.

Performance Metrics: We used a wide variety of metrics to benchmark the various tools used in our experiments. We used four well-known divergence measures for probability distributions to compare relative abundance distributions, the Jensen-Shannon divergence(JSD), Total variation distance(TVD), Hellinger distance(HD) and cumulative mass (CM). The Jensen-Shannon divergence (JSD) between two distributions P and Q, which is a smoothed and symmetric version of the KL-divergence [30], is given by $JSD(P\|Q) = KL(P\|M) + KL(Q\|M)$ where $M = \frac{1}{2}(P + Q)$. Square root of JS divergence is known to be a metric. Total variation distance (TVD) $\delta(P, Q)$ between distributions P and Q is a statistical distance metric and is given by $\frac{1}{2}\|P - Q\|_1$. Hellinger distance [23] $H(P, Q)$ between distributions P and Q is given by $\frac{1}{\sqrt{2}}\|\sqrt{P} - \sqrt{Q}\|_2$, where $\sqrt{P}$ denotes the component-wise square root of P. Hellinger distance is known to be a metric and is also related to Bhattacharya coefficient. JSD, TVD and HD are known to be special cases of f-divergences between distributions. The cumulative mass (CM) of the true positives for an estimated distribution P with respect to a true distribution Q is given by the sum $\sum_{i:Q_i \neq 0} P_i$. Additionally, we also compare the ability of the tools to correctly identify strains using the standard precision, recall and F1-score metrics. For JSD, TVD and HD, lower values indicate better performance. For CM, precision, recall and F1-scores, higher values indicate better performance.

Results on HC1, HC2 and LC Datasets: Tables 1, 2 and 3 shows the comparison of the benchmarked tools on the high complexity datasets HC1 and HC2 and the low complexity data LC. For all experiments, the best score under each metric is highlighted in bold.

As seen in the tables, for the high complexity datasets (HC1 and HC2), MAGE achieves significantly higher F1 score and lower distances. We remark that higher recall values MLE comes at the cost of low precision and significantly lower F1 due to the presence of several false positives. For the low complexity dataset (LC), MAGE achieves higher F1 score. For LC datasets, MLE achieved better distances, closely followed by MAGE.

Results on Strain Mutation Experiments: Tables 4, 5 and 6 shows the comparative performances for HC1-Mutated, HC2-Mutated and LC-Mutated datasets. MAGE exhibited superior performance on majority of the performance metrics.

Results on Strain Exclusion Experiments: We additionally performed strain exclusion (knock-out) experiments on MAGE. We considered the LC and the HC2 datasets. For the LC dataset, out of the 50 strains present in the sample, we removed 11 strains chosen at random from the MAGE reference. For the

Table 1. Strain level profiling performances on HC-1 data

	MetaPhlAn strainer	Centrifuge 1	Centrifuge 2	GOTTCHA 1	GOTTCHA 2	Kraken2 V1	Kraken2 V2	MLE	MAGE
JSD	0.8197	0.7468	0.8305	0.6960	0.7496	0.6928	0.6950	0.2829	**0.2287**
TVD	0.9714	0.8515	0.9988	0.7714	0.8523	0.8138	0.8159	0.1721	**0.0974**
HD	0.9845	0.8892	0.9923	0.8312	0.8981	0.7902	0.7931	0.3315	**0.2739**
CM	0.0331	0.1888	0.0012	0.4281	0.1488	**0.9608**	0.9495	0.8381	0.9377
Prec.	0.1111	0.0323	0.0233	0.0833	0.0606	0.0102	0.0100	0.2063	**0.9118**
Recall	0.0294	0.3235	0.2941	0.2059	0.2353	0.2353	0.2353	0.9706	**0.9118**
F1	0.0465	0.0587	0.0431	0.1186	0.0964	0.0195	0.0192	0.3402	**0.9118**

Table 2. Strain level profiling performances on HC-2 data

	MetaPhlAn strainer	Centrifuge 1	Centrifuge 2	GOTTCHA 1	GOTTCHA 2	Kraken2 V1	Kraken2 V2	MLE	MAGE
JSD	0.8139	0.7337	0.8294	0.7441	0.7508	0.7074	0.7084	0.3188	**0.3015**
TVD	0.9588	0.8061	0.9982	0.8528	0.8442	0.8585	0.8599	0.2033	**0.1560**
HD	0.9776	0.8783	0.9892	0.8889	0.9002	0.8090	0.8110	0.3706	**0.3606**
CM	0.0436	0.2167	0.0018	0.3157	0.2235	**0.9486**	0.9338	0.8308	0.8930
Prec.	0.0714	0.0342	0.0262	0.0631	0.0471	0.0095	0.0093	0.2624	**0.8772**
Recall	0.0333	0.2833	0.3000	0.1167	0.1333	0.1667	0.1667	0.9667	**0.8333**
F1	0.0455	0.0610	0.0481	0.0819	0.0696	0.0180	0.0175	0.4128	**0.8547**

Table 3. Strain level profiling performances on LC data

	MetaPhlAn strainer	Centrifuge 1	Centrifuge 2	GOTTCHA 1	GOTTCHA 2	Kraken2 V1	Kraken2 V2	MLE	MAGE
JSD	0.7877	0.7157	0.8312	0.6638	0.6638	0.6944	0.6996	**0.1666**	0.2479
TVD	0.9154	0.7843	0.9993	0.7872	0.7872	0.8349	0.8412	**0.0758**	0.1646
HD	0.9453	0.8551	0.9936	0.7755	0.7755	0.7915	0.7982	**0.1932**	0.2925
CM	0.1353	0.3016	0.0006	0.7609	0.7609	**0.9906**	0.9709	0.9389	0.9815
Prec.	0.2000	0.1552	0.1417	0.3429	0.3429	0.0136	0.0133	0.6757	**0.9487**
Recall	0.1000	0.3600	0.3400	0.2400	0.2400	0.2200	0.2200	**1.0000**	0.7400
F1	0.1333	0.2169	0.2000	0.2824	0.2824	0.0257	0.0251	0.8065	**0.8315**

Table 4. Strain level profiling performances on HC1 Mutated dataset

	MetaPhlAn strainer	Centrifuge 1	Centrifuge 2	GOTTCHA 1	GOTTCHA 2	Kraken2 V1	Kraken2 V2	MLE	MAGE
JSD	0.8170	0.7627	0.8308	0.6971	0.7527	0.6938	0.6958	0.5417	**0.4626**
TVD	0.9684	0.8805	0.9990	0.7714	0.8608	0.8158	0.8176	0.6088	**0.4674**
HD	0.9803	0.9084	0.9930	0.8328	0.9016	0.7908	0.7935	0.6276	**0.5429**
CM	0.0341	0.1475	0.0010	0.4205	0.1392	**0.9686**	0.9574	0.3912	0.5328
Prec.	0.1538	0.0437	0.0272	0.0795	0.0552	0.0101	0.0100	0.0667	**0.4923**
Recall	0.0588	0.2941	0.3235	0.2059	0.2353	0.2353	0.2353	**0.9706**	0.9412
F1	0.0851	0.0760	0.0502	0.1148	0.0894	0.0194	0.0191	0.1248	**0.6465**

Table 5. Strain level profiling performances on HC2 Mutated dataset

	MetaPhlAn strainer	Centrifuge 1	Centrifuge 2	GOTTCHA 1	GOTTCHA 2	Kraken2 V1	Kraken2 V2	MLE	MAGE
JSD	0.8029	0.7695	0.8295	0.7444	0.7532	0.7075	0.7085	0.5041	**0.4754**
TVD	0.9333	0.8771	0.9982	0.8523	0.8436	0.8598	0.8612	0.5191	**0.3984**
HD	0.9644	0.9214	0.9898	0.8896	0.9033	0.8086	0.8105	0.5869	**0.5686**
CM	0.0719	0.1328	0.0018	0.3099	0.2094	**0.9567**	0.9423	0.4954	0.6613
Prec.	0.0968	0.0286	0.0290	0.0593	0.0442	0.0083	0.0083	0.1473	**0.3895**
Recall	0.0500	0.2000	0.2667	0.1167	0.1333	0.1500	0.1500	**0.9500**	0.6167
F1	0.0659	0.0501	0.0524	0.0787	0.0664	0.0157	0.0157	0.2550	**0.4774**

Table 6. Strain level profiling performances on LC Mutated dataset

	MetaPhlAn strainer	Centrifuge 1	Centrifuge 2	GOTTCHA 1	GOTTCHA 2	Kraken2 V1	Kraken2 V2	MLE	MAGE
JSD	0.7764	0.7287	0.8315	0.6649	0.6649	0.6952	0.7003	**0.2874**	0.2990
TVD	0.8956	0.8054	0.9995	0.7872	0.7872	0.8359	0.8428	0.2148	**0.1859**
HD	0.9315	0.8710	0.9952	0.7775	0.7775	0.7924	0.7989	0.3366	**0.3541**
CM	0.1677	0.2656	0.0005	0.7482	0.7482	**0.9917**	0.9730	0.8022	0.8941
Prec.	0.1515	0.0057	0.0039	0.3077	0.3077	0.0132	0.0132	0.4673	**0.8837**
Recall	0.1000	0.3200	0.3000	0.2400	0.2400	0.2200	0.2200	**1.0000**	0.7600
F1	0.1205	0.0111	0.0077	0.2697	0.2697	0.0249	0.0250	0.6369	**0.8172**

HC2 dataset, out of the 60 strains present in the sample, we removed 24 strains chosen at random from the MAGE reference.

Table 7 shows the performance in estimating the abundances of the non-excluded ground truth strains. Here, MAGE indicates the performance using the full reference. For each dataset, MAGE_X indicates the performance of MAGE after excluding the strains from the reference.

Table 7. Strain exclusion experiments using HC2 and LC dataset. Recall is computed with respect to the non-excluded ground truth strains.

	HC2		LC	
	MAGE	MAGE_X	MAGE	MAGE_X
JSD	0.2077	0.2755	0.2170	0.2398
TVD	0.1210	0.1932	0.1433	0.2185
HD	0.2461	0.3250	0.2551	0.2749
Recall	0.8056	0.6111	0.7179	0.7179

Results on Low Abundance Datasets: For the LA1 dataset, MAGE's output was exactly the five TP strains with estimated abundances ((24.2:25.7: 24.6:25.2:0.25), which is very close to the ground truth (24.9:24.9:24.9:24.9:0.2), exhibiting MAGE's sensitivity for low abundance strains in a sample.

The real dataset consisted of four E.coli strains, namely SEC470, UT189, Sakai and 24377A, with cumulative abundance of 1% and with relative abundances 80%, 15%, 4.9% and 0.1%. Human reads accounted for the remaining 99%. MAGE predicted four strains, all from E.coli. The four detected strains consisted of one TP SEC470 (relative abundance 34%) and three FP strains WCHEC050613(27%), SF173(19.5%) and NU17(19.5%). It is noteworthy that strain FP WCHEC050613 had 99.39 ANI with TP strain SEC470. SEC470 together with WCHEC050613 accounted for 61% of the relative abundance. Also, predicted strains NU17 and SF173 had ANI of 99.99 and 99.97 respectively to the TP strain UT189. Furthermore, if we consider the ranked order of the different strains based on MAGE's iterative refinement, where the order indicates the ranked order of strain relevance to the sample, the top four ranked strains formed the MAGE's prediction and the TP strains Sakai and 24377A were very close to the top at ranks 5 and 9 respectively. This shows that, in very low abundance scenarios, MAGE exhibited promising performance even without any host read filtering, parameter tuning, and without using a tailored reference or reference preprocessing. Handling very low abundance scenarios is part of future work.

MAGE Implementation: MAGE was implemented in C++ as multi-threaded. The R-Index library [11,15] and FM-index library [25] were used for indexing level 2 sequence sub-collections. The total size of the MAGE index was 38 GB. MAGE mapper takes around 1.9 h using 64 threads for mapping a sample of 1 million reads. The strain refinement was implemented as single-threaded. Strain refinement took 374, 2931 and 1972 secs respectively for HC1, HC2 and low abundance real datasets.

4   Conclusion

Current state-of-art microbial profiling methods predominantly focus on the problem of strain level identification and species level abundance quantification. Strain level abundance estimation is challenging and remains largely unaddressed. We developed MAGE for deep taxonomic profiling and abundance estimation at strain level. For accurate profiling, MAGE makes use of read mapping information and performs a novel local search-based profiling guided by a constrained optimization based on maximum likelihood estimation. Unlike the existing approaches which often rely on strain-specific markers and homology information for deep profiling, MAGE works solely with read mapping information, which is the set of target strains from the reference collection for each mapped read. As part of MAGE, we also provide an alignment-free and k-mer-based read mapper that uses a compact and comprehensive index constructed using FM-index and R-index. We use a variety of evaluation metrics for validating abundances estimation quality. We have extensively benchmarked our results on low complexity and high complexity datasets, which demonstrates the ability of MAGE for accurate profiling. Our work shows that even in the absence of strain specific markers or marker guided read filtering, strain level profiling can be performed solely from the read mapping information. Unlike many of the existing approaches, no separate processing using various taxonomic markers is required for handling read mapping ambiguities. As a future work, it would be interesting to explore improved methods to guide the strain set refinement using more informative surrogates. We believe that our work would drive further research on novel approaches for abundance quantification at strain level resolution and various metrics for benchmarking.

References

1. Alizon, S., de Roode, J.C., Michalakis, Y.: Multiple infections and the evolution of virulence. Ecol. Lett. **16**(4), 556–567 (2013)
2. Anyansi, C., Straub, T.J., Manson, A.L., Earl, A.M., Abeel, T.: Computational methods for strain-level microbial detection in colony and metagenome sequencing data. Front. Microbiol. **11**, 1925 (2020)
3. Balmer, O., Tanner, M.: Prevalence and implications of multiple-strain infections. Lancet Infect. Dis. **11**(11), 868–878 (2011)
4. Beghini, F., et al.: Integrating taxonomic, functional, and strain-level profiling of diverse microbial communities with biobakery 3. Elife **10**, e65088 (2021)
5. Bray, N.L., Pimentel, H., Melsted, P., Pachter, L.: Near-optimal probabilistic RNA-seq quantification. Nat. Biotechnol. **34**(5), 525–527 (2016)
6. Centrifuge. https://ccb.jhu.edu/software/centrifuge/
7. Da Silva, K., Pons, N., Berland, M., Oñate, F.P., Almeida, M., Peterlongo, P.: Strainflair: Strain-level profiling of metagenomic samples using variation graphs. PeerJ **9**, e11884 (2021)
8. van Dijk, L.R., et al.: Strainge: A toolkit to track and characterize low-abundance strains in complex microbial communities. Genome Biol. **23**(1), 1–27 (2022)
9. Ferragina, P., Manzini, G.: Opportunistic data structures with applications. In: Proceedings 41st Annual Symposium on Foundations of Computer Science, pp. 390–398. IEEE (2000)
10. Freitas, T.A.K., Li, P.E., Scholz, M.B., Chain, P.S.: Accurate read-based metagenome characterization using a hierarchical suite of unique signatures. Nucl. Acids Res. **43**(10), e69–e69 (2015)
11. Gagie, T., Navarro, G., Prezza, N.: Optimal-time text indexing in BWT-runs bounded space. In: Proceedings of the Twenty-Ninth Annual ACM-SIAM Symposium on Discrete Algorithms, pp. 1459–1477. SIAM (2018)
12. Hamady, M., Knight, R.: Microbial community profiling for human microbiome projects: Tools, techniques, and challenges. Genome Res. **19**(7), 1141–1152 (2009)
13. Huang, W., Li, L., Myers, J.R., Marth, G.T.: Art: A next-generation sequencing read simulator. Bioinformatics **28**(4), 593–594 (2012)
14. Kim, D., Song, L., Breitwieser, F.P., Salzberg, S.L.: Centrifuge: Rapid and sensitive classification of metagenomic sequences. Genome Res. **26**(12), 1721–1729 (2016)
15. Kuhnle, A., Mun, T., Boucher, C., Gagie, T., Langmead, B., Manzini, G.: Efficient construction of a complete index for pan-genomics read alignment. J. Comput. Biol. **27**(4), 500–513 (2020)
16. Li, H.: WGSIM - simulating sequence reads from a reference genome. https://github.com/lh3/wgsim (2011)
17. Li, H., et al.: The sequence alignment/map format and samtools. Bioinformatics **25**(16), 2078–2079 (2009)

18. Lu, J., Breitwieser, F.P., Thielen, P., Salzberg, S.L.: Bracken: Estimating species abundance in metagenomics data. PeerJ Comput. Sci. **3**, e104 (2017)
19. McIntyre, A.B., et al.: Comprehensive benchmarking and ensemble approaches for metagenomic classifiers. Genome Biol. **18**(1), 1–19 (2017)
20. McIver, L.J., et al.: Biobakery: A meta'omic analysis environment. Bioinformatics **34**(7), 1235–1237 (2018)
21. MetaPhlAn2. https://github.com/biobakery/MetaPhlAn2
22. Neelakanta, G., Sultana, H.: The use of metagenomic approaches to analyze changes in microbial communities. Microbiol. Insights **6**, MBI-S10819 (2013)
23. Nikulin, M.S., et al.: Hellinger distance. Encyclopedia Math. **78** (2001)
24. O'Leary, N.A., et al.: Reference sequence (RefSeq) database at NCBI: Current status, taxonomic expansion, and functional annotation. Nucl. Acids Res. **44**(D1), D733–D745 (2016)
25. Petri, M.: Fm-index-compressed full-text index. https://github.com/mpetri/FM-Index (2015)
26. Roberts, A., Pachter, L.: Streaming fragment assignment for real-time analysis of sequencing experiments. Nat. Methods **10**(1), 71–73 (2013)
27. Roosaare, M., et al.: Strainseeker: Fast identification of bacterial strains from raw sequencing reads using user-provided guide trees. PeerJ **5**, e3353 (2017)
28. Scholz, M., et al.: Strain-level microbial epidemiology and population genomics from shotgun metagenomics. Nat. Methods **13**(5), 435–438 (2016)
29. Simon, H.Y., Siddle, K.J., Park, D.J., Sabeti, P.C.: Benchmarking metagenomics tools for taxonomic classification. Cell **178**(4), 779–794 (2019)
30. Sims, G.E., Jun, S.R., Wu, G.A., Kim, S.H.: Alignment-free genome comparison with feature frequency profiles (FFP) and optimal resolutions. Proc. Natl. Acad. Sci. **106**(8), 2677–2682 (2009)
31. Truong, D.T., et al.: Metaphlan2 for enhanced metagenomic taxonomic profiling. Nat. Methods **12**(10), 902–903 (2015)
32. Wood, D.E., Lu, J., Langmead, B.: Improved metagenomic analysis with kraken 2. Genome Biol. **20**(1), 1–13 (2019)
33. Wood, D.E., Salzberg, S.L.: Kraken: Ultrafast metagenomic sequence classification using exact alignments. Genome Biol. **15**(3), 1–12 (2014)
34. Walia, V., Saipradeep, V.G., Srinivasan, R., Sivadasan, N.: Supplementary Materials: MAGE (2023). https://doi.org/10.5281/zenodo.7746145

MoTERNN: Classifying the Mode of Cancer Evolution Using Recursive Neural Networks

Mohammadamin Edrisi[1]($\boxtimes$)(iD), Huw A. Ogilvie[1](iD), Meng Li[2](iD),
and Luay Nakhleh[1]($\boxtimes$)(iD)

[1] Department of Computer Science, Rice University, Houston, TX, USA
{edrisi,nakhleh}@rice.edu
[2] Department of Statistics, Rice University, Houston, TX, USA

Abstract. With the advent of single-cell DNA sequencing, it is now possible to infer the evolutionary history of thousands of tumor cells obtained from a single patient. This evolutionary history, which takes the shape of a tree, reveals the mode of evolution of the specific cancer under study and, in turn, helps with clinical diagnosis, prognosis, and therapeutic treatment. In this study we focus on the question of determining the mode of evolution of tumor cells from their inferred evolutionary history. In particular, we employ recursive neural networks that capture tree structures to classify the evolutionary history of tumor cells into one of four modes—*linear, branching, neutral,* and *punctuated.* We trained our model, MoTERNN, using simulated data in a supervised fashion and applied it to a real phylogenetic tree obtained from single-cell DNA sequencing data. MoTERNN is implemented in Python and is publicly available at https://github.com/NakhlehLab/MoTERNN.

1 Introduction

From an evolutionary perspective, clonal evolution in cancer and intra-tumor heterogeneity (ITH) are the results of an interplay between mutations and selective pressures in the tumor micro-environment [10,26,32] and can be in part some of the contributing factors in metastasis [44] and tumor drug resistance [2,12]. Aided by advances in sequencing technologies such as microarray [47], next-generation sequencing [25,27], and single-cell sequencing [31,46] that have been developed over the last three decades, the field of cancer evolution has gained attention as studies have shown supporting evidence of tumor cells being subject to selective pressures in response to their environment, which includes the immune system response as well as treatments such as chemotherapy and radiation.

As cancer cells are sampled and sequenced at a small number of time points (most often only one), understanding cancer evolution is done by inferring the evolutionary history of the sampled cells from their somatic mutations—single-nucleotide variations (SNVs) and copy number aberrations (CNAs)—obtained

This study was supported in part by the National Science Foundation, grants IIS-1812822 and IIS-2106837 (L.N.).

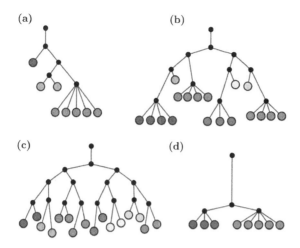

Fig. 1. The phylogenetic trees indicative of four different modes of evolution. (a) Linear evolution, where a clone takes over the cancer cell population. (b) Branching evolution, where multiple clones arise and evolve in parallel at different rates due to selective pressures. (c) Neutral evolution, where multiple clones arise and evolve at similar rates. (d) Punctuated evolution, where a burst of mutations occurs followed by the growth of clones. Ancestral cells are shown in black solid circles, whereas present-day cells are shown in colored solid circles. (Reproduced from [3])

by a variety of DNA sequencing technologies. Indeed, a wide array of cancer evolutionary tree inference methods has been developed in the last decade, which use bulk sequencing data, single-cell sequencing data or a combination thereof [5–7,14,16,20–22,40,41,52–54].

While evolution is a stochastic process, it has been shown that this process could be constrained in cancer as evidenced by different modes of evolution observed in different cancers and, sometimes, during the lifetime of the same cancer (e.g., see Table 1 in [48]). The four main modes of evolution are linear evolution (LE), branching evolution (BE), neutral evolution (NE), and punctuated evolution (PE), all of which are illustrated in Fig. 1. In linear evolution, some cells acquire somatic mutations with strong selective advantages over other cells. This selective sweep results in the tumor being dominated by a major *clone*[1] and a few persistent minor clones that survived from the previous selective sweeps. Thus, the expected phylogenetic tree would take a ladder-like shape, as illustrated in Fig. 1. In branching evolution, the clones evolve in parallel while all gaining fitness during their evolution. Consequently, multiple clones are expected to be present at the time of tissue sampling [3], and the phylogenetic tree would take an overall balanced shape as the clones do not outcompete each other. However, in each clone, one can observe fitness changes during the lifetime of the tumor (Fig. 1). Neutral evolution refers to the case where there is no selection during the lifespan of the tumor. Neutral evolution assumes that the accumulation of mutations is merely a result of tumor progression and natural selection

---

[1] A clone consists of a group of cells with similar genotypes.

does not play much of a role; thus, it provides an alternative explanation for patterns and frequencies of mutations [3,18]. The expected shape of a phylogeny following this model would be highly balanced, not only on the tree's backbone but also at the level of individual cells, as illustrated in Fig. 1. Finally, punctuated evolution, first proposed in paleontology [8], is based on the idea that tumor progression begins in a *big bang* fashion: at the earlier stages, there is a burst of a large number of mutations. Following this phase, the clones gradually grow, leading to a few dominant clones in the tumor. Since the burst of mutations occurs earlier, all the clones share a large portion of mutations yielding a long root branch in the expected phylogeny, as illustrated in Fig. 1.

Determining the mode of tumor evolution is important as these models have different diagnostic, prognostic, and therapeutic implications [48]. For example, a tumor following LE and PE models would require a simpler biopsy since a few samples are good representatives of the entire tumor. On the other hand, BE and NE indicate a high degree of ITH and thus require more biopsy samples for diagnostic purposes [3]. Common approaches to determining the mode of cancer evolution include simulations and mathematical modeling. There is a rich body of literature on mathematical modeling of tumor evolution based on stochastic processes such as the multi-type branching evolution process [4] and the Moran process [29]. These stochastic processes provide predictive statistic measurements on the population size of cancer cells, mutant allele frequencies, or mutation rates [4,51] whose agreement with the observed data determines the mode of evolution. As an example of a simulation-based approach, [11] identified punctuated evolution in triple-negative breast cancer (TNBC) by simulating CNA phylogenetic trees under gradual and punctuated modes of multi-type stochastic birth-death-mutation process, and then measured the fitness of each simulation scenario to the real data using AMOVA analysis [9]. Although this approach benefits from taking the evolutionary history of cells into account, generating realistic phylogenies is still one of the challenges in the field. Minussi *et al.* [28] assessed the fitness of two evolutionary hypotheses to eight TNBC tumors. In both models, the tumor growth starts with a punctuated burst of CNA events. In one model this punctuated phase is followed by a *"gradual accumulation of CNAs"* at a constant rate, and in the other model it leads to a *"transient instability"* in genomic evolution, then a return to the gradual evolution phase. Incorporating these two models into a likelihood framework enabled the authors to measure the fitness of the two models using Akaike Information Criterion [1]. This analysis showed punctuated evolution followed by transient instability and gradual evolution better describes the TNBC tumors. In addition to these approaches, one can use model-based approaches (e.g., in [24]) that accurately detect the speciation events that agree with the PE mode of evolution.

Outside the above categories, Phyolin [50] and the method of [37] identify the mode of evolution given the binary genotype matrices obtained from single-cell SNVs. These methods, however, are aimed at distinguishing between linear and nonlinear modes (i.e., a binary classification), which is a simpler problem than the one we address here. In this study, we tackle the problem of determining the mode of cancer evolution as a graph classification where the graphs are phylogenetic trees and their labels/classes are the corresponding evolutionary modes. Specifically, we investigate the appropriateness of recursive neural networks, or RNNs[2], first proposed in [42]. RNNs have been successfully used in natural language processing to capture semantic relationships between words in variable-sized sentences and parse trees [42,43]. Here, the *words* and their *semantic representations* are the genotypes of the present-day cells and their ancestors, respectively. It is worth mentioning that the capability of RNNs in exploiting the tree structure of phylogenies makes them a more natural candidate for our task than stand-alone multi layer perceptrons (MLPs), convolutional neural networks (CNNs), or traditional classification methods such as random forests. To train our model in a supervised manner, we modified the beta-splitting model (BSM) [38] to generate simulated phylogenetic trees and genotypes according to the four cancer evolutionary modes. Next, we applied our model, MoTERNN (Mode of Tumor Evolution using Recursive Neural Networks), on a real biological phylogenetic tree obtained from single-cell DNA sequencing (scDNAseq) data of a TNBC patient [49]. MoTERNN classified the TNBC patient's data as belonging to the punctuated mode of evolution. Our study demonstrates the suitability of RNNs for classification problems on tree structures. By developing MoTERNN, we have added to the evolutionary biology toolbox that is used to study and understand cancer biology.

2   Methods

2.1   Problem Description

The input to MoTERNN consists of a genotype matrix and a phylogenetic tree obtained from tumor samples. Let $\mathbf{G} = (g_{ij}) \in \{0,1\}^{N \times M}$ be a binary genotype matrix where 0 and 1 indicate absence and presence of mutations, respectively, N is the number of samples, and M is the number of genomic loci. Following this notation, g_i represents the genotype vector of the ith sample at the jth locus. Let $\mathcal{T} = (V, E)$ be a phylogenetic tree where V and E are the sets of nodes and edges, respectively. In this work, we assume that the phylogenetic tree is binary. The leaves of $\mathcal{T}$ are bijectively labeled by the genotype vectors (rows) in $\mathbf{G}$. Given $\mathbf{G}$ and $\mathcal{T}$, MoTERNN predicts one of the four labels in the set {LE, BE, NE, PE}. Next, we give a brief background on RNNs and then describe the model underlying MoTERNN.

2.2   Recursive Neural Networks

Dating back to the late eighties and early nineties, training neural networks on recursive data structures attracted interest from the machine learning community. Seminal works on models such as recursive autoassociative memory [34] and

---

[2] Sometimes the acronym RvNN is used to distinguish it from recurrent neural networks.

backpropagation through structure (BTS) [13] established the basis for dealing with variable-sized recursive structures and efficient computation of backpropagation. Utilizing the BTS scheme for backpropagation, Socher *et al.* [42] introduced RNNs for parsing natural scene images and natural language sentences.

The input of an RNN is a binary tree (or the corresponding adjacency matrix) and a set of vector embeddings for all the leaves. The goal is to learn and predict the labels of the internal nodes including the root of the tree. To predict the label of a node, a corresponding vector embedding for the node is required. In our case, the embeddings[3] of the leaves are generated using an encoder neural network that maps the genotype profiles into a shared lower dimension. Given a tree, the algorithm traverses it and computes the labels and embeddings of the internal nodes recursively. The embedding of a parent node is computed by a neural network which takes as input the concatenation of the two children's embeddings and generates the parent's embedding as output. Following the terminology from [43], we refer to this network as a *compositionality function*. To predict a node's label, its embedding is given to a classifier neural network that produces scores for the node's association to each class. It is worth mentioning that what separates RNNs from their predecessors is that the same compositionality function and classifier network are applied to all inputs resulting in a more flexible and computationally efficient architecture.

Although the original RNN aimed at predicting the labels of all nodes, we associate the evolutionary mode of a phylogenetic tree with its root. Thus, in our scheme, the other internal nodes do not have labels, and the RNN only predicts the root's label. Next, we describe the three components of our model including the encoder, compositionality function, and classifier.

2.3 Encoder Network

Each genotype profile g_i is mapped into a shared lower dimension using an encoder network ϕ:

$$\phi : \{0, 1\}^M \to \mathbb{R}^d, \tag{1}$$

where d is a user-specified parameter. Although the encoder network ϕ is applied to only the genotypes at the leaves of $\mathcal{T}$, we use $\phi(v)$ more generally to denote the embedding of any node v in the tree $\mathcal{T}$. In our implementation, we used a single-layer feed-forward neural network with $d = 128$ as our encoder network.

2.4 Compositionality Function

The compositionality function is a neural network that computes the embedding of a parent node given the computed embeddings of its children. Let v be an internal node and $\phi(v_1)$ and $\phi(v_2)$ be the embeddings of its children v_1 and v_2. The embedding of v based on the compositionality function is computed using the following formula:

---

[3] Hereafter, we use *embedding* instead of *vector embedding*.

$$\phi(v) = \mathscr{F}\left(\mathbf{W} \cdot \begin{bmatrix} \phi(v_1) \\ \phi(v_2) \end{bmatrix}\right), \tag{2}$$

where $\mathscr{F}$ is a non-linear activation function (such as ReLU) and $\mathbf{W} \in \mathbb{R}^{d \times 2d}$ represents the weights of the compositionality function which are multiplied by the concatenation of the two embeddings of the children. Figure 2 illustrates an example of a tree being traversed by an RNN and how the compositionality function operates recursively during this process. First, the embeddings of the leaves A, B, and C are obtained from the encoder network. The leaves' embeddings are denoted by $\phi(A)$, $\phi(B)$, and $\phi(C)$. The first internal node whose embedding is computed is D because its children's embeddings have already been computed. Next, the embedding of the root node, E, is computed by passing $\phi(C)$ and $\phi(D)$ to the compositionality function.

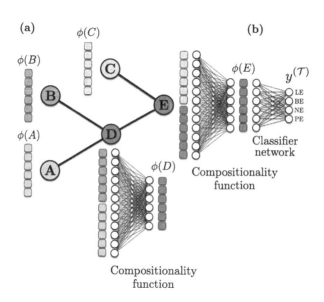

Fig. 2. An illustrative example of processing a tree by MoTERNN. (a) Given a phylogeny and the embeddings of the leaves, MoTERNN computes the embeddings of the internal nodes up to the root of the tree in a bottom up fashion. A node and its embedding are both shown in the same color. The embeddings of A and B are given to compositionality function to produce the embedding of D, $\phi(D)$. Next, the embedding of root node, E, is generated from the embeddings of C and D by compositionality function (denoted by $\phi(E)$). (b) The classifier network takes as input the embedding of the root node; the output contains the association scores for each mode of evolution based on which the label of the tree, $y^{(\mathcal{T})}$, is predicted.

2.5 Classifier Network

To predict the label/model of the phylogenetic tree, we used a classifier network. This could be a simple MLP whose weights are denoted by $\mathbf{W}^c \in \mathbb{R}^{4 \times d}$. The classifier network takes as input the embedding of the root node, $\phi(r)$, and predicts scores for all four tumor evolutionary models, LE, BE, NE, and PE (Fig. 2). The raw scores are passed through a softmax function to generate values between 0 and 1 that are treated as the probability values for each of the four models. The index of the maximum value corresponds the predicted label of $\mathcal{T}$ denoted by $y^{(\mathcal{T})} \in \{1, 2, 3, 4\}$. Let $\mathbf{z} = \mathbf{W}^c \cdot \phi(r)$, formally, we have:

$$y^{(\mathcal{T})} = \operatorname*{argmax}_{k \in \{1,2,3,4\}} \left\{ \sigma\left(\mathbf{z}\right)_k \right\}, \tag{3}$$

where $\sigma(\mathbf{z})_k$ is the kth element of the softmax activation function applied on vector $\mathbf{z}$, defined as $\sigma(\mathbf{z})_k = e^{z_k} / \sum_j e^{z_j}$.

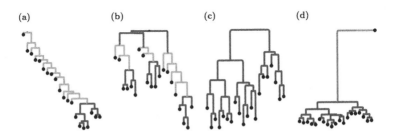

Fig. 3. Examples of simulated trees for the four modes of cancer evolution. (a) Linear evolution. (b) Branching evolution. (c) Neutral evolution. (d) Punctuated evolution. All the trees contain 20 leaves. The branch lengths are proportional to the number of mutations. Different evolutionary episodes are shown in different colors; the branches generated using BSM's unbalanced mode are shown in orange, while purple branches indicate balanced BSM mode of simulation. The particularly long branch in the PE mode is shown in pink. Since the input trees of MoTERNN must be binary, we added one extra branch connected to the root with no mutations in PE trees (colored in green) so that the root has in-degree 0 and out-degree 2.

2.6   Loss Function

We trained MoTERNN in a supervised manner using the simulated trees with their true labels (see Simulation design for more details). We used the cross-entropy loss as our objective function, which is commonly used in multi-class classification tasks. Given a tree $\mathcal{T}$ and its true label $y^{(\mathcal{T})}$ at each iteration during training, the cross-entropy between the estimated scores from the classifier network and the true labels is calculated by

$$\mathcal{H}(\mathcal{T}) = -\sum_{k=1}^{4} \mathbb{1}(y^{(\mathcal{T})} = k) \log \left\{ \mathbb{P}\left(k\right) \right\}, \tag{4}$$

where $\mathbb{1}(y^{(\mathcal{T})} = k)$ is an indicator variable that is equal to one if the true label is k and to zero otherwise. The second term is the logarithm of the probability of $\mathcal{T}$ being associated with class k. This probability is the kth element in the output of the softmax layer. In each iteration, $\mathcal{H}(\mathcal{T})$ is calculated and minimized using stochastic gradient descent or its derivations [36] such as Adam [19]. The weights of the encoder network, compositionality function, and classifier network are updated through backpropagation.

2.7 Simulation Design

The supervised learning of our model requires labeled phylogenies representing each of the four modes of cancer evolution. In general, a phylogenetic tree has two constituents, namely, a topology and branch lengths. As each mode of evolution posits conditions on the topology and branch lengths, we simulated each mode with a different scheme. To manipulate tree topologies, we used BSM, which produces binary trees with arbitrary shapes. According to BSM, the generative process of creating a tree with N cells includes the following steps.

1. **Sampling generative sequences**: sample a sequence of $N-1$ independent and identically distributed (i.i.d.) random values $B = (b_1, \cdots, b_{N-1})$ from the beta distribution $\mathcal{B}(\alpha + 1, \beta + 1)$, where $\alpha > 0$ and $\beta > 0$ are the parameters of the beta distribution. Next, sample a sequence of i.i.d. random values $U = (u_1, \cdots, u_{N-1})$ from a uniform distribution on $[0, 1]$.
2. **Initialization**: create the root of the tree and assign the interval $[0, 1]$ to it. Next, split the root into the left and right child nodes and assign the intervals $[0, b_1]$ and $[b_1, 1]$ to the left and right child nodes, respectively.
3. **Iteration**: in iteration i (considering the initialization step as the first and second iterations), among the leaves of the current tree, find the leaf whose interval $[x, y]$ contains u_i. Select the leaf and split it into the left and right child nodes. Assign $[x, x + (y - x)b_i]$ to the left child and $[x + (y - x)b_i, y]$ to the right child. Stop at iteration $N-1$.

The parameters α and β control the shape of the tree; for example, equal values of α and β generate *balanced* topologies with high probability, and this probability increases as α and β become larger. On the other hand, the difference between the values of α and β determines the *imbalance* of the tree. We used BSM in balanced and unbalanced modes. For the balanced mode, we used $(\alpha, \beta) = (10^4, 10^4)$. A balanced topology resembles the NE mode overall topology. Also, we used it to imitate the cellular evolution within a clone. For unbalanced mode, we used $(\alpha, \beta) = (10^4, 10^{-4})$. An unbalanced topology can imitate the genetic drifts as it occurs during most of a tumor's lifetime in the LE mode.

In the following sections, we detail on our simulation schemes for the four modes of cancer evolution in terms of creating the topology and sampling the mutations on the branches of the trees.

Simulation Scheme for LE. We assume the tree topology of LE model grows during two episodes. The first and second episodes occur before and after the emergence of the dominant clone, respectively. We model the first episode with the unbalanced mode of BSM. Among the total $N-1$ number of speciations—or cell divisions—required for simulating a tree with N cells, two thirds of the first speciations are done in the unbalanced mode. We model the tree growth in the second episode with the balanced mode of BSM which covers the rest of speciations. Figure 3 shows an example of an LE tree with 20 cells. The first and second episodes are shown in orange and purple, respectively. After creating a topology, we sample the number of mutations from a Poisson distribution with a mean of 5 ($\lambda = 5$) for each branch. To generate binary genotype profiles of the cells with M loci, we assign an all-zeros vector to the root, which is assumed to be a normal cell without mutations. We assume the mutations are accumulated following the infinite-sites assumption (ISA) for all cancer evolution modes. Starting from the root node's children, the tree is traversed in a breadth-first (or level-order) manner. For a branch with $\mathcal{X}$ mutations, we randomly sample $\mathcal{X}$ loci (without replacement) from the unmutated loci. Next, to create the genotype vector of the child node, we copy the genotype vector of the parent node and change the values of the entries corresponding to the randomly selected loci to 1.

Simulation Scheme for BE. We simulate BE trees in two steps. First, the total number of clones, C, is determined. We sample C uniformly from the set $\{2, 3, 4\}$. The tree grows using the balanced mode of BSM until C leaves are generated. In the second step, the number of cells associated with each clone is sampled. Note that the sum of the number of leaves under all clones must be equal to the total number of leaves, N, which is specified by the user. We sample these counts from a multinomial distribution with N trials and C categories where the success probability of each category equals $\frac{1}{C}$. Having sampled these counts, we generate each clonal lineage using the same procedure described in Simulation scheme for LE to account for the evolutionary sweeps within each clonal lineage. Figure 3 shows an example of a simulated BE tree. The number of mutations on the branches and genotype profiles of the cells are generated according to the same procedure and distributions for sampling mutations in Simulation scheme for LE.

Simulation Scheme for NE. Since there are no selection or dominant clones in the NE mode, we simulate the entire tree topology using BSM's balanced mode. The sampling of mutations and their assignment to the cells are done according to the procedure described in Simulation scheme for LE. Figure 3 shows an example tree generated according to the NE mode.

Simulation Scheme for PE. The main characteristics of the PE trees include a long root branch with the largest portion of mutations accumulated during the lifetime of a tumor followed by a few dominant clones. To simulate such trees, we first determine the number of clones, C, by sampling uniformly from $\{2, 3\}$. Next, we sample the number of cells belonging to each clone from a multinomial distribution with N trials, C categories, and the success probability of $\frac{1}{C}$ for each category (the same as in Simulation scheme for BE). Given the number of cells within each clone, we grow each clonal lineage separately using BSM's balanced mode. The number of mutations on the long root branch is determined by sampling from a Poisson distribution with $\lambda = 100$. The number of mutations on the clonal branches is sampled with $\lambda = 5$. Since MoTERNN requires binary trees as input, we attach one extra cell with no mutations to the root that represents a normal cell. Figure 3 shows an example tree generated according to this procedure.

3   Results and Discussion

3.1   Supervised Training of MoTERNN

We simulated 8100 data points (2025 for each mode) according to the schemes described in Simulation design. A data point refers to a phylogenetic tree, a genotype matrix of the leaves, and the true label of the tree. For each phylogeny, the number of cells was sampled uniformly from the set of integers ranging from 20 to 100. We set the number of loci to 3375 to match the number of candidate loci for mutation calling in our real data set that we applied our trained model later on (see Application to real data). We applied k-fold cross-validation with $k = 4$ on our model to assess its predictive ability given different randomly selected test and training subsets. First, we randomly shuffled the data, and selected 100 data points as the validation set. Next, we partitioned the 8,000 remaining data points into four equal-sized subsets. In each round of cross-validation, a single subset was chosen as the test set, and the rest retained as the training set. We trained four instances of our model, separately, on each pair of training and test subsets selected by cross-validation.

We used single layer fully-connected feed-forward neural networks for all the functions, including the encoder, compositionality function, and classifier network. In particular, the encoder network was of size 3375×128 (3375 nodes at the input and 128 nodes at the output), the compositionality function was a network of dimensions 256×128 (so that it takes concatenation of two embeddings as input), and the classifier was of size 128×4. In each iteration, one data point is processed to compute the cross-entropy loss. The networks' weights are updated via the Adam optimization algorithm implemented in PyTorch. For the Adam optimizer, we used the default parameters except for the learning rate, which was set to 10^{-4} (lower than 10^{-3}, the default value) to achieve a smoother convergence. In each cross-validation round, we trained MoTERNN for 6,000 iterations to process all the data points in the training set. We ran MoTERNN on an Nvidia A6000 GPU with 48 GB RAM, and an Intel Xeon Gold 6226R CPU with 64GB available RAM. The average training time was five minutes and 14 s. The peak of memory consumption on GPU was 1.53 GB, while on CPU, the maximum occupation of RAM was 29.6 GB.

In every iteration, we evaluated each of the four models on its entire training and validation sets. Then, we calculated the average training and validation accuracy of the four models at each iteration (Fig. 4). The average loss function values of the models are also demonstrated in Fig. 4. We observed spikes in the loss curve, especially in the early iterations. Such spikes are likely caused by the exploding gradients due to the small batch size we used. More restriction on the gradient values using the gradient clipping technique [33] could better stabilize the training. After the four models were separately trained on the cross-validation subsets, we evaluated the trained models on their corresponding test and training sets. MoTERNN achieved an average training and test accuracy of 99.98% and 99.95%, respectively.

3.2   Application to Real Data

We applied our trained models on a data set consisting of single-cell whole-exome sequencing samples from a TNBC patient [49]. The TNBC data set consists of 16 diploid cells (treated as normal control samples), eight aneuploid cells, and eight hypodiploid cells [49]. We ran Phylovar [5] to infer the SNVs and the underlying phylogeny of the single-cells. The total number of candidate loci for mutation calling was 3,375. The phylogeny inferred by Phylovar is binary and

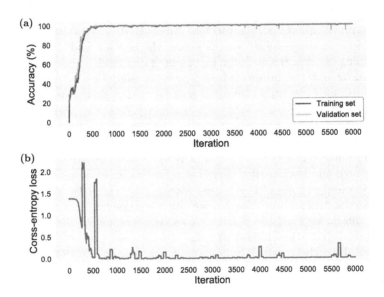

Fig. 4. MoTERNN's loss and classification accuracy during training (a) The average accuracy of the four models trained in cross-validation. For each model, the accuracy on the training and validation sets was computed in each iteration. Then, we averaged over the accuracy values of the models in each iteration. (b) The average cross-entropy loss function during training of the four cross-validation models. To further smooth the plots, we averaged the accuracy and loss values over the last ten iterations.

admits the infinite-sites assumption [5]. To apply our trained model only on the tumor cells, we detached all the diploid cells from the phylogeny except for one that was directly connected to the root. The TNBC phylogeny is illustrated in Fig. 5. Given the genotype matrix and the phylogeny, we applied the four trained models to the TNBC data and measured their average association probability scores to each mode of cancer evolution. These probability scores were approximately 0.7501, 0.2498, 0, and 0 for PE, NE, LE, and BE, respectively. Therefore MoTERNN hypothesized PE as the mode of evolution for this data set. In the original study, Wang *et al.* concluded that *"point mutations evolved gradually,"* which disagrees with our result. This difference might be attributed to the direct incorporation of phylogenetic information, as opposed to the birth-death process the authors utilized for modeling the evolution of SNVs [49]. It is worth mentioning that Wang *et al.* reported punctuated copy number evolution from the same TNBC patient. Based on the similarity between the two phylogenies—one obtained from CNA profiles by Wang *et al.* and the other from SNVs in our study—in terms of grouping the aneuploid and hypodiploid cells, and also our model's result, we hypothesize that CNAs and SNVs both followed a punctuated evolution in this patient. We leave investigating the evolutionary mode of the TNBC patient's CNA data by MoTERNN as a future direction.

We note that in order to train MoTERNN for a real data set, the number of loci in the training data set must be exactly the same as in the real data set. Also, the number of cells for training must be in a range that covers the number of cells in the real data set.

4   Summary and Future Directions

In this work, we developed MoTERNN, a supervised learning approach to classifying the mode of cancer evolution using RNNs. MoTERNN takes as input a tumor phylogeny and the binary genotype matrix of the single-cells, and associates the tumor to one of four evolutionary modes, including LE, BE, NE, and PE. To train our model, we simulated tree topologies for the four modes of evolution and generated binary genotype matrices of the leaves following an ISA model for mutation placements on the tree branches. Our model achieved 99.98% and 99.95% accuracy on the training and test sets, respectively. We applied the trained model on a single-cell DNA sequencing data set from a triple-negative breast cancer patient. MoTERNN identified a punctuated mode of evolution in this data.

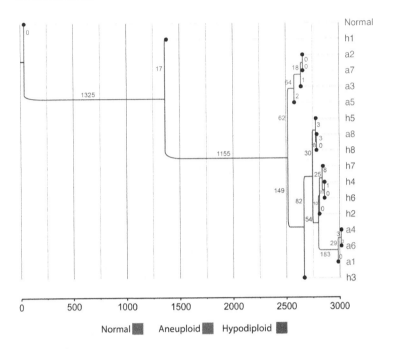

Fig. 5. The phylogeny obtained from the TNBC data set. The aneuploid cells are tagged with *a* and colored in pink. The hypodiploid cells are tagged with *h* and colored in blue. The diploid cell connected to the root is shown in green. The branch lengths indicate the number of mutations that occurred on them. The scale axis at the bottom of the figure shows the number of accumulated mutations as the tree grows. The branches are annotated from the tips for better visualization. Figure generated by FigTree (http://tree.bio.ed.ac.uk/software/figtree/).

One factor that could impact the performance of our method is sampling bias. Sampling has been shown to have an impact on the inferred tree in the field of phylogenetics [15], though the debate about this has been inconclusive so far [30]. Moreover, to extend our model to non-binary tree topologies, a potential solution is to employ N-ary Tree LSTMs (long short-term memory), a generalization of RNNs for trees with arbitrary branching factors [45]. Like any other supervised learning method, the performance of MoTERNN on real data relies on the quality of simulations used to generate the training data. The ISA that we used for simulating mutations can be violated in cancer. In this regard, simulating phylogenetic trees by incorporating more complex evolutionary models such as multi-type branching evolution process [4,11,28] or utilizing more advanced simulators for SNVs—especially, to deviate from binary genotypes and move towards ternary genotypes—and CNAs, such as [23,35], is a future research direction to explore. Extending MoTERNN to CNAs paves the way to its application to larger and more diverse data sets such as the CNA data set originally studied by Kim *et al.* [17], where the authors investigated the clonal evolution of eight TNBC patients at various stages of tumor progression.

Although learning on more advanced simulations might require more model parameters, the computational training cost would not increase dramatically by making MoTERNN more complex because such architectures train/apply, repeatedly, a few building blocks (e.g., the encoder, compositionality function, and classifier network) to the entire data structure. As the first RNN application to phylogenetics (see [39] for details on the current status of deep learning applications in phylogenetics), MoTERNN demonstrated the potential of RNN models in learning on phylogenetic trees. We believe that variations of RNN models can be suitable choices for future studies in evolutionary biology.

5 Competing Interests

The authors declare no competing interests.

References

1. Akaike, H.: A new look at the statistical model identification. IEEE Trans. Autom. Control **19**(6), 716–723 (1974)
2. Burrell, R.A., et al.: Tumour heterogeneity and the evolution of polyclonal drug resistance. Molecul. Oncol. **8**(6), 1095–1111 (2014)
3. Davis, A., et al.: Tumor evolution: Linear, branching, neutral or punctuated? Biochim. Biophys. Acta Rev. Cancer **1867**(2), 151–161 (2017)
4. Durrett, R., et al.: Intratumor heterogeneity in evolutionary models of tumor progression. Genetics **188**(2), 461–477 (2011)
5. Edrisi, M., et al.: Phylovar: Toward scalable phylogeny-aware inference of single-nucleotide variations from single-cell DNA sequencing data. Bioinformatics **38**(Supplement 1), i195–i202 (2022)
6. Edrisi, M., et al.: A combinatorial approach for single-cell variant detection via phylogenetic inference. In: Huber, K.T., Gusfield, D. (eds.) 19th International Workshop on Algorithms in Bioinformatics (WABI 2019). Leibniz International Proceedings in Informatics (LIPIcs), vol. 143, pp. 22:1–22:13. Schloss Dagstuhl-Leibniz-Zentrum fuer Informatik, Dagstuhl (2019)
7. El-Kebir, M.: SPhyR: Tumor phylogeny estimation from single-cell sequencing data under loss and error. Bioinformatics **34**(17), i671–i679 (2018)
8. Eldredge, N., et al.: Punctuated equilibria: An alternative to phyletic gradualism. In: Schopf, T.J.M. (ed.) Models in Paleobiology, pp. 82–115. Freeman Cooper (1972)
9. Excoffier, L., et al.: Analysis of molecular variance inferred from metric distances among DNA haplotypes: Application to human mitochondrial DNA restriction data. Genetics **131**(2), 479–491 (1992)
10. Fidler, I.J.: Tumor heterogeneity and the biology of cancer invasion and metastasis. Cancer Res. **38**(9), 2651–2660 (1978)
11. Gao, R., et al.: Punctuated copy number evolution and clonal stasis in triple-negative breast cancer. Nat. Genet. **48**(10), 1119–1130 (2016)
12. Gillies, R.J., et al.: Evolutionary dynamics of carcinogenesis and why targeted therapy does not work. Nat. Rev. Cancer **12**(7), 487–493 (2012)

13. Goller, C., et al.: Learning task-dependent distributed representations by backprop-agation through structure. In: Proceedings of International Conference on Neural Networks (ICNN 1996), vol. 1, pp. 347–352 (1996)
14. Hajirasouliha, I., et al.: A combinatorial approach for analyzing intra-tumor het-erogeneity from high-throughput sequencing data. Bioinformatics **30**(12), i78–i86 (2014)
15. Heath, T.A., et al.: Taxon sampling and the accuracy of phylogenetic analyses. J. System. Evol. **46**(3), 239 (2008)
16. Jahn, K., et al.: Tree inference for single-cell data. Genom. Biol. **17**(1), 86 (2016)
17. Kim, C., et al.: Chemoresistance evolution in triple-negative breast cancer delin-eated by single-cell sequencing. Cell **173**(4), 879–893 (2018)
18. Kimura, M.: Rare variant alleles in the light of the neutral theory. Molecul. Biol. Evol. **1**(1), 84–93 (1983)
19. Kingma, D.P., et al.: Adam: A method for stochastic optimization. In: Bengio, Y., LeCun, Y. (eds.) 3rd International Conference on Learning Representations, ICLR 2015, San Diego, CA, 7–9 May 2015, Conference Track Proceedings (2015)
20. Kuipers, J., et al.: Single-cell copy number calling and event history reconstruction. bioRxiv (2020)
21. Liu, Y., et al.: Nestedbd: Bayesian inference of phylogenetic trees from single-cell DNA copy number profile data under a birth-death model. bioRxiv (2022)
22. Malikic, S., et al.: PhISCS: A combinatorial approach for subperfect tumor phy-logeny reconstruction via integrative use of single-cell and bulk sequencing data. Genome Res. **29**(11), 1860–1877 (2019)
23. Mallory, X.F., et al.: Assessing the performance of methods for copy number aber-ration detection from single-cell DNA sequencing data. PLoS comput. Biol. **16**(7), e1008012 (2020)
24. Manceau, M., et al.: Model-based inference of punctuated molecular evolution. Molecul. Biol. Evol. **37**(11), 3308–3323 (2020)
25. Mardis, E.R.: A decade's perspective on DNA sequencing technology. Nature **470**(7333), 198–203 (2011)
26. Merlo, L.M.F., et al.: Cancer as an evolutionary and ecological process. Nat. Rev. Cancer **6**(12), 924–935 (2006)
27. Meyerson, M., et al.: Advances in understanding cancer genomes through second-generation sequencing. Nat. Rev. Genet. **11**(10), 685–696 (2010)
28. Minussi, D.C., et al.: Breast tumours maintain a reservoir of subclonal diversity during expansion. Nature **592**(7853), 302–308 (2021)
29. Moran, P.A.P.: Random processes in genetics. Math. Proc. Camb. Philos. Soc. **54**(1), 60–71 (1958)
30. Nabhan, A.R., et al.: The impact of taxon sampling on phylogenetic inference: A review of two decades of controversy. Brief. Bioinform. **13**(1), 122–134 (2012)
31. Navin, N., et al.: Tumour evolution inferred by single-cell sequencing. Nature **472**(7341), 90–94 (2011)
32. Nowell, P.C.: The clonal evolution of tumor cell populations. Science **194**(4260), 23–28 (1976)
33. Pascanu, R., et al.: On the difficulty of training recurrent neural networks. In: International Conference on Machine Learning, pp. 1310–1318. PMLR (2013)
34. Pollack, J.B.: Recursive distributed representations. Artif. Intell. **46**(1), 77–105 (1990)
35. Posada, D.: Cellcoal: Coalescent simulation of single-cell sequencing samples. Molecul. Biol. Evol. **37**(5), 1535–1542 (2020)

36. Ruder, S.: An overview of gradient descent optimization algorithms. arXiv (2016)
37. Sadeqi Azer, E., et al.: Tumor phylogeny topology inference via deep learning. iScience **23**(11), 101655 (2020)
38. Sainudiin, R., et al.: A beta-splitting model for evolutionary trees. Roy. Soc. Open Sci. **3**(11) (2015)
39. Sapoval, N., et al.: Current progress and open challenges for applying deep learning across the biosciences. Nat. Commun. **13**(1), 1–12 (2022)
40. Satas, G., et al.: SCARLET: Single-cell tumor phylogeny inference with copy-number constrained mutation losses. Cell Syst. **10**(4), 323–332 (2020)
41. Singer, J., et al.: Single-cell mutation identification via phylogenetic inference. Nat. Commun. **9**(1), 5144 (2018)
42. Socher, R., et al.: Parsing natural scenes and natural language with recursive neural networks. In: Proceedings of the 28th International Conference on International Conference on Machine Learning, pp. 129–136. Omnipress (2011)
43. Socher, R., et al.: Recursive deep models for semantic compositionality over a sentiment treebank. In: Proceedings of the 2013 Conference on Empirical Methods in Natural Language Processing, pp. 1631–1642. Association for Computational Linguistics (2013)
44. Swanton, C.: Intratumor heterogeneity: Evolution through space and time. Cancer Res. **72**(19), 4875–4882 (2012)
45. Tai, K.S., Socher, R., Manning, C.D.: Improved semantic representations from tree-structured long short-term memory networks. arXiv preprint arXiv:1503.00075 (2015)
46. Tang, F., et al.: mRNA-Seq whole-transcriptome analysis of a single cell. Nat. Methods **6**(5), 377–382 (2009)
47. Trevino, V., et al.: DNA microarrays: A powerful genomic tool for biomedical and clinical research. Molecul. Med. **13**(9), 527–541 (2007)
48. Venkatesan, S., et al.: Tumor evolutionary principles: How intratumor heterogeneity influences cancer treatment and outcome. Am. Soc. Clin. Oncol. Educ. Book **36**, e141–e149 (2016)
49. Wang, Y., et al.: Clonal evolution in breast cancer revealed by single nucleus genome sequencing. Nature **512**(7513), 155–160 (2014)
50. Weber, L.L., et al.: Distinguishing linear and branched evolution given single-cell DNA sequencing data of tumors. Algorithm. Molecul. Biol. **16**(1), 14 (2021)
51. Williams, M.J., et al.: Identification of neutral tumor evolution across cancer types. Nat. Genet. **48**(3), 238–244 (2016)
52. Zaccaria, S., et al.: Characterizing allele-and haplotype-specific copy numbers in single cells with chisel. Nat. Biotechnol. **39**(2), 207–214 (2021)
53. Zafar, H., et al.: SiFit: Inferring tumor trees from single-cell sequencing data under finite-sites models. Genome Biol. **18**(1), 178 (2017)
54. Zafar, H., et al.: SiCloneFit: Bayesian inference of population structure, genotype, and phylogeny of tumor clones from single-cell genome sequencing data. Genome Res. **29**, 1860–1877 (2019)

Author Index

A

Anderson, Tavis   131
Artemov, Gleb N.   84

B

Bergeron, Anne   68
Beslon, Guillaume   1
Bohnenkämper, Leonard   51
Braga, Marília D. V.   35
Brockmann, Leonie R.   35

D

Diallo, Abdoulaye Baniré   112

E

Edrisi, Mohammadamin   232
Eulenstein, Oliver   131

F

Fedorova, Valentina S.   84

G

Górecki, Paweł   131

K

Kalhor, Reza   1
Khayatian, Elahe   146
Kirilenko, Kirill M.   84
Klerx, Katharina   35
Kokhanenko, Alina A.   84

L

Lafond, Manuel   1, 68, 179
Landry, Kaari   179
Li, Meng   232

M

Markin, Alexey   131
Mirarab, Siavash   196

N

Nakhleh, Luay   232

O

Ogilvie, Huw A.   232
Ouangraoua, Aida   19
Ouedraogo, Wend Yam Donald Davy   19

R

Rachtman, Eleonora   196
Remita, Amine M.   112

S

Saipradeep, V. G.   215
Sankoff, David   100
Şapcı, Ali Osman Berk   196
Scornavacca, Celine   1
Sharakhov, Igor V.   84
Sivadasan, Naveen   215
Soboleva, Evgenia S.   84
Srinivasan, Rajgopal   215
Stoye, Jens   35
Swenson, Krister M.   68

T

Tremblay-Savard, Olivier   179

V

Valiente, Gabriel   146
Vitae, Golrokh   112

W

Wagle, Sanket   131
Walia, Vidushi   215
Wu, Yufeng   162

X

Xu, Qiaoji   100

Z

Zhang, Louxin   146, 162

K. Jahn and T. Vinař (Eds.): RECOMB-CG 2023, LNBI 13883, p. 249, 2023.
https://doi.org/10.1007/978-3-031-36911-7

Printed in the United States
by Baker & Taylor Publisher Services